ADVANCES IN MANUFACTURING AND PROCESSING OF MATERIALS

Characterization and Applications

ADVANCES IN MANUFACTURING AND PROCESSING OF MATERIALS

Characterization and Applications

Edited

Narendra Kumar, PhD

Mohit Tyagi, PhD

Dilbagh Panchal, PhD

Ravi Pratap Singh, PhD

AAP | APPLE ACADEMIC PRESS

First edition published 2025

Apple Academic Press Inc.
1265 Goldenrod Circle, NE,
Palm Bay, FL 32905 USA

760 Laurentian Drive, Unit 19,
Burlington, ON L7N 0A4, CANADA

CRC Press
2385 NW Executive Center Drive,
Suite 320, Boca Raton FL 33431

4 Park Square, Milton Park,
Abingdon, Oxon, OX14 4RN UK

Library and Archives Canada Cataloguing in Publication

Title: Advances in manufacturing and processing of materials : characterization and applications / edited, Narendra Kumar, PhD, Mohit Tyagi, PhD, Dilbagh Panchal, PhD, Ravi Pratap Singh, PhD.
Names: Kumar, Narendra (Professor of industrial and production engineering), editor. | Tyagi, Mohit, editor. | Panchal, Dilbagh, editor. | Singh, Ravi Pratap, 1987- editor.
Description: First edition. | Includes bibliographical references and index.
Identifiers: Canadiana (print) 20240315154 | Canadiana (ebook) 20240315200 | ISBN 9781774914922 (hardcover) | ISBN 9781774914939 (softcover) | ISBN 9781003408345 (PDF)
Subjects: LCSH: Manufacturing processes. | LCSH: Materials.
Classification: LCC TS183 .A38 2024 | DDC 670—dc23

Library of Congress Cataloging-in-Publication Data

CIP data on file with US Library of Congress

ISBN: 978-1-77491-492-2 (hbk)
ISBN: 978-1-77491-493-9 (pbk)
ISBN: 978-1-00340-834-5 (ebk)

About the Editors

Narendra Kumar, PhD

Assistant Professor, Industrial and Production Engineering Department, Dr. B. R. Ambedkar National Institute of Technology, Jalandhar, Punjab, India

Narendra Kumar, PhD, is an Assistant Professor in the Industrial and Production Engineering Department at Dr. B. R. Ambedkar National Institute of Technology, Jalandhar, Punjab. He obtained his PhD in Mechanical Engineering from the Indian Institute of Information Technology, Design and Manufacturing, Jabalpur. His research work focused on the "Development and Performance Evaluation of Pellet-based Additive Manufacturing Process for Flexible Parts." Prior to joining NIT Jalandhar, he worked for Bajaj Institute of Technology, Wardha, India. He also contributed to a DST/AMT-sponsored project titled "Development of Metal-Based Deposition System Using Induction Heating Method" as a research associate. Dr. Kumar is also working as a Principal Investigator (PI) of SERB-DST-funded and ISRO-YUKTI funded projects which are in progress. He has also completed two research projects as a PI. He has published over 55 research articles in SCI/Scopus-indexed journals and presented at various national/international level conferences. His research interests encompass additive manufacturing, machining methods, and material development.

Mohit Tyagi, PhD

Associate Professor, Department of Production and Industrial Engineering, Punjab Engineering College (Deemed to be University), Chandigarh, India

Mohit Tyagi, PhD, is an Associate Professor in the Department of Production and Industrial Engineering at Punjab Engineering College (Deemed to be University), Chandigarh, India. With 10 years of teaching and research experience, his areas of expertise include industrial engineering, supply chain management, corporate social responsibilities, performance measurement systems, Data Science, and Fuzzy Inference Systems. Dr. Tyagi has guided 25 postgraduate dissertations and 35 undergraduate projects, and is presently supervising seven MTech and eight PhD scholars. He has an impressive record of around 125 publications in international and national journals, as well as proceedings of international conferences. Additionally, he serves as a reviewer for many esteemed international journals such as the *International Journal of Industrial Engineering: Theory, Application and Practices, Supply Chain Management: An International Journal*, the *International Journal of Logistics System Management*, and the Journal of Manufacturing Technology Management. Dr. Tyagi has also organized four international conferences, one as a joint organizing secretary at Delhi Technological University Delhi and the remaining ones as a convener and secretary at NIT Jalandhar, and PEC Chandigarh. Dr. Tyagi has also been awarded Best Faculty on 5th September 2023 at PEC Chandigarh for his extraordinary teaching and research skills. He completed his BTech (Mechanical Engineering) with Hons. from UPTU Lucknow in 2008 and MTech (Product Design and Development) with a Gold Medal from MNNIT Allahabad in 2010. In 2015, he was awarded his PhD from the Indian Institute of Technology, Roorkee (India).

Dilbagh Panchal, PhD
Assistant Professor, (Grade-I), Department of Mechanical Engineering, National Institute of Technology, Kurukshetra, Haryana, India

Dilbagh Panchal, PhD, is an Assistant Professor (Grade-I) in the Department of Mechanical Engineering at the National Institute of Technology, Kurukshetra, Haryana, India. His research focuses on reliability and maintenance engineering, fuzzy decision-making, and operation management. He holds a Bachelor's (Hons.) in mechanical engineering from Kurukshetra University, Kurukshetra, India (2007), a Masters (Gold medalist) in Manufacturing Technology from Dr. B.R. Ambedkar National Institute of Technology Jalandhar, India (2011), and a PhD from the Indian Institute of Technology in Roorkee, India (2016). Currently, he is supervising two PhD scholars and has guided five MTech dissertations, with two in progress. He has published 56 research papers in SCI/Scopus-indexed journals and authored five book chapters with a reputed publisher. Additionally, he has edited a book on advanced multi-criteria decision-making for addressing complex sustainability issues.

Ravi Pratap Singh, PhD

Assistant Professor (Grade-I),Department of Mechanical Engineering, National Institute of Technology, Kurukshetra, Haryana, India

Ravi Pratap Singh, PhD, is presently working as an Assistant Professor (Grade-I) in the Department of Mechanical Engineering at the National Institute of Technology, Kurukshetra, Haryana, India. Prior to this, he served in the Department of Industrial & Production Engineering, NIT Jalandhar, Punjab as an Assistant Professor for about five years and the Department of Technical Education, Uttar Pradesh, India for about two years. He is a life member of the IIIE, Mumbai; The Institution of Engineers (India), Kolkata; Ultrasonic Society of India (USI), New Delhi; ISTE, New Delhi; Additive Manufacturing Society of India (AMSI), Bangalore; and Science and Engineering Institute (SCIEI), Los Angeles, USA. He has published more than 150 research articles in several SCI/Scopus indexed journals, including international/national level conferences. His research articles have received more than 5500 citations around the globe with the h-index of 40. He is also academically engaged with the editorship and reviewership with several SCI/Scopus indexed journals for the last 9–10 years. He is also an Editor of *Current Indian Science: Mechanical Engineering* (Bentham Science Publisher, UAE), and Guest Editor of *World Journal of Engineering* (Emerald Publications), *Sensors International* (KeAi Publisher), *Industrial Robot* (Emerald), *IJSMSS* (Inderscience), and *IJSSCA* (Inderscience). Dr. Singh is also working as a Principal-Investigator (PI) of SERB-GoI funded and Co-PI of ISRO-YUKTI funded projects which are under progress. He has also completed two research projects as a PI & Co-PI. He has also been granted two Indian patents in the domain of Development of Hybrid Machining Tools and Methods. Besides research excellence, he has been awarded as Best Teacher at NIT Jalandhar during 2021–2022, as Young Scientist in Mechanical Engineering in VIRA–2019 awards, a Young Faculty in Engineering in VIFA–2019 awards, and a Best Researcher Award in 4th International Scientist Awards on Engineering, Science, and Medicine during January 2020 in New Delhi. He has also been nominated as General Chair Reviewer, Technical Program Committee, National Advisory Committee, etc. during various national/international events held in China, Norway, Italy, Sweden. Dr. Singh is also a senior member of ISME, and

Senior Fellow of The Society of Innovative Educationalist & Scientific Research Professional (SFSIESRP), Chennai. In addition to this, he is also serving as an Ambassador of Bentham Open, Sharjah, UAE as well as a member of the Visor Academic Committee, Singapore. He has been awarded Young Engineer Award (Mechanical Engineering) in 37th National Convention on Mechanical Engineering and National Seminar conducted by the Telangana State Center, Khairtabad, Hyderabad under the aegis of Institution of Engineers (India), IEI held during December 2022. Dr. Singh has also been ranked among the Top 2% of Scientists in the World as per the survey conducted by Stanford University for the year 2021 and 2022.

Contents

Contributors

Israr Ahmad
Department of Mechanical Engineering, Zakir Husain College of Engineering and Technology, Aligarh Muslim University, Aligarh, Uttar Pradesh, India

Abdul Baquee Ahmed
Girijananda Chowdhury Institute of Pharmaceutical Science, Tezpur, Assam, India

Mohammed Zainul Arefeen
Department of Mechanical Engineering, Zakir Husain College of Engineering and Technology, Aligarh Muslim University, Aligarh, Uttar Pradesh, India

Pradyumn Kumar Arya
Department of Mechanical Engineering, Indian Institute of Technology, Indore, Madhya Pradesh, India

Ratnali Bania
Pratiksha Institute of Pharmaceutical Sciences, Guwahati, Assam, India

Neha Bhadauria
Department of Mechanical Engineering, KIET Group of Institutions, Ghaziabad, Uttar Pradesh, India

Suparna Bhattacharjee
Department of Applied Sciences, Gauhati University, Guwahati, Assam, India

Deepesh Bhavsar
Department of Mechanical Engineering, Ravindranath Tagore University, Raisen, Madhya Pradesh, India

Pallabi Boro
Department of Applied Sciences, Gauhati University, Guwahati, Assam, India

Suhas Chavhan
Department of Mechanical Engineering, National Institute of Technology, Hamirpur, Himachal Pradesh, India

Satish Chinchaniker
Professor, Department of Mechanical Engineering, Vishwakarma Institute of Information Technology, Pune, Maharashtra, India

Satyendra Deka
Pratiksha Institute of Pharmaceutical Sciences, Guwahati, Assam, India

Poonam S. Deshmukh
Department of Mechanical Engineering, Indian Institute of Technology, Indore, Madhya Pradesh, India

Mandeep Dhanda
Department of Production and Industrial Engineering, Punjab Engineering College, Chandigarh, India

Pulkit Dureja
Department of Production and Industrial Engineering, Punjab Engineering College, Chandigarh, India

Shelly Garg
Research Scholar, Department of Electronics and Communication, Punjabi University, Patiala, Punjab, India

Kelvii Wei Guo
Department of Mechanical and Biomedical Engineering, City University of Hong Kong, Kowloon Tong, Kowloon, Hong Kong

S. Rijuwana Haque
Department of Applied Sciences, Gauhati University, Guwahati, Assam, India

Neelesh Kumar Jain
Department of Mechanical Engineering, Indian Institute of Technology, Indore, Madhya Pradesh, India

Balkrishan Jindal
Assistant Professor, Computer Engineering Section, Punjabi University, Guru Kashi Campus, Talwandi Sabo, Punjab, India

Pratap Kalita
Assistant Professor, Pratiksha Institute of Pharmaceutical Sciences, Guwahati, Assam, India; PhD Research Scholar, Assam Science and Technology University, Guwahati, Assam, India

Swabnam Khataniar
Pratiksha Institute of Pharmaceutical Sciences, Guwahati, Assam, India

Anil Khurchania
Department of Mechanical Engineering, Ravindranath Tagore University, Raisen, Madhya Pradesh, India

Atul Kulkarni
Associate Professor, Department of Mechanical Engineering, Vishwakarma Institute of Information Technology, Pune, Maharashtra, India

Pankaj Kumar
Department of Mechanical Engineering, Indian Institute of Technology, Indore, Madhya Pradesh, India

Hussam Bin Mehare
Department of Mechanical Engineering, Zakir Husain College of Engineering and Technology, Aligarh Muslim University, Aligarh, Uttar Pradesh, India

Apurva Rane
Department of Mechanical Engineering, Vishwakarma Institute of Information Technology, Pune, Maharashtra, India

Anil Kumar Rao
Department of Mechanical Engineering, Ravindranath Tagore University, Raisen, Madhya Pradesh, India

Probin Kr. Roy
Regional Institute of Paramedical and Nursing Sciences, Zemabawk, Aizawl, Mizoram, India

Satyabrat Sarma
Faculty of Pharmaceutical Sciences, Assam Down Town University, Guwahati, Assam, India

Dan Sathiaraj
Department of Mechanical Engineering, Indian Institute of Technology, Indore, Madhya Pradesh, India

Ghanshyam Satnami
Department of Mechanical Engineering, Ravindranath Tagore University, Raisen, Madhya Pradesh, India

Afnaan Sayyed
Department of Mechanical Engineering, Vishwakarma Institute of Information Technology, Pune, Maharashtra, India

Saikat Sen
Faculty of Pharmaceutical Sciences, Assam Down Town University, Guwahati, Assam, India

Nirvedh Umesh Shirke
Department of Mechanical Engineering, Vishwakarma Institute of Information Technology, Pune, Maharashtra, India

Parnika Shrivastava
Department of Mechanical Engineering, National Institute of Technology, Jalandhar, Punjab, India; Department of Mechanical Engineering, National Institute of Technology, Hamirpur, Himachal Pradesh, India

Priyanshu Soni
Department of Mechanical Engineering, National Institute of Technology, Hamirpur, Himachal Pradesh, India

Faisal Talib
Department of Mechanical Engineering, Zakir Husain College of Engineering and Technology, Aligarh Muslim University, Aligarh, Uttar Pradesh, India

Shivank A. Tyagi
Research Scholar, Department of Mechanical Engineering, NIT, Warangal, Telangana, India

Prashant Vashishtha
Department of Mechanical Engineering, KIET Group of Institutions, Ghaziabad, Uttar Pradesh, India

Aditya Verma
Department of Production and Industrial Engineering, Punjab Engineering College, Chandigarh, India

Girish Verma
Department of Mechanical Engineering, Indian Institute of Technology, Indore, Madhya Pradesh, India

R. S. Walia
Department of Production and Industrial Engineering, Punjab Engineering College, Chandigarh, India

Abbreviations

3DP	three-dimensional-printing
ABS	acrylonitrile butadiene styrene
AFM	atomic force microscopy
Al	aluminum
AM	additive manufacturing
API	active pharmaceutical ingredients
ASTM	American Society for Testing and Materials
AWJ	abrasive water jet
BBB	blood-brain barrier
BESO	bidirectional evolutionary structural optimization
CAD	computer-aided design
CAD	computer-aided detection
CdCl2	cadmium chloride
CdS	cadmium sulfide
CF	carbon fiber
CIGS	copper–indium–gallium–selenide
CMCs	ceramic matrix composites
CMP	chemical mechanical polishing
CSL	direct ceramic stereolithography
CT	computed tomography
CVD	chemical vapor deposition
CW	continuous-wave
DDS	drug delivery system
DED	directed energy deposition
DLD	direct laser deposition
DMLS	direct metal laser sintering
DSC	differential scanning colorimetry
EBM	electron beam melting
EBSD	electron backscatter diffraction
EDS	energy dispersive X-ray spectroscopy
EEG	electroencephalogram
EFF	extrusion free-forming
ESO	evolutionary structural optimization

FDA	Food and Drug Administration
FDM	fused deposition modeling
FLC	formability limit curve
FSAM	friction stir additive manufacturing
FSISF	friction stir incremental sheet forming
FSP	friction stir processing
FSW	friction stir welding
FTC	fused tungsten carbide
FTIR	Fourier transform infrared spectroscopy
FWHM	full-width at half maximum
GA	genetic algorithm
GETS	global efficiency of tensile strength
GMO	glycerol monooleate
GTAW	gas tungsten arc welding
HEAs	high entropy alloys
HIP	hot isostatic pressing
HVOF	high-velocity oxy-fuel
IJP	inkjet printing
IMCs	intermetallic compounds
ISF	incremental sheet forming
ISO	International Standard Organization
LAM	laser additive manufacturing
LLC	lyotropic liquid crystal
LLCNPs	lipid-based liquid crystalline nanoparticles
LMD	laser melt deposition
MMCs	metal matrix composites
MMTO	multi-material topology optimization
MRI	magnetic resonance imaging
MRTO	multi-resolution topology optimization
Na2S	sodium sulfide
NF	natural fiber
Ni-P	nickel-phosphorus
OAS	optical absorption spectroscopy
OP	oxygen permeability
OTR	oxygen transmission rate
PAAM	plasma arc additive manufacturing
PL	photoluminescence
PMCs	polymer matrix composite

PPAW	plasma powder arc welding
PPT	pressure pain thresholds
QD	quantum dot
RBS	Rutherford backscattering spectrometry
RSM	surface response methods
SAXS	small angle X-ray scattering
SB	synchronous bonding
SCE	shape change effect
SEM	scanning electron microscopy
SFJ	spiral fluid jet
SHS	selective heat sintering
SiC	single-phase CoCrFeNi
SIMP	solid isotropic material with penalization
SIMS	secondary ion mass spectrometry
SKO	soft kill option
SLA	stereolithography
SLM	selective laser melting
SLS	selective laser sintering
SMC	shape memory ceramics
SME	shape memory effect
SPM	scanning probe microscopy
STL	standard tessellation language
STM	scanning tunneling microscopy
TDD	targeted drug delivery
TEM	transmission electron microscopy
TIG	tungsten inert gas
TPIF	two-point incremental forming
TWB	tailored welded blanks
UF	ultrafiltration
UFSW	underwater FSW
UTM	universal testing machine
VAM	vacuum arc melted
VED	volumetric energy density
XFEM	extended finite element method
XRD	X-ray diffraction
ZnS	zinc sulfide

Preface

The development and analysis of advanced engineering materials and their in-depth characterization in the broad domain of manufacturing are becoming an issue of paramount importance in the trendy competitive culture of industrial and production systems applications. The studies related to the fabrication of typical engineering materials, favorable processing methods, effective overall process control and optimization, etc., are further some of the spheres that reflect the importance of learning these modern manufacturing aspects at a deeper level.

The societal issues today are very complex in nature. Advancement in technology can play a vital role in dealing with these issues. Research on the advancements in the domain of materials and manufacturing is the key to solutions to current issues as well as to future innovation. Therefore, it is necessary to have a common platform where all researchers across the world discuss their innovative ideas and thoughts together to solve societal issues and challenges. Novel materials and manufacturing processes can be key to future engineering developments. The outcome of this kind of research may have great societal benefits. Therefore, researchers need to be encouraged to develop novel materials and manufacturing processes.

This book presents quality manuscripts with ideas related to materials and manufacturing processes and their relevant applications. This book provides deep insights to its readers about the current scenarios and future advancements in the domain of material and manufacturing processes with process characterization and case studies. Various areas like material characterization using modern technologies, process characterization, 3D printing, AM with high entropy alloys, natural fiber and synthetic-based composites, medical implants, 4D printing, micro-finishing, nanomaterials, turning, etc., are covered in detail.

The chapters have been made more targeted and comprehensive by including real-life case studies on additive manufacturing, composite materials, topography optimization, 4D printing and analysis, novel material, turning process, various joining processes and their applications, microstructure analysis, etc., which further makes the book more meaningful and attractive for young scholars and researchers from the sphere of advanced materials and its allied domains.

—Editors

Introduction

Advancement in technology is crucial for addressing complex societal issues. Research in the field of materials and manufacturing holds the key to finding solutions and driving future innovation. To achieve this, a global platform is needed for researchers to collaborate and exchange innovative ideas. Novel materials and manufacturing processes are essential for future engineering advancements, with the potential to greatly benefit society. It is important to encourage researchers to pursue the development of such novel approaches.

This book comprises a collection of manuscripts exploring materials and manufacturing processes, along with their applications. This book will further enlighten its readers about the current and latest material and manufacturing developments going on around the globe. The recent advancements in materials and manufacturing is covered in this book along with case studies. The ongoing research in the domains of material science, material characterization, mechanical engineering, civil engineering, and design engineering is summarized at a common platform in the form of extended manuscripts from different authors around the globe. The book will be highly valuable as it brings out ideas, knowledge, skills, and experiences shared by valuable authors, researchers, and scientists in this domain.

This book is a comprehensive resource that delves into the latest research interests and advancements in the field of advanced materials, their processing, and related domains. It offers deep insights into the current scenarios and future developments in material and manufacturing processes, along with process characterization and case studies. The book covers a wide range of topics, including material characterization using modern technologies, process characterization, 3D printing, AM with high entropy alloys, natural fiber and synthetic-based composites, medical implants, 4D printing, micro-finishing, nanomaterials, and turning, among others.

—Editors

CHAPTER 1

Additive Manufacturing of FCC-Based High Entropy Alloys

POONAM S. DESHMUKH and DAN SATHIARAJ

Department of Mechanical Engineering, Indian Institute of Technology, Indore, Madhya Pradesh, India

ABSTRACT

Among various newly emerging advanced engineering materials, high entropy alloys (HEAs) have attracted industries and academia in the past few years. Conventional alloys are usually based on one principal element, with small fractions of other elements added to enhance the properties. By contrast, HEAs have at least five or more principal elements, and the concentration of each element is between 5 and 35 wt.%. HEAs possess a combination of excellent mechanical properties, high-temperature strength, corrosion, and wear resistance. The stable solid solution phases in HEAs are achieved on account of relatively higher configurational entropy of mixing. The good microstructural stability makes it feasible to synthesize processes, analyze, and manipulate as per the application requirement. With the versatile manufacturing technique with low processing time, material, and shape design freedom, additive manufacturing (AM) possesses viability to fabricate HEAs. In recent years, FCC-based HEAs have become popular in academic research owing to their high toughness and ductility. However, the *in-situ* alloying with certain elements has been carried out to improve the mechanical strength of the FCC-based HEAs. This chapter reports previous studies on AM processing, post-processing,

Advances in Manufacturing and Processing of Materials: Characterization and Applications.
Narendra Kumar, PhD, Mohit Tyagi, PhD, Dilbagh Panchal, PhD & Ravi Pratap Singh, PhD (Eds.)

element composition design, and *in-situ* alloying of various FCC-based HEAs, along with their microstructural and mechanical properties.

1.1 INTRODUCTION

The conventional alloy system is based on one principal element, and other minor elements are added to enhance the property. Copper-based, titanium-based, iron-based, aluminum (Al)-based alloys, and nickel-based superalloys are popular alloys used in various applications (Henderson, Arrell, Heobel, Larsson, & Marchant, n.d.). However, the constraint of one principal element limits the number of useful derivable alloys. Contrary to the conventional alloy design approach, Cantor et al. & Yeh et al. in 2004, reported a novel alloy system with multiple principal elements known as high entropy alloys (HEAs) (MacDonald et al., 2017). The composition of the alloy can be equimolar or near equimolar. HEAs may contain at least five or more principal elements with 5 wt.% to 35 wt.% content of each element. However, composition definition is not a rigid criterion but a guideline (Yeh, 2013). Alloys with configurational entropy >1.5R are considered as HEA. The basis of the HEAs design is high mixing entropy (also called configurational entropy) that encourages more stable solid solutions of the elements. It makes the HEAs feasible for alloying, fabrication, analysis, and manipulation (Yeh, 2013; Yeh et al., 2004). HEAs possess an exceptional combination of high strength and ductility, strong wear, fatigue, and corrosion resistance, excellent strength at elevated temperature, and superior microstructural stability (Tsai & Yeh, 2014; Chen, Tong, & Liaw, 2018). It makes these HEAs a potential material for application in different sectors such as nuclear (Barron et al., 2020), hydrogen storage (Kao et al., 2010; Sleiman et al., 2021), precision resistors (Tsai, 2013), soft magnetic materials (Zhang, Zuo, Cheng, & Liaw, 2013), and aerospace (Ye, Wang, Lu, Liu, & Yang, 2016). Conventional casting using induction melting (Qin et al., 2020) or vacuum arc melting (Lu et al., 2018) and spark plasma sintering (Rameshbabu, Parameswaran, Vijayan, & Panneer, 2017; Zhang et al., 2019) are the techniques that have been deployed majorly to fabricate HEAs. However, the processing route is time-consuming with traditional casting processes as it requires several times remelting of the alloy and post-processing to ensure homogeneous and defect-free cast alloy. Also, higher cooling rates are preferred during the processing of some HEAs to avoid the formation of intermetallic phases (Ostovari et al., 2021). Thus, the casting proves to be time-consuming, monotonous, and infeasible to fabricate complex shaped components of some certain HEAs. In addition, mechanical alloying followed by consolidation has been widely used for fabricating FCC-based HEAs owing to its high strengthening effect. The solid-state fabrication processes such as mechanical alloying or gas atomization followed by consolidation and additive manufacturing (AM) techniques produce ultrafine grains and thus give superior strength compared to casting (Lyu, Lee, Wang, Fan, Yeh, & Liaw, 2019). Besides, AM is a versatile technique with less processing time, less material consumption and offers extraordinary freedom of material design and complex geometry design (Chen, Tong, & Liaw, 2018; Lyu, Lee, Wang, Fan, Yeh, & Liaw, 2019). Many researchers have been investigated the feasibility of the AM technique for fabricating the HEAs. AM is a layer-by-layer manufacturing process as opposed to a conventional subtractive manufacturing process. The 3D CAD model of the object is sliced into the layers and used as an input for the process (Wong & Hernandez, 2012; Frazier, 2014). Many researchers have employed AM to fabricate several conventional metals alloys such as Ti alloys (Dutta & Froes, 2017), Cu alloys (Yadav, Paul, Jinoop, Rai, & Bindra, 2020; Zeng et al., 2020), Ni-based super alloys (Jinoop, Paul, Mishra, & Bindra, 2019), Al alloys (Kotadia, Gibbons, Das, & Howes, 2021), and polyamides (Goodridge, Tuck, & Hague, 2012).

In recent years, some researchers have reported the AM processing of the HEAs. The AM of CoCrFeMnNi alloy (Dovgyy, Piglione, Hooper, & Pham, 2020; Li, Niu, Yuan, Cao, Chen, & Zhou, 2018; Li et al., 2019; Pickering, Muñoz-Moreno, Stone, & Jones, 2016; Piglione et al., 2018; Qiu, Yao, Feng, Li, & Chu, 2018; Tong et al., 2019; Zhu et al., 2018; Zhang et al., 2021) and effect of Al addition on microstructure and mechanical properties of the AM-built CoCrFeMnNi (Gao et al., 2020; Joseph, Stanford, Hodgson, & Fabijanic, 2017; Joseph et al., 2018; Kuwabara et al., 2018; Su, Luo, & Wang, 2020; Mohanty et al., 2019; Peng et al., 2021; Qiu, 2021; Sarswat, Sarkar, Murali, Huang, Tan, & Free, 2019; Shen, Kong, & Chen, 2021; Shiratori et al., 2016; Sun et al., 2021; Ocelík, Janssen, Smith, & De Hosson, 2016) has been widely investigated. Also, the phase stability (Zhang et al., 2021) and strengthening mechanism of HEAs processed through AM has been studied (Zhang et al., 2021; Shen et al., 2021; Zhang et al., 2021). This chapter mainly focuses on AM processing, post-processing, and the effect of compositional changes on the microstructure and mechanical properties of the FCC-based HEAs. AM provides an opportunity to strengthen these alloys using *in-situ* alloying with minor concentrations of the strengthening elements like-Al, Sic, TiC, WC (Zhang et al., 2021; Shen et al., 2021; Zhang et al., 2021; Luo, Zhao, Su, Liu, & Wang, 2020).

1.2 ADDITIVE MANUFACTURING (AM) OF COCRFEMNNI FCC-BASED HEA

1.2.1 SELECTIVE LASER MELTING (SLM) OF COCRFEMNNI HEA

1.2.1.1 MICROSTRUCTURE EVOLUTION

The laser powder bed fusion or selective laser melting (SLM) is one of the popular AM techniques to manufacture FCC-based HEAs. The studies show that the near-dense deposit of CoCrFeMnNi HEA with 98.79%–99.2% density has been successfully fabricated using SLM. The epitaxial grain growth along the build direction was observed. It was attributed to the solidification conditions. The grain growth was along the <001>, which is the preferred grain growth direction for FCC crystals during the solidification process (Li, Niu, Yuan, Cao, Chen, & Zhou, 2018; Piglione et al., 2018; Zhu et al., 2018; Chen, Li, Zhou, Yan, & Attallah, 2020). The microstructure of the deposit consisted of sub-micron cellular grains, columnar grains, and dislocations (Zhu et al., 2018). The rapid solidification of melt pool in the vicinity to substrate promoted fine grains. The direction of maximum heat flux is perpendicular to the fusion line (or the melt pool profile), causing the epitaxial growth from the grains of the substrate. The high intensity (200) peak in the XRD pattern also revealed <001> as a preferred grain orientation, confirming the single FCC phase as shown in Figure 1.1(a). The EBSD studies also showed <001> preferred grain orientation direction agreeing with the XRD phase analysis. The elemental mapping analysis confirmed homogenous elemental distribution of the alloying elements without any noticeable segregation, as observed from Figure 1.1(b) (Piglione et al., 2018; Zhu et al., 2018). Further, a significantly different microstructure was observed for the multilayer bulk deposit than the single-layer deposition. The epitaxial grain growth from the existing grains/cells of the substrate or previously deposited layer was observed. Several remelting events of the melt pool during the deposition of successive layers promote grain growth competition. The grains in better alignment with the maximum heat flux direction outgrew the misaligned grains resulting in the reduced number of grains (Piglione et al., 2018). Dovgyy et al. (2020) investigated the effect of process parameters like scan strategy, hatch spacing, point distance, and energy density on microstructure. The scan strategy was found to be the most dominant factor governing microstructure development. In a single-layer deposition, the growth of the cells was observed along <100> direction of maximum heat flux, which is inconsistent (Piglione et al., 2018; Zhu et al., 2018). The fish scale microstructure, along with the fusion lines, was observed in the multilayer bulk depositions. It was attributed to the local compositional variations and directional solidification along the boundaries of the melt pool. Each melt pool consisted of fine cells with a diameter ranging between 300 nm and 1,000 nm. TEM analysis revealed dislocation-free cell interiors and cell boundaries with dense dislocations. EDX analysis confirmed high oxidation content in the precipitates observed in TEM images. This suggested that the oxidation formed during SLM processing. The point distance and exposure time with values in the process window showed negligible influence on the microstructure variation. However, scan strategy significantly influences the microstructure. Figure 1.2 presents EBSD images depicting the effect of different scan strategies on grain morphologies and crystal orientations. The scan strategy is one of the factors controlling heat flux, hence possessing a profound influence on the final microstructure (Dovgyy, Piglione, Hooper, & Pham, 2020).

The densification of the HEA was observed with an increase in volumetric energy density (VED) and decreased with further increases in energy density beyond a particular value. The partial melting and balling effect with clearly visible scanning track marks were observed at very low VED. The increase in VED and certain low scan speed induced sufficient high temperature. Thus, proper melting was observed, leading to better flowability that formed better bonding in successive layers. However, further increases in the VED lead to increased porosity due to an unstable melt pool causing gas entrapment. The highest density of 98.2% was observed at optimum VED and scan speed. The visible scan marks with micro-cracks and pores on the top surface and multi-layer overlapped weld tracks along build direction were observed in the as-built sample. The micro-cracks and pores disappeared after hot isostatic pressing (HIP) and densified the deposit to 99.1%. The microstructure comprised ultrafine cellular or columnar grains exhibiting epitaxial grain growth along <100> (Li, Niu, Yuan, Cao, Chen, & Zhou, 2018). It was observed that the microstructure is a function of thermal gradient and solidification rate, and both are governed by the scan strategy (Dovgyy, Piglione, Hooper, & Pham, 2020). The as-built and HIP samples show texture in <001>, which is inconsistent (Dovgyy, Piglione, Hooper, & Pham, 2020; Piglione et al., 2018; Zhu et al., 2018). The EDS mapping revealed a homogeneous distribution of all the elements except segregation of Mn at grain boundaries which is inconsistent (Zhu et al., 2018). Moreover, the Mn segregation disappeared after HIP, and a more homogeneous elemental distribution was obtained. The phase analyses ensured that the FCC phase of the powder had been retained in the SLM-built depositions. However, the right-side shift of the (111) XRD peak was observed due to the grain refinement and decreased (111) interplanar spacing, which was attributed to the Mn volatilization (Li, Niu, Yuan, Cao, Chen, & Zhou, 2018).

The study on SLM deposition (Chen, Li, Zhou, Yan, & Attallah, 2020) of CoCrFeNi HEA with *in-situ* alloying with Mn showed homogeneous elemental distribution and a single FCC phase. The effect of VED on the density and defects such as micro-cracks and micro-pores presented the same trend as observed by Li, Niu, Yuan, Cao, Chen, & Zhou (2018); Dovgyy, Piglione, Hooper, & Pham (2020). The irregular-shaped Mn powder consumed energy through evaporation and absorption of laser energy. This aided fabrication of the highly-dense structures at high VED. The peak of α-Mn phase observed in powder XRD was disappeared in the SLM-built alloy giving a single FCC phase. The microstructure consisted of epitaxial grain growth along the build direction and near-equiaxed grains along the scan direction. The EBSD investigation showed texture in <001> direction, i.e., the preferred direction of grain growth in FCC crystals, which was along the build direction, and it is inconsistent (Zhu et al., 2018; Piglione et al., 2018; Li, Niu, Yuan, Cao, Chen, & Zhou, 2018; Dovgyy, Piglione, Hooper, & Pham, 2020).

1.2.1.2 MECHANICAL PROPERTY STUDIES

The studies on the mechanical property of the AM-built CoCrFeMnNi HEA are required to investigate the direct application ability of the as-built components. The experimental investigation by Zhu et al. (2018) showed that as-built CoCrFeMnNi HEA possessed a higher yield strength of 510

Mpa. The as-built HEA exhibited superior strain hardening ability that resulted in uniform elongation. The strength enhancement was observed through dislocation hardening. The cellular structures played an important role in trapping and storing the dislocations and thereby maintaining the dislocation hardening capability. Also, the microstructural heterogeneity may cause the generation of back stresses during deformation that assists the strain hardening (Zhu et al., 2018). The average hardness value of 212 HV_1 without any significant variation was observed along the deposition direction (Piglione et al., 2018). The samples fabricated with different scan strategies gave the average hardness values with deviations <10%. The hardness deviation was attributed to the variations in grain size. The superior mechanical properties were observed with ultimate tensile strength, yield strength, and strain rate of 530±20 Mpa, 783±7 Mpa, and 23±1%, respectively. The development of crystallographic textures (Figure 1.2) and variations in grain size accounted for the plastic anisotropy (Dovgyy, Piglione, Hooper, & Pham, 2020).

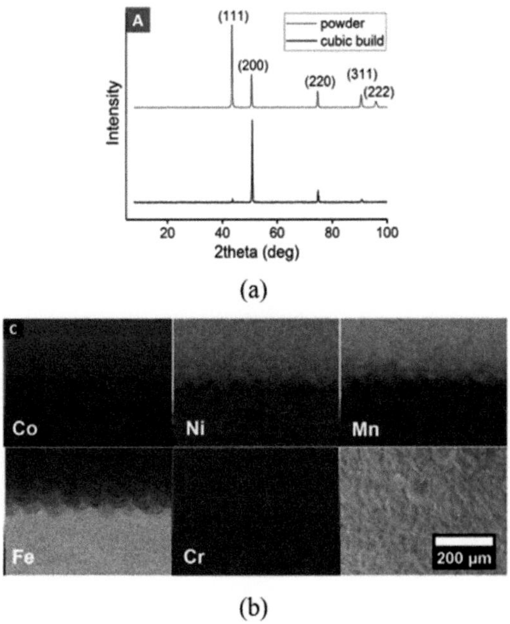

FIGURE 1.1 (a) XRD plots of CoCrFeMnNi powder and SLM-built bulk sample; (b) EDS maps at the interface of deposit and substrate.

Source: Reprinted with permission from Piglione et al., 2018. Copyright 1969, Elsevier. https://creativecommons.org/licenses/by/4.0/

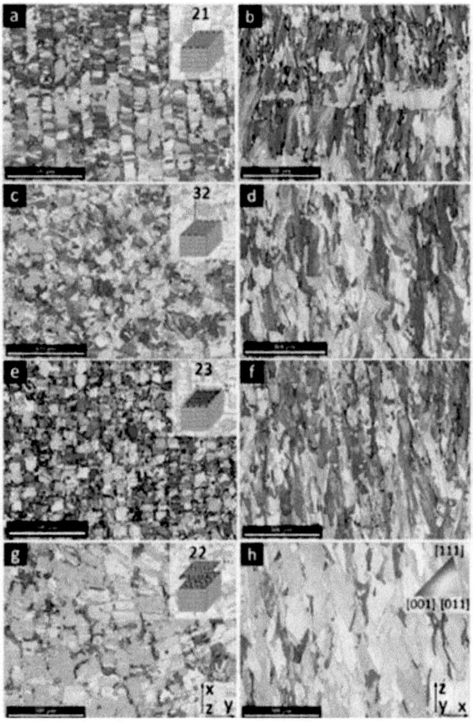

FIGURE 1.2 EBSD of the SLM-built CoCrFeMnNi HEA with different scan strategies. Top view in left column and side view in right column.

Source: Reprinted with permission from Dovgyy et al., 2020. Copyright Elsevier 2020.

The tensile strength increased with an increase in VED because the densification occurs with an increase in VED. However, beyond a particular value of the VED, the tensile strength remained constant. After HIP, enhancement in tensile strength and reduction in elongation from 601 MPa to 649 MPa and 35% to 18%, respectively, was noticed. The fracture analysis of the as-built and HIP-processed samples showed ductile failure with a more uniform distribution of void mode in the HIP-processed samples because of the high density and absence of micro-cracks and pores (Li, Niu, Yuan, Cao, Chen, & Zhou, 2018). The *in-situ* alloyed SLM-built HEA samples possessed 261±7 HV hardness, 681±14 MPa tensile strength, and 12.5±0.5% elongation. The strength and hardness are superior to the SLM-built samples using pre-alloyed HEAs powder (Chen, Li, Zhou, Yan, & Attallah, 2020).

1.2.2 LASER MELT DEPOSITION (LMD) OF COCRFEMNNI

1.2.2.1 MICROSTRUCTURE EVOLUTION

The cryogenic deformation studies of laser additive manufacturing (LAM) built HEA showed that the single FCC phase was retained even after deformation. The as-built structure was highly dense without any significant cracks and voids. The microstructure showed a typical coarse dendritic structure with grain growth normal to the laser scan direction. The FCC phase was highly stable since there were no phase transformations in any deformation stage. The EDS analysis revealed a homogeneous distribution of alloying elements except for minor precipitation of Mn and Cr (Qiu, Yao, Feng, Li, & Chu, 2018).

Tong et al. (2019) performed an experimental study to investigate the influence of heat treatment on the microstructure, phase, residual stress, and mechanical properties. The high-density deposit was observed at certain high laser power. The defects such as micro-cracks, pores, and voids were observed at low and very high power. A rational reason behind this is inconsistent (Li, Niu, Yuan, Cao, Chen, & Zhou, 2018). The high-temperature gradient formed during layer-by-layer deposition provided the driving force for epitaxial growth of the columnar grains. The different directions of the grain growth (Figure 1.3) were attributed to the Marangoni effect. However, equiaxed grains were observed outside the melt pool owing to the low thermal gradient of the previously deposited layer. The XRD studies confirmed the single FCC phase in the as-built LAM sample. Moreover, the higher angle side shift of the (111) and (200) was observed compared to the powder XRD peaks. It was attributed to the decrease in lattice parameter and inter-planar spacing due to Mn volatilization observed during AM processing in agreement (Li, Niu, Yuan, Cao, Chen, & Zhou, 2018). Moreover, the FCC phase was dominant in heat-treated samples, with an additional peak of the bcc phase in the samples heat-treated at 700C. The bcc peak weakens after 900°C temperature and completely dissolves at 1,100°C. The EDS mapping analysis revealed minor segregation of Mn along grain boundaries which is inconsistent with the findings by Li, Niu, Yuan, Cao, Chen, & Zhou (2018); Zhu et al. (2018). Further, the Mn segregation was eliminated, and a homogeneous recrystallized grain structure was obtained after the samples were treated at 1,100C. The EBSD analysis confirmed the crystal orientation of as-built

HEA in <001> and <101> directions and that of heat-treated samples in <111> and <101> directions (Tong et al., 2019).

An equiaxed grain structure without any micro-cracks and pores was obtained with the addition of 5 wt.% WC (Li et al., 2019; Zhang et al., 2021) and TiC (Zhang et al., 2021). As depicted in Figure 1.4(b), the laser melt deposition (LMD) built HEA comprised of long columnar grains, while the HEA+TiC (Figure 1.4(c)) and HEA+WC (Figure 1.4(d)) showed the equiaxed crystalline structure. The grain size reduced significantly with the addition of these compounds. The phase analysis studies (Figure 1.5) confirmed the single FCC phase of all the samples with TiC phase in HEA+TiC samples and $M_{23}C_6$ phase in HEA+WC samples (Li et al., 2019; Zhang et al., 2021). The TEM analysis revealed that the FCC phase accompanied several reinforced phases of the size 200 nm-1 μm in the HEA+TiC samples, as depicted in Figure 1.6(a)–(c). In contrast, the HEA+WC samples consisted of FCC phase accompanied with the $M_{23}C_6$ cubical reinforced phases of the size 50 nm-100 nm as observed from Figure 1.6(d)–(f). The TiC phases were retained owing to their high chemical stability, while the WC phases dissolved completely and formed $M_{23}C_6$ reinforcement (Zhang et al., 2021).

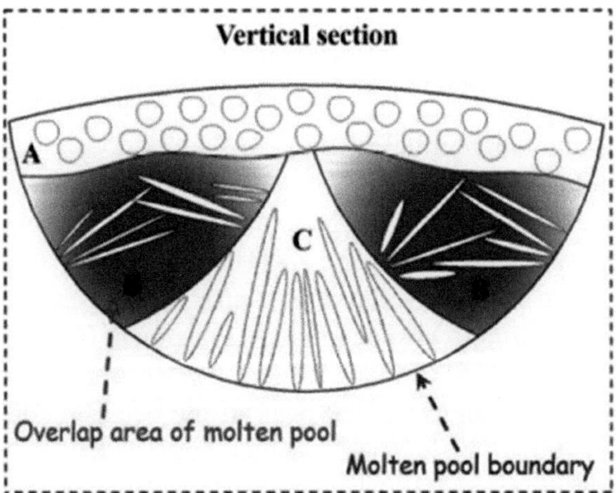

[Areas (A) Equiaxed grain; (B) Columnar grains with multiple directions; and (C): Columnar grains with single direction.

FIGURE 1.3 Schematic illustration of the microstructural features in the vertical section of single molten pool.

Source: Reprinted with permission from Tong et al., 2019. Copyright 2019, Elsevier

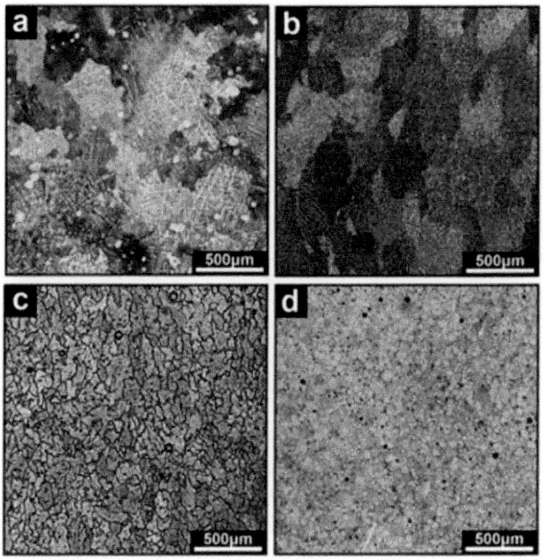

FIGURE 1.4 Microstructure of the (a) as-cast CoCrFeMnNi; (b) LMD-build CoCrFeMnNi; (c) CoCrFeMnNi + TiC; and (d) CoCrFeMnNi + WC.

Source: Reprinted with permission from Zhang et al., 2021. Copyright 2021 Elsevier.

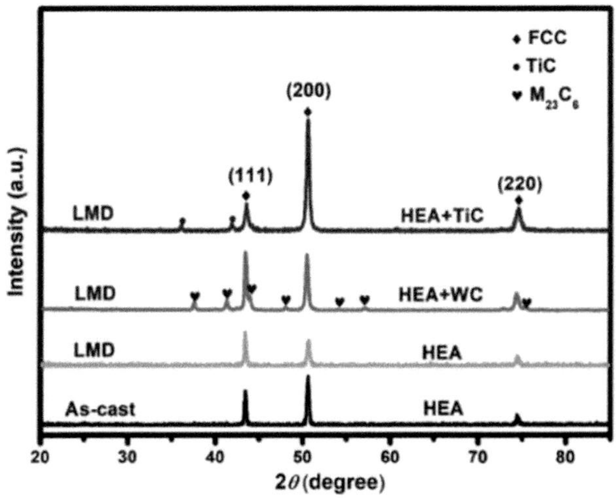

FIGURE 1.5 XRD spectrums of the as-cast and LMD-built CoCrFeMnNi HEA and HEA with TiC and WC.

Source: Reprinted with permission from Zhang et al., 2021. Copyright 2021. Elsevier.

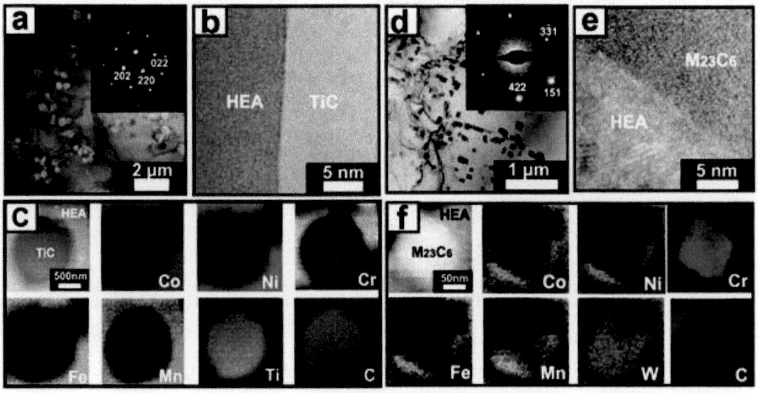

FIGURE 1.6 (a) The TEM image of the CoCrFeMnNi+TiC sample; (b) the interface between HEA matrix and TiC phase; (c) elemental mapping near the TiC phase; (d) the TEM image of the CoCrFeMnNi + WC sample; (e) the interface between HEA matrix and $M_{23}C_6$ phase; and (f) elemental mapping near the $M_{23}C_6$ phase.

Source: Reprinted with permission from Zhang et al., 2021. Copyright 2021. Elsevier.

1.2.2.2 MECHANICAL PROPERTY STUDIES

The high-density dislocations and dislocation motion yield the superior mechanical strength of the AM-built structures. In addition, twinning during the cryogenic deformation hindered dislocation motion and increased work hardening rate, resulting in the bulk HEA's further strengthening. As the temperature during tensile testing decreased to 77 K from 298 K, the yield strength, the ultimate tensile strength, and the ductility, respectively, increased by 60% to 650 Mpa, 65% to 891 Mpa, and 38% to 0.36 (Qiu, Yao, Feng, Li, & Chu, 2018).

The average micro-hardness of the as-built HEA was ~ 200 HV and reduced gradually with an increase in the heat treatment temperature. The highest value of yield strength (346 MPa) and ultimate strength (566 MPa) was observed at lower laser power because the grain size increases with increased laser power. However, a significant increase in elongation was observed due to increased grain size. Nevertheless, the retained strength of the HEA after heat treatment could be attributed to the residual stress relaxation after heat treatment (Tong et al., 2019). After alloying with five wt.% WC, the yield strength and ultimate tensile strength at room temperature increased from 300 MPa to 502 MPa and 550 MPa to 776 MPa, respectively, and elasticity decreased from 50% to 37% (Li, Niu, Yuan,

Cao, Chen, & Zhou, 2018; Zhang et al., 2021). Moreover, tensile strength and an elongation of 405 Mpa and 45% were observed at a temperature of 873 K. The fracture surfaces showed ductile fracture with abundant dimples (Li et al., 2019). While the tensile strength improved to 610 Mpa and elongation reduced to 47% with additions of 2.5 wt.% of TiC (Zhang et al., 2021). The enhancement in mechanical properties was attributed to the combined effect of the grain size refinement and the homogeneously distributed $M_{23}C_6$ reinforcement precipitates (Li, Niu, Yuan, Cao, Chen, & Zhou, 2018; Zhang et al., 2021).

1.2.3 PLASMA ARC ADDITIVE MANUFACTURING (AM) OF COCRFENIX

1.2.3.1 MICROSTRUCTURE EVOLUTION

Shen et al. (2021) systematically investigated the effect of SiC addition in various proportions on the microstructure and phases of the CoCrFeNi HEA built using plasma arc additive manufacturing (PAAM). It has been observed that the microstructure of the single-phase CoCrFeNi $(SiC)_x$ HEA changes with the concentration of SiC. The microstructure with SiC_0 was composed of single-phase FCC and transformed to FCC+Cr_7C_3 dual phase with increasing content of SiC. The formation of the Cr_7C_3 phase was attributed to the largest negative enthalpy of the Cr and C among all other alloying elements. As shown in Figure 1.7, $SiC_{0.1}$, $SiC_{0.3}$, and $Sic_{0.5}$ are composed of precipitation of Cr_7C_3 along grain boundaries, uniformly distributed Cr_7C_3 in the FCC, and a considerable amount of columnar Cr_7C_3, respectively. The XRD studies also confirmed the increase in the Cr_7C_3 phase with an increase in the SiC content. The higher angle side peak shift was observed due to the decrease in lattice parameters on account of the dissolution of Si and C in the FCC phase, and the same was confirmed through elemental mapping analysis as shown in Figure 1.7 (Shen et al., 2021).

The experimental study on microstructure evolution (Figure 1.8) PAAM built of $CoCrFeNiX_{0.4}$ (X_Al, Nb, Ta) has been performed by Zhang et al. (2021). The microstructure studies showed that the Al-Ni rich bcc phase without any micro-voids formed with $Al_{0.4}$ (Figure 1.8(b)–(f)). In addition to the FCC+bcc phase, XRD and EDS analyses confirmed the formation of a bright σ phase. Very high heat and rapid cooling rate are the driving factors in the formation of the σ phase. The microstructure of $CoCrFeNiTa_{0.4}$ HEA, as seen from Figure 1.8(c)–(g), is composed of FCC matrix with minor laves phase structure and eutectic phase with needle-like and dot-like structure. However, Figure 1.8(d)–(h) shows the microstructure of $CoCrFeNiNb_{0.4}$ HEA composed of hypo-eutectic dendrites and layered eutectic matrix consisting of FCC and uniformly distributed laves phase (Zhang et al., 2021).

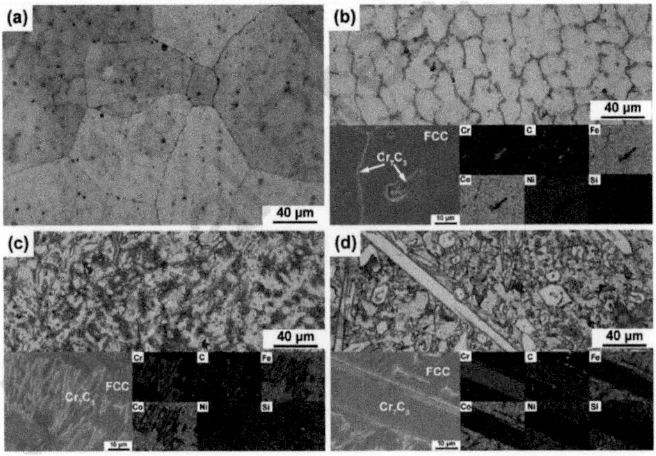

FIGURE 1.7 Optical microscope images and elemental maps of PAAM-built (a) $CoCrFeNi(SiC)_0$; (b) $CoCrFeNi(SiC)_{0.1}$; (c) $CoCrFeNi(SiC)_{0.3}$; and (d) $CoCrFeNi(SiC)_{0.5}$.

Source: Reprinted with permission from Zhang et al., 2021. Copyright 2021. Elsevier.

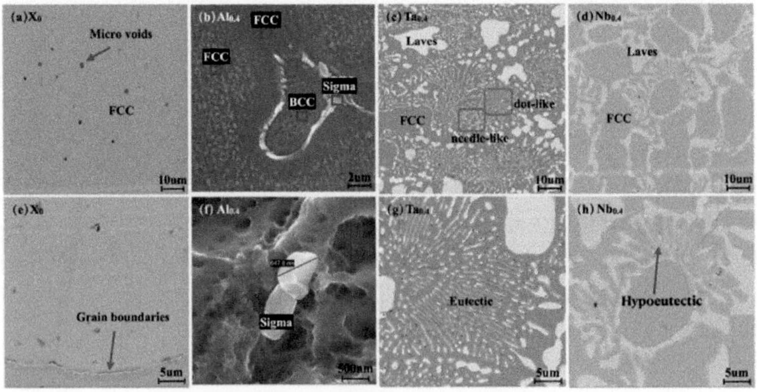

FIGURE 1.8 SEM images – (a) & (e) CoCrFeNi; (b) & (f) $CoCrFeNiAl_{0.4}$; (c) & (g) $CoCrFeNiTa_{0.4}$; and (d) & (h) $CoCrFeNiNb_{0.4}$.

Source: Reprinted with permission from Zhang et al., 2021. Copyright 2021. Elsevier.

1.2.3.2 MECHANICAL PROPERTY STUDIES

The investigation on the effect of SiC addition in CoCrFeNi$(SiC)_x$ alloy came up with a significant enhancement in the mechanical properties of the HEA. This strengthening was ascribed to the solid solution strengthening accompanied with the second phase strengthening. The microhardness and yield strength, respectively increased from ~139 HV to ~310 HV and ~141 Mpa to ~712 Mpa with an increase in SiC content from SiC_0 to $SiC_{0.5}$. However, the fracture strength and ductility were reduced from 2,499 MPa and 48.5% at $SiC_{0.3}$ to 2,155 MPa and 36.5% at $SiC_{0.5}$ (Shen et al., 2021).

The study had shown that with the addition of $Al_{0.4}$ in CoCrFeNi, the hardness increases slightly (4.578 GPa to 4.674 GPa) and retained ductility. On the other hand, the addition of $Ta_{0.4}$ and $Nb_{0.4}$ considerably increases hardness to 5.672 GPa and 5.617 GPa, respectively, and reduces ductility. Reduction in ductility was ascribed to the laves phases, and an increase in strength was attributed to the solid solution strengthening and precipitation strengthening (Zhang et al., 2021).

3. Additive Manufacturing (AM) of Al_xCoCrFeNi FCC-based HEA

Microstructure Evolution

Joseph et al. (2015) conducted the comparative study of Al_xCoCrFeNi FCC-based HEA fabricated using DLF and arc melting with x=0.3, 0.6, and 0.85. Both the alloys possessed single FCC, FCC+bcc, and single bcc phases with $Al_{0.3}$, $Al_{0.6}$, and $Al_{0.85}$, respectively. The Al addition stabilized the bcc solid solutions in the alloy, tending to form FCC+bcc dual-phase at $Al_{0.6}$ and further Al addition resulting in a single-bcc phase at $Al_{0.85}$. The DLF sample with $Al_{0.3}$ exhibited columnar grains elongated along the build direction and the equiaxed grains along the scan direction (Figure 1.9(a)–(c)). Likewise, the arc-melted sample consisted of columnar grains along the solidification direction, and equiaxed grains traversed to solidification. The FCC phase in either sample displayed a firm <001> texture. The EDS studies revealed Al segregation along very few grain boundaries in DLF-built $Al_{0.3}$ HEA. Figure 1.9(d)–(f) presents the $Al_{0.6}$ DLF-built sample was composed of Widmanstatten grain structure with random grain orientation, and the arc melted sample was composed of coarse dendrites due to lower cooling rates, with a firm <001> texture. The inter-plates of DLF-built sample and interdendritic of arc melted sample with $Al_{0.6}$ comprised Ni and Al-rich ordered B_2 phase. The spinodal decomposed bcc phase was also observed in some regions of the bcc phase at $Al_{0.6}$ in both samples. Further, as seen from Figure 1.9(g)–(i), the $Al_{0.85}$ DLF-built sample exhibited columnar grains with a typical <001> texture, and the arc melted sample exhibited transition from columnar grains to randomly oriented equiaxed grains towards the center (Joseph et al., 2015). Further, detailed studies on the microstructure of DLF-built AlxCoCrFeNi have been carried out to understand the mechanical behavior of HEA at different Al content. Figure 1.9(a)–(c) presents the microstructure of $Al_{0.3}$ HEA, and it shows that there was no significant elemental segregation. The $Al_{0.6}$ HEA, as shown in Figure 1.9(d)–(f), is composed of a B_2 phase with 30% volume fraction, disordered BCC phase with 7% volume fraction, and FCC-widmanstatten plates with 63% volume fraction. In the $Al_{0.85}$ HEA, 48% and 50% volume fraction of the B_2 phase and a disordered bcc phase were observed, as depicted in Figure 1.9(g)–(i). The B_2 phase was enriched in Ni and Al, while the disordered bcc phase was a Cr and Fe rich phase (Joseph, Stanford, Hodgson, & Fabijanic, 2017). The HIP resulted in the homogenization and grain coarsening of $Al_{0.3}$ HEA. However, there were no other significant changes in microstructure and no phase transformations, as

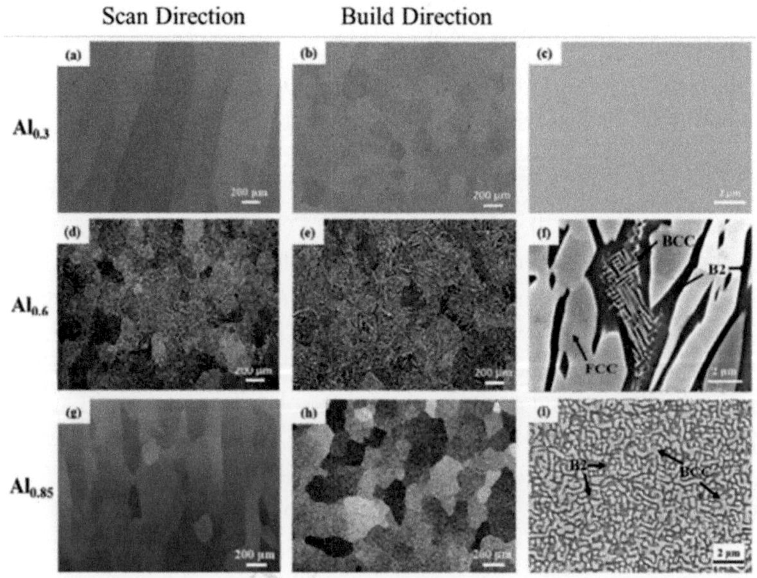

FIGURE 1.9 Microstructure of DLF-built HEA (a)–(c) $Al_{0.3}$; (d)–(f) $Al_{0.6}$; and (g)–(i) $Al_{0.85}$ along scan direction and deposition direction along with micrographs at high magnification.

Source: Reprinted with permission from Joseph et al., 2017. Copyright Elsevier 2017

can be seen from Figure 1.10(a) and (d). The porosity was reduced after HIP, and densification occurred in all the samples irrespective of Al content. The coarsening of the widmanstatten plates and development of second phase precipitates in the FCC phase of $Al_{0.6}$ occurred after HIP, as shown in Figure 1.10(b) and (e). As confirmed in XRD studies, the HIP processed $Al_{0.85}$ alloy composed of $B2^{+}$ bcc phase + σ phase. The σ phase formation was ascribed to the slower cooling rates with the HIP process. Also, the coarsening of the particles enriched in Fe and Cr was observed (Figure 1.10(c) and (f)) in the B_2 matrix (Joseph et al., 2018).

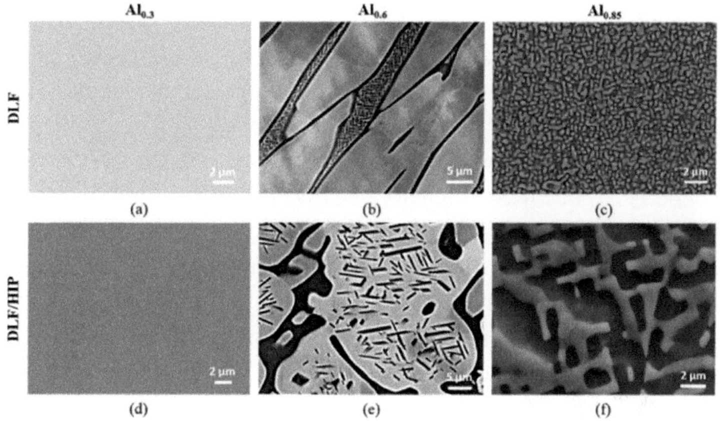

FIGURE 1.10 The SEM images of (a)–(c) DLF-built HEA; and (d)–(f) DLF/HIP processed HEA.

Source: Reprinted with permission from Joseph et al., 2018. Copyright Elsevier 2018

The direct laser deposition (DLD) built $Al_{0.3}CoCrFeMn$ possessed the same microstructure, grain orientation, and phases observed by Joseph et al. (2015). The homogeneous distribution of the alloying elements in $Al_{0.3}$ alloy was confirmed in the EDS analysis. The microstructure of DLD-built $Al_{0.7}$ alloy exhibited dendritic coarser grains with growth direction along build direction. The dendritic structure is composed of FCC-phase in the region of dendrites and bcc-phase in the region of inter-dendritic. The minor elemental segregation of Al and Cr in the FCC matrix was confirmed in the EDS analysis. The EBSD studies revealed the columnar grains oriented equally along <001>, <111>, and <101> direction. The higher temperature oxidation resistance studies showed that the $Al_{0.7}$ possessed superior oxidation resistance at 1,100°C elevated temperature in air for 200 hrs. As seen from Figure 1.11(a)–(c), the oxidized $Al_{0.3}$ HEA exhibited an external layer of Cr_2O_3 accompanied by an internal subscale of discontinuous Al_2O_3. The increase in thickness of Cr_2O_3 and coarsening of Al_2O_3 occurred with the duration. However, Figure 1.12(a)–(c) shows that the oxidized $Al_{0.7}$ HEA consisted of a Cr_2O_3 thin discontinuous layer accompanied by a few Al_2O_3 precipitates. Further oxidation for 200 hrs. leads to the coarsening of the Al_2O_3 subscale and formation of Al_2O_3 thick layer, leading to better oxidation resistance than $Al_{0.3}$ HEA. The EDS analysis revealed that the top layer was enriched in Cr and O, while the subscale was enriched in Al and O in both HEAs. The high Al content in $Al_{0.7}$ HEA (Mohanty et al., 2019). The experimental investigation on the effect of the addition of Al in CoCrFeNi was performed by Gao et al. (2020). The XRD phase analysis showed that with an increase in Al content from 0 wt.% to 8 wt.%, the single-phase FCC was observed up to 2 wt.% of Al content. At 5 wt.% of Al content, a small (110) bcc peak was observed, indicating the phase change to FCC+bcc dual-phase. The (110) bcc peak intensity became stronger with the increase in Al content to 8 wt.%. The lattice distortion energy increased with the Al addition and relaxed through the FCC to bcc crystal structure transition. Thus, precipitation of the bcc phase in the FCC matrix was observed at higher Al content. The phase analysis results are inconsistent with the previous studies by Joseph, Stanford, Hodgson, & Fabijanic (2017); Joseph et al. (2015).

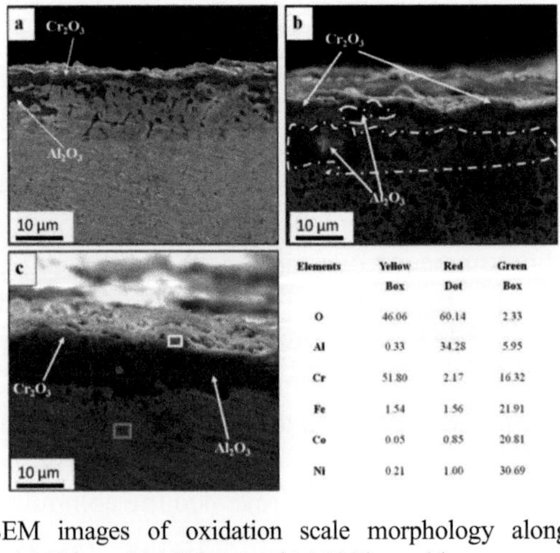

Elements	Yellow Box	Red Dot	Green Box
O	46.06	60.14	2.33
Al	0.33	34.28	5.95
Cr	51.80	2.17	16.32
Fe	1.54	1.56	21.91
Co	0.05	0.85	20.81
Ni	0.21	1.00	30.69

FIGURE 1.11 SEM images of oxidation scale morphology along cross-section of $Al_{0.3}$CoCrFeNi after (a) 20 hrs.; (b) 100 hrs.; and (c) 200 hrs. with corresponding EDS mapping.

Source: Reprinted with permission from Mohanty et al., 2019. Copyright. 2019. Elsevier.

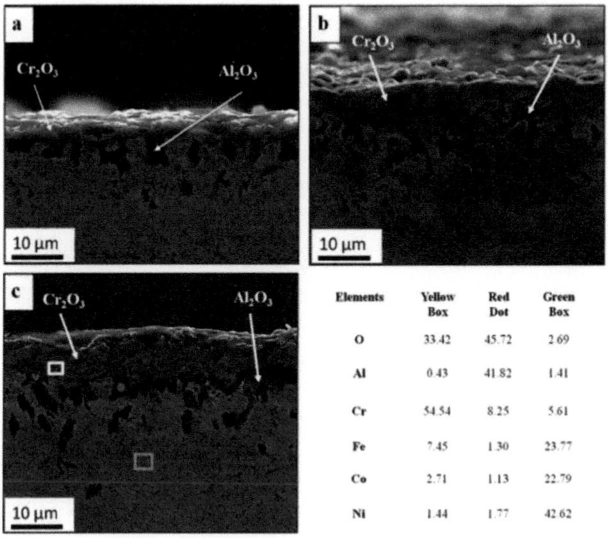

Elements	Yellow Box	Red Dot	Green Box
O	33.42	45.72	2.69
Al	0.43	41.82	1.41
Cr	54.54	8.25	5.61
Fe	7.45	1.30	23.77
Co	2.71	1.13	22.79
Ni	1.44	1.77	42.62

FIGURE 1.12 SEM images of oxidation scale morphology along cross-section of $Al_{0.7}$CoCrFeNi after (a) 20 hrs.; (b) 100 hrs.; and (c) 200 hrs. with corresponding EDS mapping.

Source: Reprinted with permission from Mohanty et al., 2019. Copyright. 2019. Elsevier.

Shen et al. (2021) deployed combined cable wire AM to fabricate the AlCoCrFeNi structure with the wire arrangement as shown in Figure 1.13. The cold metal transfer technique with argon gas shielding was used. The bulk HEA was deposited without any macroscopic defects. The deposit was composed of 20.13, 17.08, 3.18, 27.18, and 32.42% (approx.) of Al, Co, Cr, Fe, and Ni, respectively. The microstructure of the as-deposited HEA comprises equiaxed and columnar grains with 32.67% of FCC phase and 67.33% of bcc phase. In agreement with the results obtained by Joseph, Stanford, Hodgson, & Fabijanic (2017) and Joseph et al. (2018), the bcc and FCC phase was observed to be enriched in Al-Ni and Fe-Cr, respectively. The fine grain structure was observed with an increase in torch travel speed, resulting in faster cooling rates. The grain orientation of FCC grains was along the <001> direction, which agrees with the results obtained by Zhu et al. (Li, Niu, Yuan, Cao, Chen, & Zhou, 2018; Piglione et al., 2018; Zhu et al., 2018; Chen, Li, Zhou, Yan, & Attallah, 2020). The XRD analysis revealed the intense (110) and (200) diffraction peaks. However, the (110) peak became intense, and the (200) peak weakened with an increase in the torch travel speed. This indicated that the faster cooling rates are favorable for the growth of bcc grains in (110) orientation (Shen, Kong, & Chen, 2021). The studies on SLM deposition of $Al_{0.3}$CoCrFeNi HEA with 98.1% density followed by annealing heat treatment have been carried out recently by Peng et al. (2021). The microstructure of the as-built sample is composed of columnar grains grown along the build direction. The interior of the grains contained a subgrain cellular structure. The SEM images of the as-built HEA along the scan direction and deposition direction revealed a mesoscale crack, as seen from Figure 1.14(a)–(b). The region near to crack was enriched in Co, Cr, and Fe, and the region away from the crack showed higher content of Al and Ni. The annealed sample exhibited FCC phase and few B2 and σ phase precipitates, as confirmed by XRD and EBSD analyses. The segregation of B2 precipitates was observed near the cracks and grain boundaries, as shown in Figure 1.14. The <110> texture was observed along the build direction in the as-built HEA. The <110> texture was attributed to the 0° laser scanning direction during the SLM that led to grain nucleation at the edges of the melt pool. It suggested that the texture variation is strongly influenced by the remelting process and angle of the laser scanning (Peng et al., 2021). A detailed investigation of Al addition during AM of CoCrFeNi alloy to reduce the hot tearing tendency has been reported. As observed from Figure 1.15, the crack density increased in the $Al_{0.1}$ alloy, then decreased drastically in the $Al_{0.5}$ alloy, and again increased in the $Al_{1.0}$ alloy (Sun et al., 2021). The low Al content alloy exhibited a single FCC-phase crystal structure. Further increase in the Al content, alloy showed FCC+bcc+B2 phase structure. After a certain threshold Al content, the alloy possessed a single bcc phase structure. The $Al_{0.5}$ alloy showed <001> crystallographic structure along the build direction. The reported phase formation and texture are inconsistent with the previous studies (Gao et al., 2020; Joseph, Stanford, Hodgson, & Fabijanic, 2017). In the $Al_{0.5}$ alloy, the Al from the FCC phase segregated into the interdendritic region during solidification followed by spinodal decomposition caused to form ordered NiAl rich B_2 phase and Cr rich disordered bcc phase. It led to the occurrence of high tensile residual strains owing to thermal contraction. The B_2/BCC phases relived residual thermal stresses under the combined effect of molar volume expansion and plastic deformation. A higher percentage of B_2/bcc phases may transform the significant residual thermal stresses into small compressive stresses (Figure 1.16). Thus, the $Al_{0.5}$ alloy exhibited lower residual stresses and thus the least hot cracking tendency. However, the B_2/bcc phases are very brittle. The presence of a continuous film of these phases may encourage crack initiation and propagation. Thus, the Al concentration slightly lower than 0.5 might potentially eliminate all the hot cracks (Sun et al., 2021).

FIGURE 1.13 3D image of wires arrangement for CCW-AM.

Source: Reprinted with permission from Shen et al., 2021. Copyright. 2021. Elsevier.

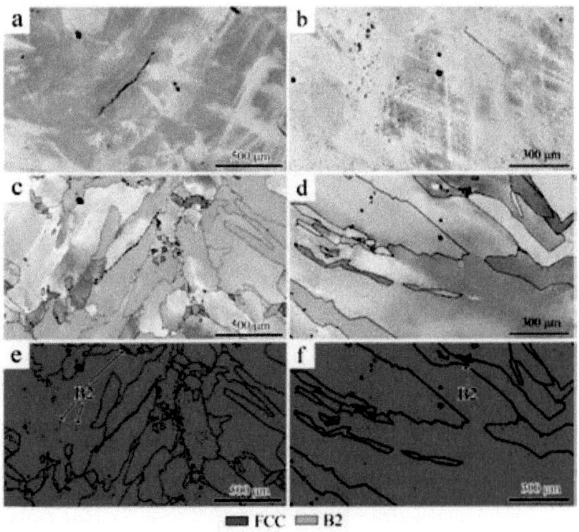

FIGURE 1.14 SEM image of as-built $Al_{0.3}CoCrFeNi$ (a) scan direction; (b) build direction; and IPF orientation distribution maps along; (c) scan direction; (d) build direction; corresponding phase distribution of the FCC and B2 phases in annealed $Al_{0.3}CoCrFeNi$ HEA along; (e) scan direction; and (f) build direction.

Source: Reprinted with permission from Shen et al., 2021. Copyright. 2021. Elsevier.

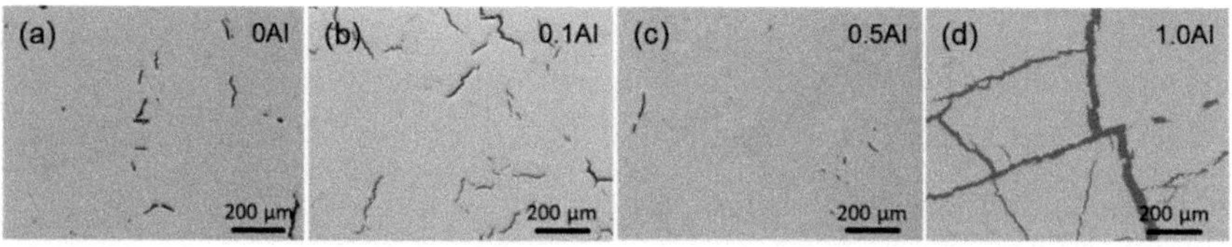

FIGURE 1.15 Representative optical images of SLM-built (a) CoCrFeNi; (b) $Al_{0.1}$CoCrFeNi; (c) $Al_{0.5}$CoCrFeNi; and (d) $Al_{1.0}$CoCrFeNi (Sun et al., 2021).

Source: Reprinted with permission from Sun et al., 2021. Copyright. 2021. Elsevier.

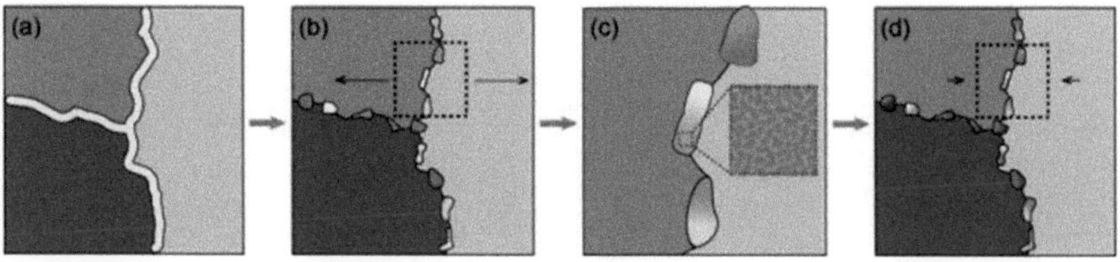

FIGURE 1.16 Schematics of residual stress minimization for $Al_{0.5}$CoCrFeNi sample.

Source: Reprinted with permission from Sun et al., 2021. Copyright. 2021. Elsevier.

1.2.3.3 MECHANICAL PROPERTY STUDIES

The studies have shown that the AlxCoCrFeNi HEA's strength increases with Al content at the expense of ductility. The excellent strength was attributed to the high-density spinodal precipitates resulting in the precipitation hardening (Joseph et al., 2015). The HIP processed $Al_{0.3}$ HEA exhibited a minor reduction in the yield strength owing to microstructural coarsening. However, the HIP processed $Al_{0.6}$ HEA showed brittle fracture on account of the development of the coarse B_2 phase precipitates. Moreover, the microstructure was coarsening, and σ-phase formation led to a significant expense of ductility in $Al_{0.85}$ alloy (Joseph et al., 2018).

The increase in Al content led to an increase in the hardness. The $Al_{0.3}$ HEA possessed the average micro-hardness of 170 HV due to the lower resistance of the FCC crystal structure to plastic deformation. The $Al_{0.7}$ showed the average micro-hardness of 380 HV because of the presence of the bcc phase that imparts hardness because of the lower dislocation slip systems it possesses (Mohanty et al., 2019). The tensile strength of

CoCrFeMn increases from 370 MPa to 978 MPa with the addition of 8 wt.% of Al. However, tensile elongation rapidly reduced from 49.5% to 7.5%. The HEA with Al concentration less than 8 wt.% showed ductile fracture with several dimples and few cavities observed on the fracture surface. Contrary to this, the fracture surfaces of the HEA with 8 wt.% of Al showed a quasi-cleavage fracture. The HEA becomes brittle beyond this concentration of Al, as observed by Joseph, Stanford, Hodgson, & Fabijanic (2017), thus 5 wt.% of Al addition showed a better combination of mechanical strength and ductility (Gao et al., 2020). In the case of CCW-AAM-built HEA, there was a slight increase in the hardness, strength, and elongation with the increase in torch travel speed was observed. The as-deposited HEA sample exhibited yield strength, ultimate strength, and elongation of 816 Mpa, 2,835 Mpa, and 41.8%. The CCW-WAAM process can help overcome the limitations of powder-based processes and can feasibly be deployed to fabricate HEA structures (Shen, Kong, & Chen, 2021).

The precipitation of B2 and σ-phase in annealed $Al_{0.3}CoCrFeNi$ sample caused an increase in hardness and wear resistance. However, annealing caused a decrease in yield strength on account of dislocation recovery induced during annealing. The micro-hardness and elongation of as-built HEA increased from 208.8 MPa and 0.60% to 221 MPa and 2.96% after annealing. In contrast, the annealing reduced yield strength and ultimate strength from 476.9 MPa and 501.8 MPa to 373.5 MPa and 473.7 MPa. The fracture surfaces of the as-built alloy samples showed dimples and cleavage steps forming river patterns. The poor ductility of the as-built sample might be attributed to these cleavage steps originating from the initial cracks observed in the microstructure studies, as shown in Figure 1.15(a)–(b).

Furthermore, the cleavage steps were displaced by equiaxed grains on the fracture surface of the annealed HEA sample. In addition, the dimples on the surface indicated a mixture the combination of brittle and ductile fracture (Peng et al., 2021). The study on the hot cracking tendency of $Al_xCoCrFeNi$ postulated that the reduction in the density of hot cracks could not be attributed to the variations in the load-bearing ability of the HEA. It was observed that both Al_0 and $Al_{0.5}$ alloys showed a 300 Mpa increase in the ultimate tensile strength. The elongation of $Al_{0.5}$ dropped significantly as a consequence of the crack minimization. It evidenced that porosity and cracks have a negligible effect on strength but a considerable influence on ductility. The fracture surfaces of the $Al_{0.5}$ and $Al_{1.0}$ showed ductile fracture with dimples and brittle fracture, respectively (Sun et al., 2021).

1.3 ADDITIVE MANUFACTURING (AM) OF MISCELLANEOUS FCC-BASED HEA

1.3.1 MICROSTRUCTURE EVOLUTION

Ocelik et al. (2016) deployed AM to clad layers of HEA on the vacuum arc melted (VAM) HEA sample with varied Al content. In the $Al_{0.4}$CoCrFeNi HEA, there was no significant change in the phase and grain orientation in the remelted region. However, phase fractions and elemental compositions of the phases varied. The remelted zone of the $Al_{0.7}$CoCrFeNi sample exhibited the FCC phase with dendritic grains and the bcc phase inside the interdendritic region. The fraction of the bcc phase was observed to increase in the remelted region compared to the non-remelted $Al_{0.7}$ alloy sample. It was attributed to the rapid cooling in the AM process. The $Al_{1.0}$ alloy possessed a microstructure and composition similar to the $Al_{0.7}$ alloy. Further, the cladding of AlCoCrFeNi and AlCrFeNiTa compositions was performed on the SS bar substrate. In the AlCoCrFeNi alloy, the dilution was observed in the first layer, while the successive layers exhibited uniform elemental distribution with bcc phase and columnar grains having grain growth in <001> direction. Besides, the Fe dilution was observed in the first two layers of the AlCrFeNiTa alloy. Also, the observed less content of Al at the top layers was attributed to the evaporation of Al due to induced high temperatures during deposition of this Ta containing refractory alloy. The cladding was composed of sporadic bright spots throughout the coating and equiaxed dendrites at the top (Ocelík, Janssen, Smith, & De Hosson, 2016).

The AlCoFeNiSmTiVZr HEA alloy with varying concentrations of V and Sm has been fabricated by Sarswat et al. (2019) using the SLM technique. The EDS analysis confirmed the uniform distribution of alloying elements in all compositions. As seen from Figure 1.17, The homogenous deposition without significant cracks and irregular molten tracks was obtained. The etched samples showed flat interconnected flower type (Figure 1.17(b)) or spherical cauliflower type grains (Figure 1.17(f)–(g)) depending on the composition. The ESBD analysis revealed uniformly distributed equiaxed grains in the AlNiCoFeTi$V_{0.9}Sm_{0.1}$ composition. The

SEM analysis revealed dendritic and interdendritic type regions, and the Kikuchi pattern confirmed the single FCC phase in this composition. The XRD plot of $AlNiCoFeTiV_{0.9}Sm_{0.1}$ HEA exhibited many different peaks corresponding to Al-Sm, Al_3Zr, $(Fe, Al)_2Zr$, and Al_3V intermetallic phases. These intermetallic phases introduce brittleness in the alloy. The XRD pattern of equiatomic AlCoFeNiTiVZr showed a perfect match with FCC crystal structure XRD peaks. Besides, the XRD peaks of Zr-free alloys also evidenced FCC crystalline structure (Sarswat et al., 2019).

Yang et al. (2019) fabricated $Ni_6Cr_4WFe_9$ HEA using the SLM process and compared the microstructure and mechanical properties with the VAM HEA. The XRD pattern of both the VAM-HEA and the SLM-built HEA showed the γ-FCC phase accompanied by minor precipitates of the unknown phase. The microstructure studies of the SLM-built alloy revealed the laser scanning track marks in the scan direction and melt pool overlaps along the build direction. The microcracks were observed in sections along with both directions. The appearance of the cracks was attributed to the distinct coefficients of thermal expansion of the matrix and the precipitates. The microstructure was composed of a combination of columnar dendritic structure and equiaxed dendritic structure. Also, the size of precipitates and γ-phase was more refined in the SLM-built alloy than in the VAM-HEA. The fine grain structure was ascribed to the high thermal gradient and rapid solidification in the SLM process. Conversely, the VAM-alloy was composed of a coarse-grained γ phase (Yang et al., 2019).

The Experimental investigation on the effect of Ni concentration in SLM built $AlCrCuFeNi_x$ showed that the micro-cracks reduced with an increase in Ni content from x=0 to x=2.75. Moreover, cracks were disappeared completely at x=3.0, as can be seen from Figure 1.18(a)–(d). The FCC phase started appearing at x=2, and the fraction of the FCC phase increased with the increase in Ni content at the expense of a reduction in the bcc phase. The higher angle side shift of the peak (111) and (110) suggested the increase in interplanar spacing of the FCC and bcc phase, respectively, and it was ascribed to Ni's relatively small atomic radius. The microstructure of $AlCrCuFeNi_3$ is composed of near-equiaxed grains comprising ultrafine lamellar and cellular structures. A considerable percentage of low-angle grain boundaries were observed interior to the near-equiaxed FCC grains, while bcc grains hardly comprised any low-angle grain boundaries. The single grain consisted of FCC (89.2%) + bcc (10.4) dual-phase as confirmed in EBSD studies. The EDS analysis revealed that the B_2 lamellas and FCC lamellas were Al-Cu and Cr-Fe enriched, respectively, and Ni content was the same in both FCC and B_2 lamellas. Minor segregation of Al, Cu, and Ni along FCC and B_2 phase boundaries was observed. In addition, profuse nano-precipitates and Cr enriched A_2 precipitates were present in the B_2 phase, as confirmed in the TEM analysis. Thus, the $AlCrCuFeNi_3$ HEA is composed of heterogeneous microstructure and phase composition (Luo, Zhao, Su, Liu, & Wang, 2020).

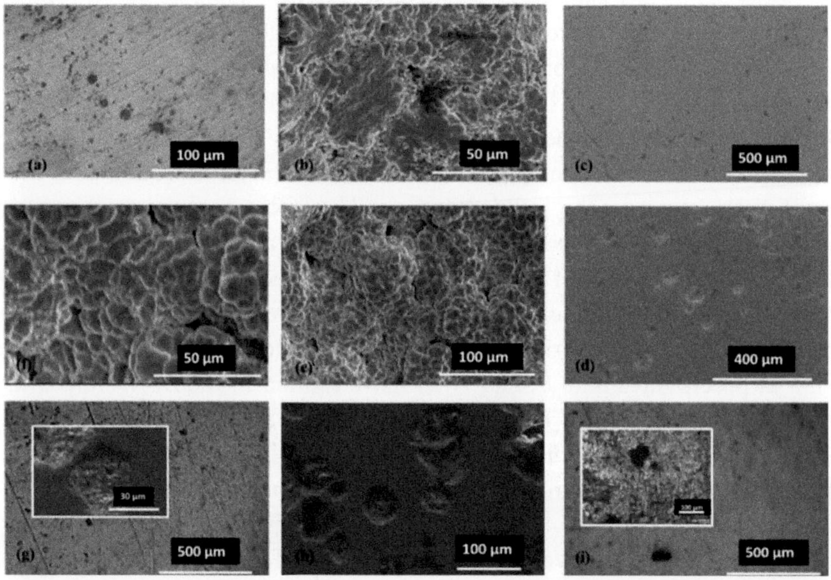

FIGURE 1.17 (a) and (b) SEM images of AlCoFeNiTiVZ HEA without etching and after etching; (c) OM of $AlCoFeNiSm_{0.1}TiV_{0.9}$; (d) SEM image of $AlCoFeNiSm_{0.1}TiV_{0.9}$ corroded HEA sample; (e) and (f) SEM image of $AlCoFeNiV_{0.9}Sm_{0.1}$ samples; (g) $AlCoFeNiV_{0.9}Sm_{0.1}$ sample and corroded portion (inset); (h) SEM image of corroded $AlCoFeNiV_{0.9}Sm_{0.1}$ sample after prolonged exposure; and (i) $AlCoFeNiSm_{0.05}TiV_{0.95}Zr$ sample before and after corrosion (inset).

Source: Reprinted with permission from Sarswat et al., 2019. Copyright. 2019. Elsevier.

Conversely to the studies by Luo et al. (2020); Su et al. (2020) studied the effect of Al addition on $Al_xCrCuFeNi_2$. The coarse columnar grains grown epitaxially along build direction were observed in the Al_0 sample. With the increase in Al content to x=0.5, the coarse columnar grains were refined to fine columnar grains and few equiaxed grains due to constitutional supercooling. The Al also triggered the formation of the bcc/B_2 phase in the interdendritic regions of the FCC phase. The near-equiaxed grain structure comprising cellular and lamellar structure and a few nano-precipitates were observed at x=0.75, 1.0. However, the $Al_{0.75}$ and $Al_{1.0}$ alloy exhibited the FCC+bcc/B_2 and single bcc phase, respectively. It was observed that a considerable number of cracks were present in the Al_0 alloy, which reduced significantly in the $Al_{0.5}$ alloy. The massive and long cracks in Al_0 and $Al_{0.5}$ alloy were propagated along the build direction and distributed along the grain boundaries, as confirmed in the EBSD analysis. The cracks observed along the high grain boundaries suggested the inter-granular hot cracking tendency of the Al_0 and $Al_{0.75}$ alloy. It was attributed to the combined effect of low molten material feeding rate (due to sluggish diffusion in HEAs) along the channels between adjacent long columnar grains and rapid solidification. The cracks were disappeared entirely in the $Al_{0.75}$ alloy but reappeared in the $Al_{1.0}$ alloy. The elimination of hot cracks in the $Al_{0.75}$ alloy was attributed to the finer and equiaxed grains, which provide better feeding of molten material. However, cold cracks were observed in the single bcc phase $Al_{1.0}$ alloy. The cold cracks were ascribed to the combined effect of induced thermal stresses during the SLM process and the bcc/B_2 phase (Su, Luo, & Wang, 2020).

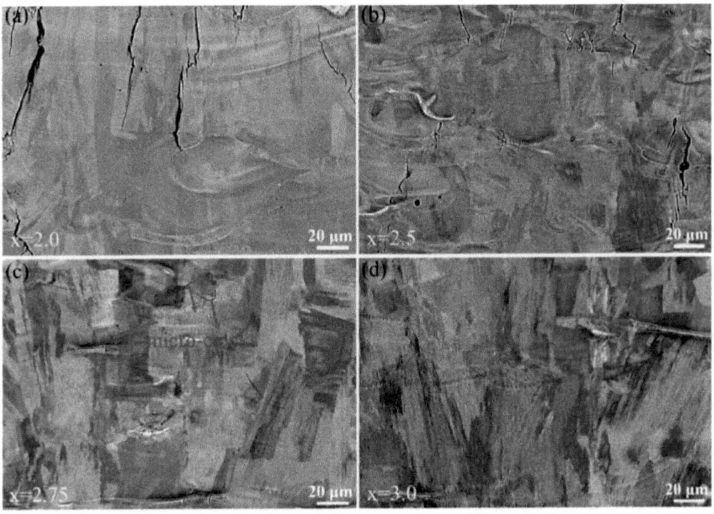

FIGURE 1.18 BSE images of the as-built AlCrCuFeNix HEAs: (a) x = 2.0; (b) x = 2.5; (c) x = 2.75; and (d) x = 3.0.

Source: Reprinted with permission from Luo et al., 2020. Copyright. 2020. Elsevier.

Qiu (2021) cladded the $Al_2CrFeCo_xCuNiTi$ HEA with the addition of 2% Si on the Q235 steel plate. The Co_0 HEA possessed the equiaxed crystal structure, with the cracks extending along the edge of the crystal due to the residual thermal stresses inside the clad layer. At the bonding zone of the $Co_{2.0}$ alloy, the diffusion of the elements of the HEA and Fe from the substrate was observed. It led to a good bond between the clad and substrate. The microstructure of the $Al_2CrFeCo_xCuNiTi$ HEA with 2% Si mainly exhibited equiaxed grains and columnar grains near the bonding region. The corrosion resistance of the coating degraded with the increase in Co content. It was ascribed to the tendency of Co to promote the formation of intermetallic compounds (IMCs), which increases the interfacial energy and results in reduced corrosion resistance of the HEA (Qiu, 2021).

Saraswat et al. (2019) fabricated AlCoCrFeNiTi and $AlCo_{0.8}CrFeNiTi$ using the powder bed fusion AM technique and compared the HEA with Ti free HEAs. The AlCoCrFeNiTi and $AlCo_{0.9}CrFeNiMn_{0.9}$ HEAs exhibited a single FCC crystal structure. EDS analysis revealed uniform distribution of the alloying elements with minor segregation (Sarswat et al., 2019).

1.3.2 MECHANICAL PROPERTY STUDIES

The increased fraction of bcc phase in the laser cladding of $Al_{0.7}CoCrFeNi$ and $Al_{1.0}CoCrFeNi$ alloy resulted in higher hardness than the VAM samples of the same composition. Moreover, the laser cladding of equimolar AlCrFeNiTa showed a three times higher hardness (860 HV) than arc melted $Al_{1.0}CoCrFeNi$ (270 HV). The study postulated that the solidification rate is crucial in determining phases and phase distribution in $Al_xCoCrFeNi$ HEAs (Ocelík, Janssen, Smith, & De Hosson, 2016). The additively manufactured Zr containing hybrid HEAs AlCoFeNiTiVZr and $AlCoFeNiSm_{0.05}TiV_{0.95}Zr$ showed superior corrosion resistance than Zr free $AlCoFeNiV_{0.9}Sm_{0.1}$; and $AlCoFeNiSm_{0.1}TiV_{0.9}$ HEA. The Rockwell A hardness testing of hybrid HEAs showed that the $AlCoFeNiV_{0.9}Sm_{0.1}$ possessed a maximum hardness of ~86.7 while the AlCoFeNiTiVZr possessed the lowest hardness of ~42.8 (Sarswat et al., 2019).

The SLM-built $Ni_6Cr_4WFe_9$ HEA exhibited excellent yield strength and tensile strength of 742 MPa and 972 MPa, 93% and 50% more than the corresponding strengths of the VAM-HEA. The elongation of 12.2% was observed, which was 77% higher than the VAM-HEA. The strengthening of HEA was attributed to the solid solution strengthening and grain refinement. The SLM-built HEA showed intergranular failure with the presence of small-sized dimples on fracture surfaces. Also, the interwoven grains along with the precipitates were observed. In contrast, fracture surfaces of the VAM-HEA possessed noticeably large-sized dimples (Yang et al., 2019).

The absence of cracks and pores in the $AlCrCuFeNi_3$ provided the excellent ultimate tensile strength of 957 MPa and ductility of 14.3%. The severe distortion of B_2/A_2 interfaces resulted in deformation twinning and stacking fault arrays in the B_2 phase. The different severe deformation modes and back-stress strengthening due to heterogeneous microstructure contributed to the excellent tensile properties of the as-built HEA (Luo, Zhao, Su, Liu, & Wang, 2020).

The study had showed that the AlCoCrFeNiTi possessed good elastic properties with similar elastic constants as Al-based AlCoCrFeNiMn, $AlCoCrFeNi_{1.3}$ HEAs, and $CoCr_{1.3}FeMnNi_{0.7}$ (Sarswat, Smith, Sarkar, Murali, & Free, 2020).

Table 1.1 summarizes the mechanical properties of different FCC-based HEAs fabricated with various AM processes and after post-processing.

TABLE 1.1 Mechanical Properties of Different FCC-based HEAs Fabricated with Various AM Processes

Author and Year	Material	Process	σ_y (Mpa)	σ_{ut} (Mpa)	εl (%)	Hardness
Zhu et al. (2018)	CoCrFeMnNi	SLM	510	–	–	–
Piglione et al. (2018)			–	–	–	212HV
Dovgyy et al. (2020)			530±20	783±7	23±1	–
Li et al.			601		35	–
(2018)		SLM/HIP	649	–	18	–
Chen et al. (2020)		SLM	681±14	–	12.5±0.5	261±7 HV
Qiu et al. (2018)		DED (cryogenic deformation at 77 K)	650	891	38	–
Tong et al. (2019)		–	346	566	–	200 HV

TABLE 1.1 *(Continued)*

Author and Year	Material	Process	σ_y (Mpa)	σ_{ut} (Mpa)	εl (%)	Hardness
Li et al. (2019)	CoCrFeMnNi	DED	300	550	50	
	CoCrFeMnNi + 5% WC	DED	502	776	37	–
		DED (High temperature deformation at 873 K)	–	405	45	–
Zhang et al. (2021)	CoCrFeMnNi + 2.5% TiC	DED	–	610	47	–
Shen et al. (2020)	CoCrFeNi	PAAM	141	–	–	139 HV
	CoCrFeNi+$SiC_{0.5}$		712	–	36.5	310 HV
	CoCrFeNi+$SiC_{0.3}$		–	–	48.5	
Zhang et al. (2020)	CoCrFeNi	PAAM	–	–	–	4.578 Gpa
	$Al_{0.4}$CoCrFeNi		–	–	–	4.674 Gpa
	CoCrFeNi$Ta_{0.4}$		–	–	–	5.672 Gpa
	CoCrFeNi$Nb_{0.4}$		–	–	–	5.617 Gpa
Joseph et al. (2015)	$Al_{0.3}$CoCrFeNi	DLF	200	–	–	–
	$Al_{0.6}$CoCrFeNi		400	–	–	–
	$Al_{0.85}$CoCrFeNi		1,400	2,250	–	–
Mohanty et al. (2019)	$Al_{0.3}$CoCrFeNi	DLD	–	–	–	170 HV
	$Al_{0.7}$CoCrFeNi		–	–	–	380 HV
Gao et al. (2020)	CoCrFeNi	DLF	370	–	49.5	–
	CoCrFeNi+8 wt.% Al		978	–	7.5	–
Shen et al. (2020)	Al based CoCrFeNi	WAAM	816	2,835	41.8	–
Peng et al. (2021)	CoCrFeNi	SLM	208.8	–	0.6	–
	$Al_{0.3}$CoCrFeNi		–	–	2.98	–
Ocelik et al. (2016)	AlCrFeNiTa	Laser Cladding	–	–	–	870 HV
	$Al_{1.0}$CoCrFeNi		–	–	–	270 HV
Sarswat et al. (2019)	AlCoFeNi$V_{0.9}Sm_{0.1}$	SLM	–	–	–	Rockwell A 86.7
	AlCoFeNiTiVZr		–	–	–	Rockwell A 42.8

TABLE 1.1 *(Continued)*

Author and Year	Material	Process	σ_y (Mpa)	σ_{ut} (Mpa)	ɛl (%)	Hardness
Yang et al. (2019)	$Ni_6Cr_4WFe_9$	SLM	742	972	12.2%	–
Luo et al. (2020)	$AlCrCuFeNi_3$	SLM	–	957	14.3	–

1.4 SUMMARY

The present study has shown that the AM is a feasible and more suitable technique for fabricating FCC-HEAs than the conventional arc melting technique. The AM processing of CoCrFeMnNi HEA (well known as cantor alloy) and its derivable HEAs have been widely investigated by researchers so far. The *in-situ* alloying capability of the AM process makes it compatible with changing the composition during the processing also. Al has proven the most popular strengthening element which promotes the bcc-phase formation on account of rapid solidification in the AM process. The strengthening of the FCC-based HEAs can be achieved easily by *in-situ* adding compounds such as Al, SiC, WC, and TiC. However, very little was reported and published in this direction and needed to be investigated in detail for various HEAs.

KEYWORDS

- **additive manufacturing**
- **FCC crystal structure**
- ***in-situ* adding compounds**
- **laser melt deposition**
- **microstructure**
- **solidification**
- **tensile properties**

REFERENCES

Barron, P. J., Carruthers, A. W., Fellowes, J. W., Jones, N. G., Dawson, H., & Pickering, E. J., (2020). Towards V-based high-entropy alloys for nuclear fusion applications. *Scr. Mater., 176*, 12–16. https://doi.org/10.1016/j.scriptamat.2019.09.028.

Cantor, B., Chang, I. T. H., Knight, P., & Vincent, A. J. B., (2004). *Mater. Sci. Eng. A., 375–377*, 213–218.

Chen, P., Li, S., Zhou, Y., Yan, M., & Attallah, M. M., (2020). Fabricating CoCrFeMnNi high entropy alloy via selective laser melting *in-situ* alloyin., *J. Mater. Sci. Technol., 43,* 40–43. https://doi.org/10.1016/j.jmst.2020.01.002.

Chen, S., Tong, Y., & Liaw, P. K., (2018). Additive manufacturing of high-entropy alloys: A review. *Entropy, 20*. https://doi.org/10.3390/e20120937.

Dovgyy, B., Piglione, A., Hooper, P. A., & Pham, M. S., (2020). Comprehensive assessment of the printability of CoNiCrFeMn in laser powder bed fusion. *Mater. Des., 194*, 108845. https://doi.org/10.1016/j.matdes.2020.108845.

Dutta, B., & (Sam). Froes, F. M., (2017). The additive manufacturing (AM) of titanium alloys. *Met. Powder Rep., 72*, 96–106. https://doi.org/10.1016/j.mprp.2016.12.062.

Frazier, W. E., (2014). Metal additive manufacturing: A review. *J. Mater. Eng. Perform., 23*, 1917–1928. https://doi.org/10.1007/s11665-014-0958-z.

Gao, X., Yu, Z., Hu, W., Lu, Y., Zhu, Z., Ji, Y., Lu, Y., Qin, Z., & Lu, X., (2020). In situ strengthening of CrMnFeCoNi high-entropy alloy with Al realized by laser additive manufacturing. *J. Alloys Compd., 847*. https://doi.org/10.1016/j.jallcom.2020.156563.

Goodridge, R. D., Tuck, C. J., & Hague, R. J. M., (2012). Laser sintering of polyamides and other polymers. *Prog. Mater. Sci., 57*, 229–267. https://doi.org/10.1016/j.pmatsci.2011.04.001.

Henderson, M. B., Arrell, D., Larsson, R., Heobel, M., & Marchant, G., (2004). Nickel-based superalloy welding practices for industrial gas turbine applications. *Science and Technology of Welding and Joining, 9*(1), 13–21.

Jinoop, A. N., Paul, C. P., Mishra, S. K., & Bindra, K. S., (2019). Laser additive manufacturing using directed energy deposition of inconel-718 wall structures with tailored characteristics. *Vacuum, 166*, 270–278. https://doi.org/10.1016/j.vacuum.2019.05.027.

Joseph, J., Hodgson, P., Jarvis, T., Wu, X., Stanford, N., & Fabijanic, D. M., (2018). Effect of hot isostatic pressing on the microstructure and mechanical properties of additive manufactured Al_xCoCrFeNi high entropy alloys. *Mater. Sci. Eng. A., 733*, 59–70. https://doi.org/10.1016/j.msea.2018.07.036.

Joseph, J., Jarvis, T., Wu, X., Stanford, N., Hodgson, P., & Fabijanic, D. M., (2015). Comparative study of the microstructures and mechanical properties of direct laser fabricated and arc-melted Al_xCoCrFeNi high entropy alloys. *Mater. Sci. Eng. A., 633*, 184–193. https://doi.org/10.1016/j.msea.2015.02.072.

Joseph, J., Stanford, N., Hodgson, P., & Fabijanic, D. M., (2017). Understanding the mechanical behaviour and the large strength/ductility differences between FCC and BCC Al_xCoCrFeNi high entropy alloys. *J. Alloys Compd., 726*, 885–895. https://doi.org/10.1016/j.jallcom.2017.08.067.

Kao, Y. F., Chen, S. K., Sheu, J. H., Lin, J. T., Lin, W. E., Yeh, J. W., Lin, S. J., et al., (2010). Hydrogen storage properties of multi-principal-component CoFeMnTi xVyZrz alloys. *Int. J. Hydrogen Energy, 35*, 9046–9059. https://doi.org/10.1016/j.ijhydene.2010.06.012.

Kotadia, H. R., Gibbons, G., Das, A., & Howes, P. D., (2021). A review of laser powder bed fusion additive manufacturing of aluminum alloys: Microstructure and properties. *Addit. Manuf., 46*, 102155. https://doi.org/10.1016/j.addma.2021.102155.

Kuwabara, K., Shiratori, H., Fujieda, T., Yamanaka, K., Koizumi, Y., & Chiba, A., (2018). Mechanical and corrosion properties of AlCoCrFeNi high-entropy alloy fabricated with selective electron beam melting. *Addit. Manuf., 23*, 264–271. https://doi.org/10.1016/j.addma.2018.06.006.

Li, J., Xiang, S., Luan, H., Amar, A., Liu, X., Lu, S., Zeng, Y., et al., (2019). Additive manufacturing of high-strength CrMnFeCoNi high-entropy alloys-based composites with WC addition. *J. Mater. Sci. Technol., 35*, 2430–2434. https://doi.org/10.1016/j.jmst.2019.05.062.

Li, R., Niu, P., Yuan, T., Cao, P., Chen, C., & Zhou, K., (2018). Selective laser melting of an equiatomic CoCrFeMnNi high-entropy alloy: Processability, non-equilibrium microstructure and mechanical property. *J. Alloys Compd., 746*, 125–134. https://doi.org/10.1016/j.jallcom.2018.02.298.

Lu, Y., Gao, X., Dong, Y., Wang, T., Chen, H. L., Mao, H., Zhao, Y., et al., (2018). Preparing bulk ultrafine-microstructure high-entropy alloys: Via direct solidification. *Nanoscale, 10*, 1912–1919. https://doi.org/10.1039/c7nr07281c.

Luo, S., Zhao, C., Su, Y., Liu, Q., & Wang, Z., (2020). Selective laser melting of dual phase $AlCrCuFeNi_x$ high entropy alloys: Formability, heterogeneous microstructures, and deformation mechanisms. *Addit. Manuf., 31*, 100925. https://doi.org/10.1016/j.addma.2019.100925.

Lyu, Z., Lee, C., Wang, S. Y., Fan, X., Yeh, J. W., & Liaw, P. K., (2019). Effects of constituent elements and fabrication methods on mechanical behavior of high-entropy alloys: A review. *Metall. Mater. Trans. A Phys. Metall. Mater. Sci., 50*. https://doi.org/10.1007/s11661-018-4970-z.

MacDonald, B. E., Fu, Z., Zheng, B., Chen, W., Lin, Y., Chen, F., Zhang, L., et al., (2017). Recent progress in high entropy alloy research. *JOM, 69*, 2024–2031. https://doi.org/10.1007/s11837-017-2484-6.

Mohanty, A., Sampreeth, J. K., Bembalge, O., Hascoet, J. Y., Marya, S., Immanuel, R. J., & Panigrahi, S. K., (2019). High temperature oxidation study of direct laser deposited $Al_xCoCrFeNi$ (X=0.3,0.7) high entropy alloys. *Surf. Coatings Technol., 380*, 125028. https://doi.org/10.1016/j.surfcoat.2019.125028.

Ocelík, V., Janssen, N., Smith, S. N., & De Hosson, J. T. M., (2016). Additive manufacturing of high-entropy alloys by laser processing. *JOM, 68*, 1810–1818. https://doi.org/10.1007/s11837-016-1888-z.

Ostovari, M. A., Shaburova, N. A., Samodurova, M. N., Abdollahzadeh, A., & Trofimov, E. A., (2021). Additive manufacturing of high entropy alloys: A practical review. *J. Mater. Sci. Technol., 77*, 131–162. https://doi.org/10.1016/j.jmst.2020.11.029.

Peng, H., Xie, S., Niu, P., Zhang, Z., Yuan, T., Ren, Z., Wang, X., et al., (2021). Additive manufacturing of $Al_{0.3}CoCrFeNi$ high-entropy alloy by powder feeding laser melting deposition. *J. Alloys Compd., 862*, 158286. https://doi.org/10.1016/j.jallcom.2020.158286.

Pickering, E. J., Muñoz-Moreno, R., Stone, H. J., & Jones, N. G., (2016). Precipitation in the equiatomic high-entropy alloy CrMnFeCoNi. *Scr. Mater., 113*, 106–109. https://doi.org/10.1016/j.scriptamat.2015.10.025.

Piglione, A., Dovgyy, B., Liu, C., Gourlay, C. M., Hooper, P. A., & Pham, M. S., (2018). Printability and microstructure of the CoCrFeMnNi high-entropy alloy fabricated by laser powder bed fusion. *Mater. Lett., 224*, 22–25. https://doi.org/10.1016/j.matlet.2018.04.052.

Qin, G., Chen, R., Liaw, P. K., Gao, Y., Wang, L., Su, Y., Ding, H., Guo, J., & Li, X., (2020). An as-cast high-entropy alloy with remarkable mechanical properties strengthened by nanometer precipitates. *Nanoscale, 12*, 3965–3976. https://doi.org/10.1039/c9nr08338c.

Qiu, X., (2021). Microstructure and corrosion properties of Al_2CrFeCoxCuNiTi high entropy alloys prepared by additive manufacturing. *J. Alloys Compd., 887*, 161422. https://doi.org/10.1016/j.jallcom.2021.161422.

Qiu, Z., Yao, C., Feng, K., Li, Z., & Chu, P. K., (2018). Cryogenic deformation mechanism of CrMnFeCoNi high-entropy alloy fabricated by laser additive manufacturing process. *Int. J. Light. Mater. Manuf., 1*, 33–39. https://doi.org/10.1016/j.ijlmm.2018.02.001.

Rameshbabu, A. M., Parameswaran, P., Vijayan, V., & Panneer, R., (2017). Diffraction, microstructure and thermal stability analysis in a double phase nanocrystalline $Al_{20}Mg_{20}Ni_{20}Cr_{20}Ti_{20}$ high entropy alloy. *J. Mech. Behav. Mater., 26*, 127–132. https://doi.org/10.1515/jmbm-2017-0021.

Sarswat, P. K., Sarkar, S., Murali, A., Huang, W., Tan, W., & Free, M. L., (2019). Additive manufactured new hybrid high entropy alloys derived from the AlCoFeNiSmTiVZr system. *Appl. Surf. Sci., 476*, 242–258. https://doi.org/10.1016/j.apsusc.2018.12.300.

Sarswat, P., Smith, T., Sarkar, S., Murali, A., & Free, M., (2020). Design and fabrication of new high entropy alloys for evaluating titanium replacements in additive manufacturing. *Materials (Basel), 13*, 1–37. https://doi.org/10.3390/ma13133001.

Shen, Q., Kong, X., & Chen, X., (2021). Fabrication of bulk Al-Co-Cr-Fe-Ni high-entropy alloy using combined cable wire arc additive manufacturing (CCW-AAM): Microstructure and mechanical properties. *J. Mater. Sci. Technol., 74*, 136–142. https://doi.org/10.1016/j.jmst.2020.10.037.

Shen, Q., Kong, X., Chen, X., Yao, X., Deev, V. B., & Prusov, E. S., (2021). Powder plasma arc additive manufactured CoCrFeNi(SiC)x high-entropy alloys: Microstructure and mechanical properties. *Mater. Lett., 282*, 128736. https://doi.org/10.1016/j.matlet.2020.128736.

Shiratori, H., Fujieda, T., Yamanaka, K., Koizumi, Y., Kuwabara, K., Kato, T., & Chiba, A., (2016). Relationship between the microstructure and mechanical properties of an equiatomic AlCoCrFeNi high-entropy alloy fabricated by selective electron beam melting. *Mater. Sci. Eng. A., 656*, 39–46. https://doi.org/10.1016/j.msea.2016.01.019.

Sleiman, S., Moussa, M., & Huot, J., (2021). Microstructure and hydrogen storage properties of the multiphase Ti0. 3V0. 3Mn0. 2Fe0. 1Ni0. *1 Alloy. Reactions, 2*(3), 287–300.

Su, Y., Luo, S., & Wang, Z., (2020). Microstructure evolution and cracking behaviors of additively manufactured $Al_x CrCuFeNi_2$ high entropy alloys via selective laser melting. *J. Alloys Compd., 842*, 155823. https://doi.org/10.1016/j.jallcom.2020.155823.

Sun, Z., Tan, X., Wang, C., Descoins, M., Mangelinck, D., Tor, S. B., Jägle, E. A., et al., (2021). Reducing hot tearing by grain boundary segregation engineering in additive manufacturing: Example of an $Al_xCoCrFeNi$ high-entropy alloy. *Acta Mater., 204*, 116505. https://doi.org/10.1016/j.actamat.2020.116505.

Tong, Z., Ren, X., Jiao, J., Zhou, W., Ren, Y., Ye, Y., Larson, E. A., & Gu, J., (2019). Laser additive manufacturing of FeCrCoMnNi high-entropy alloy: Effect of heat treatment on microstructure, residual stress and mechanical property. *J. Alloys Compd., 785*, 1144–1159. https://doi.org/10.1016/j.jallcom.2019.01.213.

Tsai, M. H., & Yeh, J. W., (2014). High-entropy alloys: A critical review. *Mater. Res. Lett., 2*, 107–123. https://doi.org/10.1080/21663831.2014.912690.

Tsai, M. H., (2013). Physical properties of high entropy alloys. *Entropy, 15*, 5338–5345. https://doi.org/10.3390/e15125338.

Wong, K. V., & Hernandez, A., (2012). A review of additive manufacturing. *ISRN Mech. Eng., 2012*, 1–10. https://doi.org/10.5402/2012/208760.

Yadav, S., Paul, C. P., Jinoop, A. N., Rai, A. K., & Bindra, K. S., (2020). Laser directed energy deposition based additive manufacturing of copper: Process development and material characterizations. *J. Manuf. Process., 58*, 984–997. https://doi.org/10.1016/j.jmapro.2020.09.008.

Yang, X., Zhou, Y., Xi, S., Chen, Z., Wei, P., He, C., Li, T., et al., (2019). Additively manufactured fine-grained $Ni_6Cr_4WFe_9Ti$ high entropy alloys with high strength and ductility. *Mater. Sci. Eng. A., 767*, 138394. https://doi.org/10.1016/j.msea.2019.138394.

Ye, Y. F., Wang, Q., Lu, J., Liu, C. T., & Yang, Y., (2016). High-entropy alloy: Challenges and prospects. *Mater. Today, 19*, 349–362. https://doi.org/10.1016/j.mattod.2015.11.026.

Yeh, J. W., (2013). Alloy design strategies and future trends in high-entropy alloys. *JOM, 65*, 1759–1771. https://doi.org/10.1007/s11837-013-0761-6.

Yeh, J. W., Chen, S. K., Lin, S. J., Gan, J. Y., Chin, T. S., Shun, T. T., Tsau, C. H., & Chang, S. Y., (2004). Nanostructured high-entropy alloys with multiple principal elements: Novel alloy design concepts and outcomes. *Adv. Eng. Mater., 6*, 299–303. https://doi.org/10.1002/adem.200300567.

Zeng, C., Zhang, B., Hemmasian, E. A., Wen, H., Yao, H., Meng, W. J., & Guo, S., (2020). Mechanical, thermal, and corrosion properties of Cu-10Sn alloy prepared by laser-powder-bed-fusion additive manufacturing. *Addit. Manuf., 35*, 101411. https://doi.org/10.1016/j.addma.2020.101411.

Zhang, M., Peng, Y., Zhang, W., Liu, Y., Wang, L., Hu, S., & Hu, Y., (2019). Gradient distribution of microstructures and mechanical properties in a fecocrnimo high-entropy alloy during spark plasma sintering. *Metals (Basel), 9*. https://doi.org/10.3390/met9030351.

Zhang, X., Li, R., Huang, L., Amar, A., Wu, C., Le, G., Liu, X., et al., (2021). Influence of in-situ and ex-situ precipitations on microstructure and mechanical properties of additive manufacturing CoCrFeMnNi high-entropy alloys. *Vacuum, 187*, 110111. https://doi.org/10.1016/j.vacuum.2021.110111.

Zhang, Y., Chen, X., Jayalakshmi, S., Singh, R. A., Deev, V. B., & Prusov, E. S., (2021). Factors determining solid solution phase formation and stability in $CoCrFeNiX_{0.4}$ (X=Al, Nb, Ta) high entropy alloys fabricated by powder plasma arc additive manufacturing. *J. Alloys Compd., 857,* 157625. https://doi.org/10.1016/j.jallcom.2020.157625.

Zhang, Y., Shen, Q., Chen, X., Jayalakshmi, S., Singh, R. A., Konovalov, S., Deev, V. B., & Prusov, E. S., (2021). Strengthening mechanisms in $cocrfenix_{0.4}$ (Al, Nb, Ta) high entropy alloys fabricated by powder plasma arc additive manufacturing. *Nanomaterials, 11*, 1–14. https://doi.org/10.3390/nano11030721.

Zhang, Y., Zuo, T., Cheng, Y., & Liaw, P. K., (2013). High-entropy alloys with high saturation magnetization, electrical resistivity, and malleability. *Sci. Rep., 3*, 1–7. https://doi.org/10.1038/srep01455.

Zhu, Z. G., Nguyen, Q. B., Ng, F. L., An, X. H., Liao, X. Z., Liaw, P. K., Nai, S. M. L., & Wei, J., (2018). Hierarchical microstructure and strengthening mechanisms of a CoCrFeNiMn high entropy alloy additively manufactured by selective laser melting. *Scr. Mater., 154*, 20–24. https://doi.org/10.1016/j.scriptamat.2018.05.015.

CHAPTER 2

Characterization of Composite Materials Using Natural and Synthetic Fiber: A Review

GHANSHYAM SATNAMI, DEEPESH BHAVSAR, ANIL KUMAR RAO, and ANIL KHURCHANIA

Department of Mechanical Engineering, Ravindranath Tagore University, Raisen, Madhya Pradesh, India

ABSTRACT

Natural fiber (NF) composites are now a day's play a big important role in aero-space, medical automotive and construction industries. In present the demand for high strength materials is low weighted, low cost, and also economical and environment friendly. Hybrid composite has become increasing in various field such as automotive industry, aeronautical industry, manufacturing industry, army field, medical industry, etc. As a result, we have seen that various researchers use cellulose fiber in the automotive and structural industries for the benefit, but some problems have been connected during the uses of some fibers, such as moisture content, low strength, and water absorption than the inorganic fiber. These types of problem have been reduced by the various researchers with the enhancements in mechanical properties through hybridization of NF with synthetic fiber. This review is focused on the use of hybrid composite, which can enhance the mechanical properties and structural properties.

Advances in Manufacturing and Processing of Materials: Characterization and Applications.
Narendra Kumar, PhD, Mohit Tyagi, PhD, Dilbagh Panchal, PhD & Ravi Pratap Singh, PhD (Eds.)

2.1 WHY A COMPOSITE?

The knowledge of the composite is very necessary for every engineer. We have seen that various types of polymer fiber are available in an open environment. The knowledge of the composite is good but understanding of the composite with their behavior on the properties, geometry, amounts, and properties of the constituent phases of materials is also compulsory. We are seen that a polymer fiber property is better than the ceramics, metal alloys, etc. In order to develop new materials, particularly in research, it is imperative (essential) to know the various materials. Now a day's analytically researcher prefer the application of the materials which has the light weight comparatively materials in vogue like iron, steel, copper, etc. Research shows that a number of materials are available having light weight able to replace the materials which has the heavy weight.

2.2 INTRODUCTION

The demand of raw materials is increasing in present era for the reinforcement of composite and fulfill the world market demand (Valadez-Gonzalez & Harera, 2005). In the last decades, we have seen that researchers have been notified of the manufacturing of new composites by using synthetic and natural fibers (NFs).

Generally, composite materials are produced from two or more materials. These components may have similar or dissimilar materials (Figure 2.1).

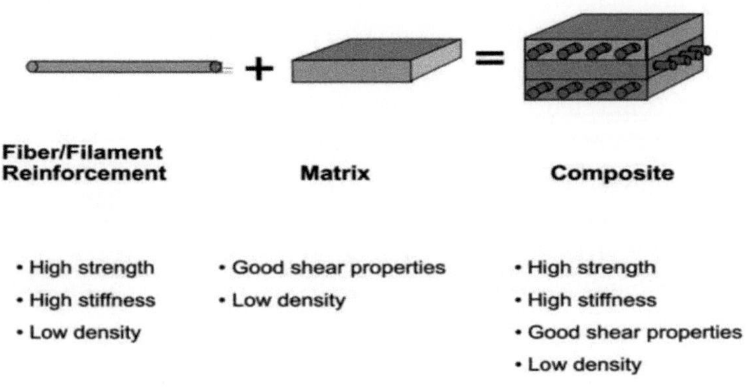

FIGURE 2.1 Composite.

Source: Adapted and modified from Mrazova (2013).

There are two types of constituent materials: matrix and reinforcement. The reinforcement is a special part of mechanical and enhance the matrix properties. Here are various types of reinforcement and matrix materials are available.

According to Donaldson & Miracle (2001), composites are classified into two different categories. The first category is based on the matrix constituent such as:

1. **Polymer Matrix Composite (PMCs):** These are constituents of a matrix from thermoplastic (polycarbonate, nylon, polyvinylchloride, polystyrene) or thermoset (Epoxy, Unsaturated polyester) and embedded Kevlar, glass, steel.
2. **Metal Matrix Composites (MMCs):** These matrices are constituent of a metallic matrix such as iron, Al, magnesium, cobalt copper and metallic phase (tungsten, lead, molybdenum) or dispersed ceramic (carbides, oxides) phase.
3. **Ceramic Matrix Composites (CMCs):** These matrixes are constituent of ceramics and ceramic materials embedded over it (in dispersed phase).

Second classification is based on reinforcing materials structure such as:

1. **Particulate Reinforced Composite:** In this composite consists of a particle in dispersed phase with matrix reinforced.
2. **Flake Composite:** These consist of flat reinforcement of flat reinforcements of matrices, such as Al, mica, silver, etc.
3. **Fiber Reinforced Composite:** This is a combination of different physical properties.
4. **Laminated Composite:** These fibers constituent of different layers of fiber.

2.3 HYBRID COMPOSITE

Generally, we see that presently, the industry is growing in the field of hybrid composite. These are attracting in industrial area. Here this area combines the synthetic fiber with NF and generates a hybrid composite. Hybridization means fulfill the requirements of industry with balance of performance cost and enhancement of load bearing capacity (Harbi, 2001).

Here, investigated the hybridization of natural with synthetic and then the enhancement of mechanical properties (Desilva et al., 2008).

Here are studies that between kenaf and flax, these are provide very good mechanical properties with the reinforcement of NF (Akil et al., 2011).

These are investigated that if the combine of NF with the synthetic fiber, then these boosted demand for NF composite materials (Mohanty & Kline, 2002).

Murugan et al. (2020) observed that a mudguard made up of a composite of banana fiber and glass fiber-reinforced polymer. The mudguard weight could be reduced. The researcher work fabricated a mudguard by using composite material. This mudguard hardness and weight percentage is calculated, also analyze its water observation capacity and impact resistance test.

Karthik et al. (2020) in this research work was taken the materials in the stacking sequences of c/k/c/k/k/c shows the highest value of materials tensile strength 385.09 MPa. It is also done by scanning to find the fracture defects in the composite.

Chethan et al. (2020) observed hybrid composites laminate which prepared from glass fiber, Kenaf and Jute fiber as a reinforcement for epoxy resin under the process of vacuum bagging. The fabricated laminated composite was tested for tensile and flexural as per ASTMD 3039 and ASTMD 790, respectively. The results of tensile strength and flexural strength is 88.08 MPa of S6 plate and 243.856 MPa of S9 plate, respectively. These results proved that the hybrid composite laminates having the amalgamation of glass and kenaf fibers can develop mechanical properties than that of jute and glass fibers with rise the number of glass plies can again enhance the various properties of mechanical.

Ramasamy et al. (2020) this research work was deal with mechanical representation of Kenaf and Kevlar fiber reinforced epoxy based on composites materials. The flexural, tensile, double shear, impact and hardness test were conducted on laminates of composite developed by using hand layup making process. The investigation is based on experimental result show the mechanical property of epoxy was increase by the hybridization of Kevlar and Kenaf fiber.

Chandramohan et al. (2020) studied the hybrid composite reinforced by NFs over Kevlar/epoxy were manufactured using hand-layup making process. Here use the SCM machine for study exposed that the primary failure mechanism of composites was fiber breakage, delimitation and matrix cracking.

Srivastava et al. (2020) analyzed the fiber reinforced epoxy composites (carbon–Kevlar, hybrid and sandwich configurations) dynamic mechanical properties were compared to the current study of fiber composite. Which is manufactured with hardener using a hand layup technique. This researcher shows that hybrid fiber sandwich composites exhibit higher damping properties at high frequencies.

Ramesh et al. (2020) in this work fiber woven composite in the mat form was added to matrix element with varying fiber stacking sequences of symmetric laminates composite. They were using various hybrid composite laminates having Basalt and Kevlar fibers are the reinforcements in varying stacking sequences are produced by hand lay-up technique after that compression molding. The mechanical strength is evaluated of the produced composite. The experimental results recorded that composite having stacking sequences of Kevlar/Basalt/Kevlar/Basalt/Kevlar/Basalt/Kevlar (K/B/K/B/K/B/K) shows the highest value of flexural strength and tensile strength of 188 MPa and 174 MPa, respectively.

Tiwari et al. (2021) worked on NF, which was blended with carbon fiber (CF) to fabricate jute to make CF-reinforced polypropylene composite laminates. The composite laminate was fabricated by compression molding process. Jute fibers are arranged in a bi-directional pattern to add strength and a good surface finish to the new composite laminates. For characterization, the absorption test done. It had been recorded that the stacking sequence of C/J/C reinforced polymer composite has resulted in a maximum tensile strength of 92.336 MPa, Rockwell hardness number of 103.65, a flexural strength of 44.771 MPa, and impact strength of 2.65 (±0.5) Jules as compared to J/C/J reinforced polymer composite laminates.

Rana et al. (2021) found that hybrid epoxy composite is reinforced by using nano-silica particles and Kevlar fiber. In this research, they found that the incorporation of nano-silica particles into the epoxy matrix improved the mechanical properties of the composite materials. However, Kevlar fibers as an internal reinforcement layer enhanced the effect of mechanical properties of the epoxy composites materials.

1. **Kelvar Fiber:** Kevlar is the trade name of DuPont's para-aramid fiber. Kevlar fiber is basically a synthetic fiber material with corrosion resistance and heat resistance. This is firstly in 1970 for

replacement of steel in place of racing tiers. The uses of Kevlar fiber include the manufacturing of fabric sheets, ropes, airplanes, ships, and various composite materials.

2. **Kevlar Fiber Grade:**

SL. No.	Name of Kevlar Fiber	Uses
1.	K-29	The use of this fiber in brake lining, cables, tiers, etc.
2.	K-49	Manufacturing of rope product.
3.	K-100	This is used in the form of Kevlar fiber color.
4.	K-119	For fatigue resistance, flexible and more elongation.
5.	K-129	Uses in ballistic applications.
6.	K-149	For aerospace and armor.
7.	K-AP	In sports.
8.	XP	Lighter weight with fiber combination.
9.	KM2	Ballistic applications such as armor applications.

➢ **The Properties of the Kevlar Fiber:**

- High tenacity;
- For maximum modulus of rigidity;
- For lower ductility;
- Low electrical conductivity for lower electrical conductivity;
- For maximum toughness;
- For lower corrosion.

➢ **The Application of Kevlar Fiber is:**

- This use in the field of cryogenics because of higher strength;
- Use in armor industry such as ballistic vests, helmets, etc.;
- Use in sports industry such as string of bows;
- Uses in manufacturing of belts and hoses;
- This is also used in music, such as loudspeaker cones;
- Manufacturing of electromechanical cables;
- Uses in aircrafts;
- Uses in motor vehicles;
- Uses in fiber optic;
- Uses in automobile such as manufacturing of tires.

3. **Natural Fiber (NF):** These are produced from geological process and also produced from the body of plants and animals. By the use of NF may be manufactured of composite materials. This is also used in mat form and manufacturing of rope, sheets and paper, etc. They can also be matted into sheets, paper, etc. This is also used in research works such as the development of composite materials. Classification of NF is given in Figure 2.2.

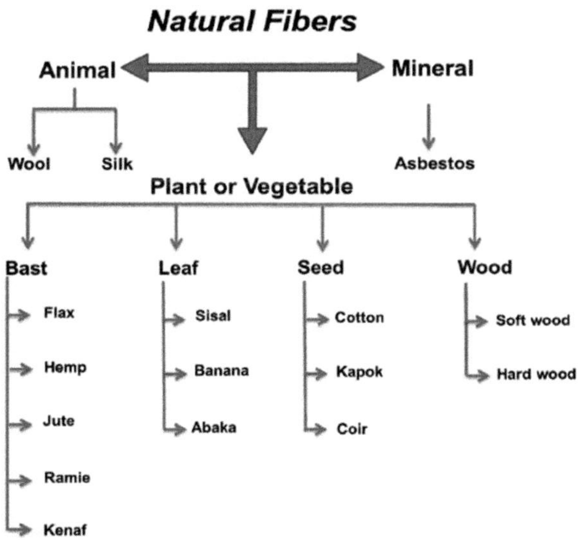

FIGURE 2.2 Classification of natural fiber.

Source: Adapted and modified from Hassan et al., 2016.

- **Properties of Natural Fiber (NF):**
  - Natural fibers have high strength and stiffness;
  - It is having high flexural strength;
  - Lightweight;
  - Low cost;
  - Low pollution;
  - Easily available.

4. **Kenaf Fiber:** It is a spring crop. Kenaf was first used in northern Africa. Kenaf is used in eating and also used in the form of fiber.

- **Chemical Composition of Kenaf Fiber:**

Chemical Composition	Weight (%)
Ash	2–3
Moisture	6.2–12
Cellulose	45–57
Lignin	21.5
Hemicelluloses	8–13
Pectin	0.6
Wax	0.8

- **Kenaf Fiber Mechanical Property:**

Mechanical Properties	Characteristics
Density (g/cm^3)	1.2
Elastic modulus (GPa)	23
Breaking strength (MPa)	100.64
Elongation (%)	8.3
Tensile strength (MPa)	283–800
Moisture absorption (%)	8.3
Yarn breaking (N)	79

As shown in Figure 2.3, it was seen that in tensile test results, it was found that if the orientation of fiber strength was decreased and if it increased, the orientation of kenaf fiber polyester strength of fiber was increased.

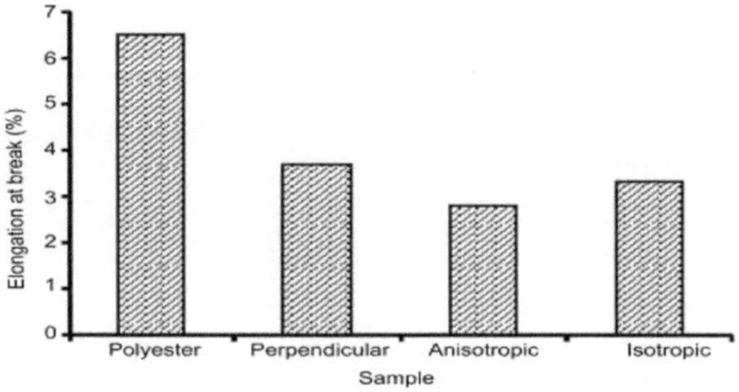

FIGURE 2.3 Fracture and elongation of polyester and kenaf polyester sandwich composite.

Source: Adapted from Rahman and Putra, 2019.

Cikala et al. (2009) – study that when reinforced of kenaf mat with woven glass mat and find the tensile and flexural strength result, from this result we can manufactured of the pipe by the use of kenaf and woven glass fiber.

Davoodi et al. (2010) – here was manufactured of car bumper beam by the use of glass fiber and kenaf fiber materials. This fiber was taken in the form of fiber mat. Here investigated that yielding and mechanical properties was good.

From Figure 2.4, we see that for the properties of hybrid composite materials like Young's modulus, the strength of tensile and strength of flexural is higher for GMT, but here, we found that impact strength was very low. From the above result that we can use of this hybrid composite for manufacturing of bumper beams.

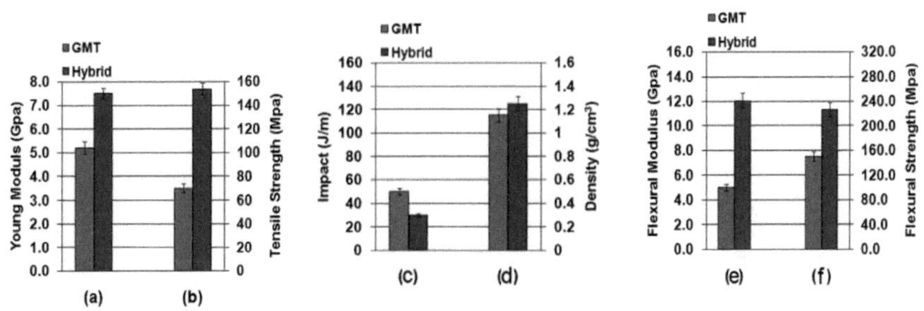

FIGURE 2.4 (a) Youngs modulus (Gpa); (b) tensile strength (Mpa); (c) impact strength (J/m); (d) density (g/cm^3); (e) flexural modulus (Gpa); and (f) flexural strength (Mpa).

2.4 CONCLUSIONS

In this review chapter, studies that the mechanical, physical, thermal and dynamic mechanical properties of Kevlar (synthetic) and kenaf fiber (natural fiber). From the above study we can be summarized as below:

- The main objective of review of the NF hybrid composite is the reduction of use of synthetic fiber.
- There is always a challenge to the researchers to achieve and enhance properties in the fiber to the required standard which needs suitable techniques. Hence, in order to reinforce polymer composites,

the thermal, mechanical, and physical properties should be totally found out.

- To develop hybrid composites for the application of industrial fields should be environment friendly and also more energetic for hybridization of natural and synthetic fiber.
- To develop a composite material with better mechanical performance by using comparatively NF (kenaf) and synthetic fiber (Kevlar) is simple as considered development techniques.
- Natural fiber should not be moisture absorb.
- Butea monosperma fiber may can also use in hybrid composite.
- The research were as a hybrid composite has a tremendous scope in the field of medical equipment, armor industries, aircraft equipment, etc.

KEYWORDS

- **hybrid composite**
- **kenaf fiber**
- **Kevlar fiber**
- **natural fiber**
- **reinforcement**
- **synthetic fiber**

REFERENCES

Davoodi, M., Sapuan, S. M., Ahmad, Ali, Khalina, & Jonoobi (2012). "Effect of polybutylene terephthalate (PBT) on impact property improvement of hybrid kenaf/glass epoxy composite. *Materials Letters, 67*(1), 5–7. doi: 10.1016/j.matlet.2011.08.101.

Davoodi, M., Sapuan, S. M., Ahmad, Ali, Khalina, & Jonoobi, (2010). Mechanical properties of hybrid kenaf /glass reinforced epoxy composite for passenger car bumper beam. *Materials and Design, 31*(10), 4927–4932. doi: 10.1016/j.matdes.2010.05.021.

Davoodi, M., Sapuan, S. M., Ahmad, Ali, Khalina, & Jonoobi, (2011). Concept selection of car bumper beam with developed hybrid bio-composite materials. *Materials and Design, 32*(10), 4857–4865. doi: 10.1016/jmatdes,2011.06.011.

Dipa, & Jogeswari, R., (2005). Thermoset biocomposites. In: Drzal, L. T., (ed.), *Natural Fibers, Biopolymers, and Biocomposites.* CRC Press, Boca Raton, FL. doi: 10.1201/978020358206,ch9.

El-Tayeb, N. S. M., (2009). Development and characterizations of low–cost polymeric composite materials. *Materials and Design, 30*(4), 1151–1160. doi: 10,1016/j. matdes.2008.06.024.

Elasiad, Dawood, Seracino, R., & Bobko, C., (2011). Mechanical properties of kenaf fiber reinforced concrete. *Construction and Building Materials, 25*(4), 1991–2001. doi: 10,1016/j conbuidmat,2010.11,052.

Faruk, O., Bledzki, A. K., Fink, H. P. N., & Sain, M., (2012). Biocomposites reinforced with natural fibers: 2000-2010. *Progress in Polymer Science, 37*(11), 1552–1596. doi: 10,1016 /jprogpolymsci,2012,04.

Hassan, E., Wei, Y., Jiao, H., Mu Huo, Y. (2016). Plant Fibers Reinforced Poly (Lactic Acid) (PLA) as a Green Composites: Review. *International Journal of Engineering Science and Technology*.

Ishak, M. R., Leman, Sapuan, Edeerozey & Othman, (2010). Mechanical properties of kenaf bast and core fiber reinforced unsaturated polyester composites. In: *9th National Symposium on Polymeric Materials (NSPM 2009),* 012006. doi: 10.1088/1757-899X/11/1/012006.

Ismail, H., Omar, N. F., & Othman, N., (2011). The effect of kenaf fiber loading on curing characteristics and mechanical properties of waste tire dust/kenaf fiber hybrid filler filled natural rubber compounds. *Bioresearches, 6*(4), 3742–3756. doi: 10.15376/ biores.6.4.3742-3756.

Mrazovia, M. (2013). Advanced composite materials of the future inaerospace industry. *INCAS Bulletin 5*(3):139-150. DOI: 10.13111/2066-8201.2013.5.3.14

Rahman, R., Putra, S.Z.F.S. (2019). Tensile properties of natural and synthetic fiber-reinforced polymer composites. Editor(s): Mohammad Jawaid, Mohamed Thariq, Naheed Saba, In Woodhead Publishing Series in Composites Science and Engineering, Mechanical and Physical Testing of Biocomposites, Fibre-Reinforced Composites and Hybrid Composites, Woodhead Publishing, 2019,81-102. ISBN 9780081022924,doi. org/10.1016/B978-0-08-102292-4.00005-9

Rosa, I. M., Santulli, C., Sarasini, F., & Valente, M., (2009). Post-impact damage characterization of hybrid configurations of jute/glass polyester laminates using acoustic emission and IR thermographs. *Composite Science and Technology, 69*(7, 8), 1142–1150. doi: 10.1016/j.compscitech.2009.02.011.

CHAPTER 3

Topology Optimization Techniques Employed for Additively Manufactured Medical Implants

PRIYANSHU SONI[1] and PARNIKA SHRIVASTAVA[2]

[1]*Department of Mechanical Engineering, National Institute of Technology, Hamirpur, Himachal Pradesh, India*

[2]*Department of Mechanical Engineering, National Institute of Technology, Jalandhar, Punjab, India*

ABSTRACT

Additive manufacturing (AM) has taken a huge transformation into the medical industry through its potential of turning raw materials into complex and usable products. Additively manufactured parts have higher accuracy and precision than parts produced from subtractive manufacturing techniques. It is evident that the share of AM in the medical and health-care industries is large and increasing day by day. In the medical field, implants and prosthetics play a significant role to give superior esthetics and preferable life. In the modern era, there are many genres of medical implants such as facial implants, head, and orthopedics, etc. During the manufacturing, weight reduction of medical implants is the most critical factor as heavily weighted implants tend to lose their stability and lead to Osseo-integration. Therefore, topology optimization of the implants produced from AM comes into picture as it helps in saving material as well as weight reduction of the same. In recent years, many techniques are

Advances in Manufacturing and Processing of Materials: Characterization and Applications.
Narendra Kumar, PhD, Mohit Tyagi, PhD, Dilbagh Panchal, PhD & Ravi Pratap Singh, PhD (Eds.)

used for topology optimization in many fields such as medical, aerospace, mechanical and civil, electronics, etc. However, still there is a lack of understanding regarding this emerging field as the information and the reported results are scattered. Therefore, to review topology optimization techniques that are applied in the course of time to produce more efficient 3D printed implants is of the prime importance. The chapter provides an overview of conventional topology optimization methods and reviews the recent techniques used for topology optimization of medical implants manufactured by 3D printing. This will help in preparing a database of the different employed techniques as well as their outcomes. This will also help in establishing the techniques that enable novel ways for the modification of medical implants.

3.1 INTRODUCTION

Additive manufacturing (AM), also known as 3D printing, was developed in 1980 for models and prototype fabrication. It is commonly termed as "Rapid prototyping" in the early stage of the AM process. The process creates 3D complex objects by depositing material in a layer-by-layer manner (Wong & Hernandez, 2012; Shannon & Sullivan, 2021). The process claims its merit as the customized and complex parts that are difficult or merely impossible to manufacture using subtractive manufacturing techniques such as milling, grinding, and boring, etc., can be easily manufactured by the process (Pontevedra et al., 2019). The application of AM in the modern era is enormous in different fields such as medical, aerospace, automobile, defense, oil and gas industries (Munk et al., 2015). The growth of AM is rapid as the prerequisites of users and clients are more distinctive (Nguyen & Vignat, 2017). In today's time along with customized and complex structures, the requirement is also to manufacture parts with reduced weight without compromising with the shape, strength, and efficiency of the part.

Martin & Noboru (1988) developed topology optimization as a computational technique that optimizes shapes for a prescribed design domain based on a set of design considerations such as load and boundary conditions. Nguyen & Vignat (2017) highlighted that topology optimization is a subset of structural optimization. Applications of topology optimization are in many fields such as aerospace, medical, automobile, etc., to create lightweight structures. Topology optimization was first used in the medical field for modeling implants to match the human skeletal structure. Salmi et al. (2021) and Abdulaziz et al. (2020) reported that medical industry employed topology optimization because of the possibility of creating lightweight implants. It upgrades the mechanical performance and fatigue behavior and other problems such as stress shielding or tissue bonding. Topology optimization is used for optimizing a variety of implants. Some popular implants are dental, facial, orthopedic, and hip bone implants, as well as prosthetics. Tesk & Widera (1973) are the first who implemented a finite element method in medical implants to optimize the topology of the same. Henceforth, it became a growing tool to forecast stress distribution at the implant-bone contact region.

In recent years, researchers have introduced various topology optimization techniques for different types of implants. These include Homogenization SIMP, SKO, ESO, and BESO. Additionally, newer techniques like Multi-Material Topology and Multi-Resolution Optimization are used for reconstructing the faces of patients with facial injuries or who have undergone tumor surgery (Park et al., 2021; Sutradhar et al., 2016; Zhu et al., 2021). However, there is a lack of understanding in this emerging field due to scattered information and reported results. Therefore, it is crucial to review topology optimization techniques applied in 3D-printed implants to enhance efficiency. This chapter provides an overview of conventional methods and reviews recent techniques used in medical implants manufactured by 3D printing. The aim is to create a database of employed techniques and their outcomes, facilitating the development of novel modification approaches for medical implants. The upcoming sections of this chapter serve as a reference for further research in biomedical applications.

3.2 ADDITIVE MANUFACTURING (AM) OVERVIEW

Additive manufacturing (AM) is also known as "3DP (three-dimensional-printing) or Solid Freeform Fabrication Technology." In which 3D complex object is created by deposition of material in layer-by-layer formation directly as per the CAD model. AM is introduced by Charles Hull (1986). The reason behind the popularity of AM these days is its advantages over the conventional manufacturing process. AM has unique advantages such as low material wastage, reduce energy consumption, ease to recreate optimized legacy parts, free of complexity with higher accuracy, and so on (Ngo et al., 2018).

The applications of the AM process in various sectors, especially in the medical sector, are discussed in subsections.

3.2.1 ADDITIVE MANUFACTURING (AM) APPLICATIONS

In recent years, the use of AM in various industries has increased. Applications of AM in various industries such as medical and health care, consumer products, industrial, aerospace, automobile, oil and gas industries, defense, etc., as shown in Figure 3.1 (Ngo et al., 2018; Shannon & Sullivan, 2021).

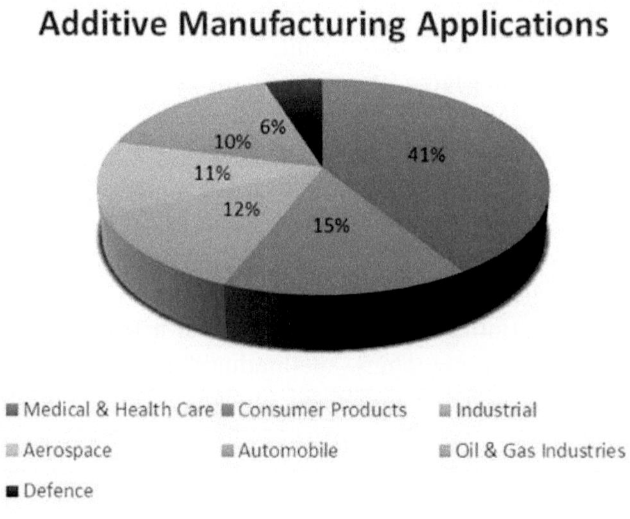

FIGURE 3.1 Additive manufacturing in different industries.

Source: Adapted from Ngo et al., 2018

3.2.2 ADDITIVE MANUFACTURING (AM) IN THE MEDICAL SECTOR

The use of AM in the medical and healthcare sector is huge because of its various advantages over conventional manufacturing technologies (Singh & Ramakrishna, 2017; Series, 2021). According to Figure 3.1, the medical and healthcare industries hold a 41% share of the AM market additive.

This is because nowadays researchers are interested in designing a medical implant that has a lightweight design with suitable mechanical properties and preferable esthetics (Javaid & Haleem, 2018). AM performs a decent role in the fabrication of complex medical implants such as dental, facial orthopedic, and as well as prosthetics. Medical implants esthetics is an important parameter for patients to live a preferable life (Bhargav et al., 2018; Wu et al., 2021; Salmi, 2021).

3.3 TOPOLOGY OPTIMIZATION

Topology is a major area of mathematics concerned with properties that are preserved under the continuous deformation of an object. Topology optimization is a very useful method that proposes optimized shapes within the design domain based on a set of design considerations, such as load and boundary conditions. It has been well-developed and applied to a variety of applications, structural, mechanical, and material systems (Rozvany, 2001; Song & Le, 2010).

3.3.1 ADDITIVE MANUFACTURING (AM) AND TOPOLOGY OPTIMIZATION

Topology optimization is a growing field of modern engineering technologies. The lightweight structures are developed by topology optimization. The shape and structure of topology optimized parts are complex (Abdulaziz et al., 2020). These shapes are manufactured by AM because of their advantages, such as being free of complexity, reduction in material wastage, energy, cost, production time, etc. (Patel & Gohil, 2021; Liu & Ma, 2016).

3.3.2 METHODOLOGY OF TOPOLOGY OPTIMIZATION FOR MEDICAL IMPLANTS

To perform topology optimization on medical implants, various steps are involved, which are shown in Figure 3.2. The first step involves patient data acquisition. In this step, the data of a patient who is suffering from any kind of physical defect is collected using reverse engineering technology such as CT (Computer Tomography) or MRI (magnetic resonance imaging; Javaid et al., 2019). These 2D images are converted into 3D STL files. Further, a 3D CAD model is created using STL Files. The defective region geometry is extracted into the 3D CAD model. Finite element analysis is performed on the defective geometry using finite element software such as ANSYS, ABQUAS, and Hyper mesh. After getting a successful analysis result, topology optimization is performed on the implant geometry. At last, finite element analysis on the new optimized geometry is performed to check the feasibility and also to compare its analysis result with old geometry (Cholkar et al., 2021b; Gómez Pérez et al., 2017; Li et al., 2020).

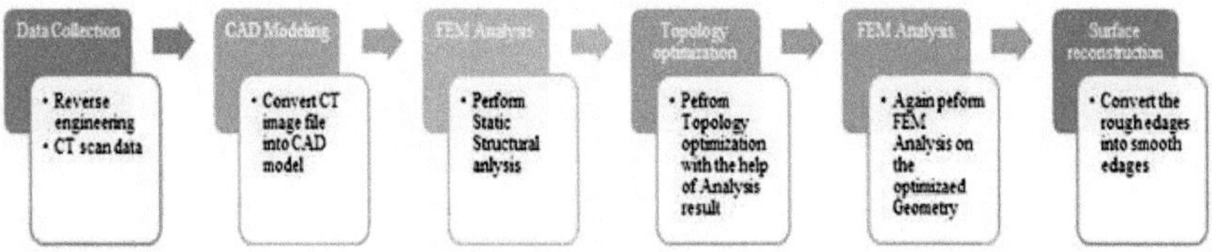

FIGURE 3.2 Flow chart of topology optimization.

3.4 TOPOLOGY OPTIMIZATION TECHNIQUES EMPLOYED FOR MEDICAL IMPLANTS

Various topology optimization techniques were developed to optimize the topology of structures. Some of the techniques which are used to optimize the topology of medical implants are discussed here. This chapter reviews SKO, SIMP, evolutionary structural optimization (ESO) and bidirectional evolutionary structural optimization (BESO) techniques and reviews two more updated techniques which are multi-material topology optimization (MMTO) and multi-resolution topology optimization (MRTO) (Rozvany, 2009) (Figure 3.3).

3.4.1 SOFT KILL OPTION (SKO)

Baumgartner et al. (1992) developed a technology which is based on adaptive bone mineralization and known as SKO. It is used for an optimal solution by adaption in young modulus in accordance with stress distribution. In the initial step, discretization in the design domain is done by finite elements. In the next step, elastic modulus is varied in the design domain according to stress distribution, and then the loading state is used to obtain the new stress distribution for the new/optimized design region, which has variable elastic modulus. In this technique young modulus is variable and element density is not involved (Baumgartner et al., 1992; Yoon et al., 2013; Sigmund & Maute, 2013).

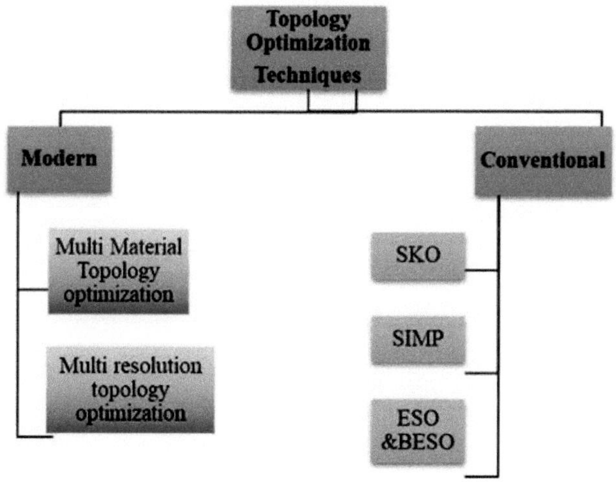

FIGURE 3.3 Topology optimization techniques.

3.4.2 SOLID ISOTROPIC MATERIAL WITH PENALIZATION (SIMP)

Martin & Noboru (1988) developed a density-based approach for topology optimization in further development to homogenization method. SIMP involves numerical finite-element-based topology optimization method. It has many other names such as "material interpolation," "power law" or "density method" but SIMP is the most common term (Sigmund & Maute, 2013). This technique aims to distribute the material, and this distribution is considered as a main variable for the optimization procedure. Element density behaves like the design parameter in the SIMP technique. The entire design region is discretized into small N finite elements some assumptions are considered for material. There is a relationship between the relative density and element stiffness. It behaves as an optimization variable and can take values between 0 (void) and 1 (solid). In SIMP approach, the intermediate relative density values are eliminated so that a practical solution is determined which has either void or material in the design domain and fractional densities are penalized. This is done using the penalization factor. Eqn. (1) is the basic objective function for the SIMP approach to minimize the compliance c(x), where "x" denotes the design variable vector and Eqn. (2) relates the global vector "F." the displacement vector "U" and the global stiffness matrix "K." Eqn. (3) deals with the prescribed volume fraction "f" which is the ratio of material volume "V(x)" to the design domain "v_0." Eqn. (4) is a mathematical constraint on design variables as they cannot be equal to 0 or a null vector.

$$\min_x : c(\mathrm{x}) = \mathrm{U^T K U} = \sum_{e=1}^{N} (\mathrm{X}^e)^p u_e^T k_0 u_e \tag{1}$$

$$\mathrm{KU} = \mathrm{F} \tag{2}$$

$$\frac{v(\mathrm{x})}{v_0} = f \tag{3}$$

$$0 < x_{\min} \leq X \leq 1 \tag{4}$$

3.4.3 ESO AND BESO

Xie & Steven (1993) developed a new technique for Structural Optimization, ESO, which gives a new approach to structural optimization. However, it is better than the traditional approaches. Optimized structure was gained by the rejection of ineffective material from the discretized design domain. There are numerous parameters, such as stiffness, frequency, and buckling load, that can be used as design control limits. The material removal is based on the stress levels or von-mises stress in the design region (Razvan, 2014; Munk et al., 2015). The equation for von Mises stress is given by Eqn. (5) (Xie & Steven, 1993; Tanskanen, 2002).

$$\sigma_e^y = \sqrt{\sigma_{11}^2 + \sigma_{22}^2 - \sigma_{11}\sigma_{22} + 3\sigma_{12}^2} \tag{5}$$

$$\frac{\sigma_e^y}{\sigma_{\max}^y} < RR^i \tag{6}$$

At the end of finite element analysis, elements followed Eqn. (6) (Xie & Steven, 1993) from the design region where RR is known as rejection ratio. To improve the capability of ESO it's extended version is introduced which is called Bidirectional structural optimization (Young et al., 1999; Radman & Sc, 2013; Sigmund & Maute, 2013). BESO can add elements to the design region as well as can remove the elements from the design region.

3.4.4 *MULTI-MATERIAL TOPOLOGY OPTIMIZATION (MMTO)*

In density-based methods, topology optimization is formulated as a material distribution problem. Usually, the formulation of topology optimization reduces compliance. According to Li & Kim (2018), in solid material with penalization (SIMP) has a limited amount of material for distribution with reference to input boundary conditions. In a design region Ω, R^2 or R^3, a MMTO finds optimal locations of 'm' number of different materials that reduce the compliance under a volume fraction constraint V_s (amount of material that can be utilized by the algorithm). The finite element that discretizes the design region is used to find the performance and visualize the optimized solution. The normalized density is considered as the design variables. Since one element can have combinations of different materials in MMTO, Park et al. (2021) comply the density at a given location and represented with a group of density values for different material phases.

3.4.5 *MULTI-RESOLUTION TOPOLOGY OPTIMIZATION (MRTO)*

Multi-Resolution topology optimization (MRTO) assumes decoupling of the design and analysis of discretization. It can achieve high-resolution designs at relatively low computational analysis. Recent literature has shown that MRTO is 3 to 30 times faster than the conventional topology optimization techniques for two-dimensional and three-dimensional problems (Keulen, 2020).

3.5 TOPOLOGY OPTIMIZATION APPLICATION IN MEDICAL IMPLANT FIELD

Various implant design is created using topology optimization and manufactured through AM. Some of the recent work which are done on different types of the medical implants are discussed in subsections.

3.5.1 HEAD IMPLANTS

Gómez Pérez et al. (2017) designed and developed a customized craniofacial implant using topology optimization. The optimization process starts with the patient's CT scan data acquisition. These 2D images of patients are converted into STL (standard tessellation language) file format. Basically, STL files are standard transmission format in CAD system. A solid CAD model of patient's head CT scan is converted and a model surgery is performed in some problems such as cancer or tumors. FEA analysis is performed to analyze the stress distribution on these models using suitable boundary conditions. Then, SIMP technique is used for topology optimization. Topology optimization resulted in 40% overall reduction in the volume of the head implant. Finally optimized customized implant is manufactured by AM.

3.5.2 DENTAL IMPLANTS

Chang et al. (2012) generated a computerized model of the edentulous maxillary first molar, which contained buccal bone irregularities. 3D CAD model. As the model was balanced, instead of considering the entire FE model, only half was enough for the analysis and optimization procedure. The implant support was created by titanium and the crown was of gold alloy and these materials were isotropic, homogenous, and linearly elastic. The have employed SIMP technique which was applied on threaded implant support. The loading conditions included an oblique load of 100 N in vertical and 20 N in horizontal from the buccal direction of the crown surface. The new implant design volume was reduced by 9.8 mm^3, which is almost equal to an 18% decrease in volume. Due to a reduction in volume, reduction in material cost, improved sustainability, and less displacement in mastication.

Yoon et al. (2013) performed the refinement and optimization of a 3D dental implant with the complex root shapes of natural teeth. Due to the complex shape of roots, it is difficult to drill manually current commercial implants. For designing such types of roots, a conductive robotic milling algorithm has been used. Due to the presence of sharp curves and undercuts, anatomically correct models must be refined for 3D robotic milling and these refined shapes must show to be optimized for load bearing. Refinement of the anatomically correct natural tooth-shaped models for robotic milling was accomplished using CAS tools for smoothing the sharp curves and undercuts. The load-bearing optimization is done by SKO.

Chen et al. (2015) performed a contact shape optimization on removable partial dentures. The contact shape optimization is based on BESO (bi-directional evolutionary structural optimization). Both initial and optimized dentures were manufactured by AM and checked their strength with the *vitro* testes. With the help of optimization techniques, the contact pressure is reduced by 70%, and uniformity is improved by 63%. The *vitro* tests verified that hydrostatic pressure which is in the mucosa is sufficient below the pressure pain thresholds (PPT) and due to it there are less risk of residual ridge resorption and minimizing pain or discomfort. The BESO and AM fabrication provided a novel method for rapid denture design at a low cost for a specific patient.

Ziervogel et al. (2008) optimized the dental bridge framework (three-unit and four-unit) based on the distribution of two materials which are porcelain and zirconia, the region with relatively more stress was assigned zirconia and the remaining region was porcelain. The extended finite element method (XFEM) was used for stress analysis, and the BESO algorithm was used for topological optimization. Models with an optimum material distribution for various volume fractions (20%, 30%, 40%, 50% of the metal) for both the four-unit and three-unit dental bridge design were determined. The fracture initiation began near the gingival connector at the first molar and the crack was fully eliminated when the model consisted of 40% zirconia and 60% porcelain for the four-unit and three-unit dental bridges design. Seitz et al. (2018) obtained the root of the lower mandibular incisor through using SKO. SKO is the most suitable technique of topology optimization for dental roots. They also compare the SKO technique with the SIMP techniques and found some comparable results. According to the comparison results, it is found that the SKO technique gives a better result compared to SIMP as far as topology optimization is concerned.

Park et al. (2019) optimized a complete denture framework as shown in Figure 3.4. This optimization aims to ensure the efficient transference of the masticatory load acting on the denture to auxiliary fixtures like implants and screws also reducing the weight of the model by reduction of inefficient material repeatedly. They also analyzed commercially available denture framework to identify any remaining area for material removal. MRTO techniques were utilized to achieve the final optimized version with 16% less stress.

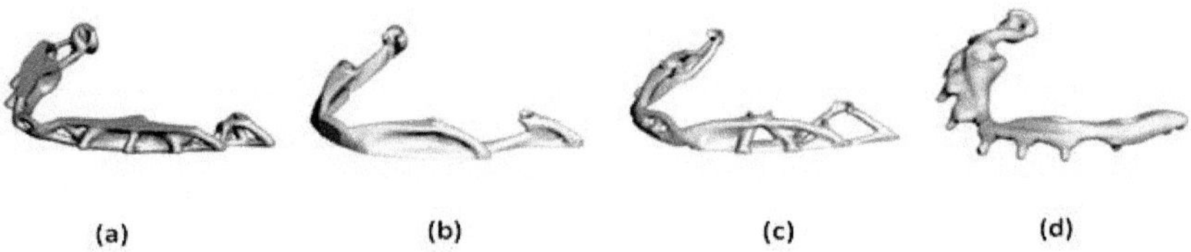

FIGURE 3.4 (a–c) Topology optimized version of specific design domain; and (d) commercially available framework.

Source: Reprinted/Adapted with permission from Park et al., 2019. Copyright © 2019 John Wiley & Sons, Ltd.

Li et al. (2020) accomplished a focused work on the optimization of patient-specific mandible reconstruction as shown in Figure 3.5. Numerous parts of the prosthesis, like the fixation wing and the internal structure, were analyzed and optimized for the mandible bone. The defective region was recognized from the first premolar to the second premolar to the second molar in the unilateral mandibular body. There are some parameters such as prosthesis weight, maximum and minimum stress, and von-misses stress which are considered for topology optimization. The two-load condition on premolar and molar were described as axial and oblique with a load of 100 N and 200 N. The first step was the optimization of the fixation wing and then two different models are prepared, and these are based on load condition. The optimized wing structure of the WOS model and the internal structures were 60% WOSA and 40% WOSO.

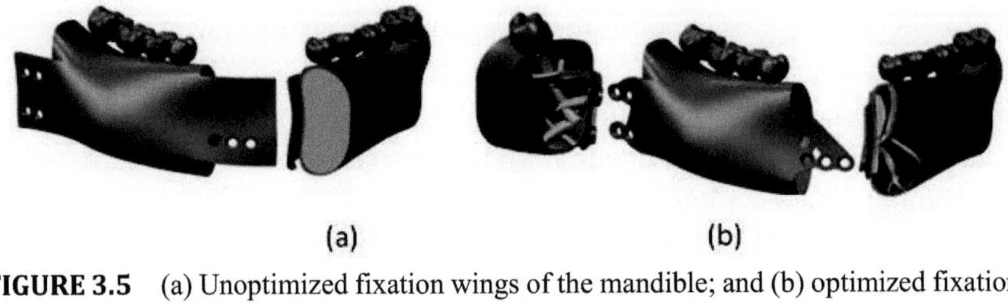

FIGURE 3.5 (a) Unoptimized fixation wings of the mandible; and (b) optimized fixation wing of the mandible.

Source: Reprinted with permission from Li et al., 2020. Copyright © 2020 Elsevier Ltd.

3.5.3 FACIAL IMPLANTS

Dai et al. (2018) proposed a new design of maxillofacial (upper jawbone) to achieve a lightweight custom medical implant that is based on topology optimization. SIMP technique is used for topology optimization. Firstly, parametric predesign of the optimized model was carried out to ensure that the optimized model conformed to the main physiological anatomical structure and recover basic function structure. The design region is defined under the given load conditions. SIMP technique with the penalization factor helps to get the final optimized 3D model. Finally, optimization results verified that the proposed method of topology optimization for the maxillofacial implant is efficient.

Park, Zobaer, & Sutradhar (2021) proposed a topology optimization approach that provides multi-material designs for bone replacement implants in complicated craniofacial segmental defects. A two-scale multi-material topology optimization approach that can be used for both macrostructure and microstructure levels. In the first step of optimization, a multi-resolution topology optimization approach is used to produce a multi-material design with maximum stiffness. This approach is beneficial for bone implant design since, in addition to imparting the desired functionality property to the design of facial implant it is also showing the porosity. To check the efficacy of the technique, four different large craniofacial defects due to maxillectomy are considered with their respective implant design are shown. This optimized design. A good potential in developing patient's specific optimized designs suitable for AM is shown in Figure 3.6.

Cholkar et al. (2021a) researched and evaluated modeling method for the zygoma bone (human jawbone) also found the most suitable material and optimized design for this critical bone implant. The same is shown in Figure 3.7. The methodology of this process is very effective and innovative, which includes modeling, design validation, topological optimization, and numerical analysis. It included the reverse engineering of CT scan data to generate the 3D model, and the topology optimization technique is used to generate the lightweight implant without causing extra stress. For choosing the best material for implant, three materials (TiAl6l4V, stainless Steel316L, and CoCr alloys) are tested. SIMP is used for Topology optimization. After optimization, Taguchi method is applied to the designs to find the optimum design. The optimum design is found with the highest mass reduction of 25%.

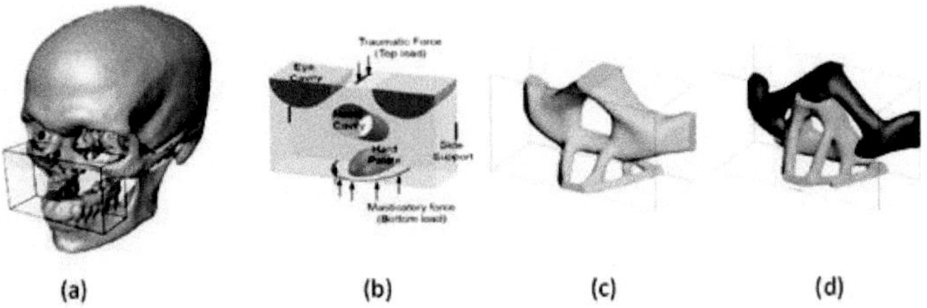

FIGURE 3.6 (a) Design domain extraction from CT scan data; (b) boundary conditions; (c) bone replacement shape using single material topology optimization; and (d) bone replacement shape using multi-material topology optimization.

Source: Reprinted with permission from Park et al., 2021. Copyright © 2021 by the authors. Licensee MDPI, Basel, Switzerland. https://creativecommons.org/licenses/by/4.0/

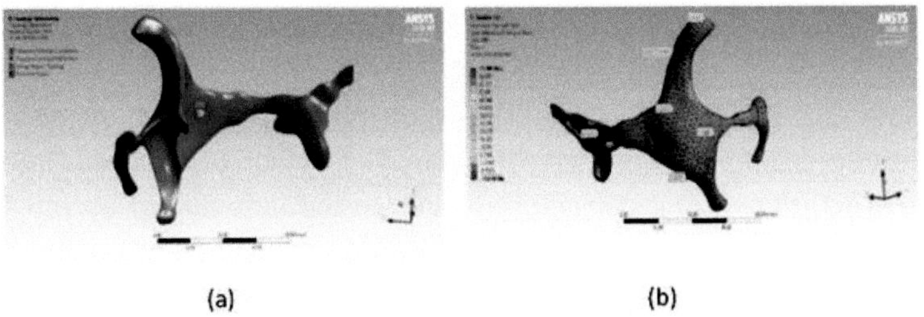

FIGURE 3.7 (a) Zygoma part 3D CAD model before optimization; and (b) 3D CAD model after optimization (Cholkar et al., 2021a).

Source: Reprinted with permission from Cholkar et al., 2021a. Copyright 2021 Elsevier.

3.5.4 *ORTHOPEDICS IMPLANTS*

Iqbal et al. (2019) proposed a workflow of a customized pelvis prosthesis using multi-objective topology optimization. They carried optimization via various sectioning types with a minimization of a weighted sum of the compliance (maximize stiffness) under six daily routine activities as the objective function and volume reduction as a constraint. The geometry with low weight and high strength has a remarkable match for pelvis bone. The finite element method is used to check the strength of prosthesis. This analysis showed that the initial geometry could withstand the static loading condition for various daily activities with less concentration.

3.6 CONCLUSION

In this chapter, the authors presented the state of art on topology optimization of medical implants to be produced by AM technologies. The chapter discusses topological optimization techniques such as SKO, SIMP, ESO and BESO (bidirectional evolutionary structural optimization), MMTO, and MRTO techniques. The chapter also discusses the aspects like implant position and geometric parameters. The peculiar research and their outcomes related to topology optimization on various implants like Head, Facial, Dental, and Orthopedic are also presented. It can be concluded that the most used technique, in terms of research and commercial software application, is SIMP. BESO is another preferred technique by the researchers. Results were summarized, keeping in the record the research accomplishments in the past few years and effective 3D printings of the topology-optimized medical implants.

KEYWORDS

- **additive manufacturing**
- **facial implants**
- **medical implant field**
- **medical sector**
- **orthopedics implants**
- **penalization**
- **soft kill option**
- **topology optimization**

REFERENCES

Abdulaziz, A., Tamimi, A., Almeida, H., & Bartolo, P., (2020). Structural optimization for medical implants through additive manufacturing. *Progress in Additive Manufacturing, 0123456789*. https://doi.org/10.1007/s40964-020-00109-7.

Baumgartner, A., Harzheim, L., & Mattheck, C., (1992). SKO (soft kill option): The biological way to find an optimum structure topology. *International Journal of Fatigue, 14*(6), 387–393. https://doi.org/10.1016/0142-1123(92)90226-3.

Bhargav, A., Sanjairaj, V., Rosa, V., Feng, L. W., & Fuh, Y. H. J., (2018). Applications of additive manufacturing in dentistry: A review. *Journal of Biomedical Materials Research – Part B Applied Biomaterials, 106*(5), 2058–2064. https://doi.org/10.1002/jbm.b.33961.

Chang, C. L., Chen, C. S., Huang, C. H., & Hsu, M. L., (2012). Finite element analysis of the dental implant using a topology optimization method. *Medical Engineering and Physics*, *34*(7), 999–1008. https://doi.org/10.1016/j.medengphy.2012.06.004.

Chen, J., Ahmad, R., Suenaga, H., Li, W., Sasaki, K., Swain, M., & Li, Q., (2015). Shape optimization for additive manufacturing of removable partial dentures: A new paradigm for prosthetic CAD/CAM. *PLoS One*, *10*(7), 1–17. https://doi.org/10.1371/journal.pone.0132552.

Cholkar, A., Kinahan, D., & Brabazon, D., (2021a). *Topology and FEA Modeling and Optimization of a Patient-Specific Zygoma Implant*.

Cholkar, A., Kinahan, D., & Brabazon, D., (2021b). Topology and FEA modeling and optimization of a patient-specific zygoma implant. *ESAFORM 2021*. https://doi.org/10.25518/esaform21.756.

Dai, N., Zhu, J. F., Zhang, M., Meng, L. Y., Yu, X. L., Zhang, Y. H., Liu, B. Y., & Zhang, S. L., (2018). Design of a maxillofacial prosthesis based on topology optimization. *Journal of Mechanics in Medicine and Biology*, *18*(3). https://doi.org/10.1142/S0219519418500240.

Gómez, P. C. A., Medellín-Castillo, H. I., & Espinosa-Castañeda, R., (2017). *Computer Assisted Design and Structural Topology Optimization of Customized Craniofacial Implants*. http://www.asme.org/about-asme/terms-of-use (accessed on 11 September 2023).

Iqbal, T., Wang, L., Li, D., Dong, E., Fan, H., Fu, J., & Hu, C., (2019). A general multi-objective topology optimization methodology developed for customized design of pelvic prostheses. *Medical Engineering and Physics*, *69*, 8–16. https://doi.org/10.1016/j.medengphy.2019.06.008.

Javaid, M., Haleem, A., & Kumar, L., (2019). Current status and applications of 3D scanning in dentistry. *Clinical Epidemiology and Global Health*, *7*(2), 228–233. https://doi.org/10.1016/j.cegh.2018.07.005.

Keulen, F. V., (2020). *Design and Analysis Adaptivity in Multiresolution Topology Optimization*, 450–476. https://doi.org/10.1002/nme.6217.

Li, C. H., Wu, C. H., & Lin, C. L., (2020). Design of a patient-specific mandible reconstruction implant with dental prosthesis for metal 3D printing using integrated weighted topology optimization and finite element analysis. *Journal of the Mechanical Behavior of Biomedical Materials*, *105*, 103700. https://doi.org/10.1016/j.jmbbm.2020.103700.

Li, D., & Kim, I. Y., (2018). Multi-material topology optimization for practical lightweight design. *Structural and Multidisciplinary Optimization*, *58*(3), 1081–1094. https://doi.org/10.1007/s00158-018-1953-z.

Liu, J., & Ma, Y., (2016). A survey of manufacturing-oriented topology optimization methods. *Advances in Engineering Software*, *100*, 161–175. https://doi.org/10.1016/j.advengsoft.2016.07.017.

Martin, P. B., & Noboru, K., (1988). Generating optimal topologies in structural design using a homogenization method. *Computer Methods in Applied Mechanics and Engineering*, *71*, 197–224.

Munk, D. J., Vio, G. A., & Steven, G. P., (2015). Topology and shape optimization methods using evolutionary algorithms: A review. *Structural and Multidisciplinary Optimization*, *52*(3), 613–631. https://doi.org/10.1007/s00158-015-1261-9.

Ngo, T. D., Kashani, A., Imbalzano, G., Nguyen, K. T. Q., & Hui, D., (2018). Additive manufacturing (3D printing): A review of materials, methods, applications and challenges. *Composites Part B*, *143*(2017), 172–196. https://doi.org/10.1016/j.compositesb.2018.02.012.

Park, J., Zobaer, T., & Sutradhar, A., (2021). A two-scale multi-resolution topologically optimized multi-material design of 3D printed craniofacial bone implants. *Micromachines*, *12*(2), 1–19. https://doi.org/10.3390/mi12020101.

Patel, P., & Gohil, P., (2021). Role of additive manufacturing in medical application COVID-19 scenario: India case study. *Journal of Manufacturing Systems*, *60*, 811–822. https://doi.org/10.1016/j.jmsy.2020.11.006.

Radman, A., & Sc, B., (2013). *Bi-directional Evolutionary Structural Optimization (BESO) for Topology Optimization of Material's Microstructure. 1*, 1–244. https://researchrepository.rmit.edu.au/esploro/outputs/9921861503701341 (accessed on 11 September 2023).

Razvan, C., (2014). *Overview of Structural Topology Optimization Methods for Plane and Solid Structures.* Annals of the Oradea University. Fascicle of Management and Technological Engineering, *XXIII (XII*(3). https://doi.org/10.15660/auofmte.2014-3.3043.

Rozvany, G. I. N., (2009). A critical review of established methods of structural topology optimization. *Structural and Multidisciplinary Optimization*, *37*(3), 217–237. https://doi.org/10.1007/s00158-007-0217-0.

Salmi, M., (2021). Additive manufacturing processes in medical applications. *Materials*, *14*(1), 1–16. https://doi.org/10.3390/ma14010191.

Seitz, K. F., Grabe, J., & Köhne, T., (2018). A three-dimensional topology optimization model for tooth-root morphology. *Computer Methods in Biomechanics and Biomedical Engineering*, *21*(2), 177–185. https://doi.org/10.1080/10255842.2018.1431778.

Series, C., (2021). *Progress on Different Topology Optimization Approaches and Optimization for Additive Manufacturing: A Review*. https://doi.org/10.1088/1742-6596/1939/1/012101.

Shannon, A., & Sullivan, L. W. O., (2021). *The Application of Additive Manufacturing / 3D Printing in Ergonomic Aspects of Product Design: A Systematic Review, 97*(2020). https://doi.org/10.1016/j.apergo.2021.103528.

Sigmund, O., & Maute, K., (2013). Topology optimization approaches: A comparative review. *Structural and Multidisciplinary Optimization*, *48*(6), 1031–1055. https://doi.org/10.1007/s00158-013-0978-6.

Singh, S., & Ramakrishna, S., (2017). Biomedical applications of additive manufacturing: Present and future. *Current Opinion in Biomedical Engineering*, *2*, 105–115. https://doi.org/10.1016/j.cobme.2017.05.006.

Tanskanen, P., (2002). The evolutionary structural optimization method: Theoretical aspects. *Computer Methods in Applied Mechanics and Engineering*, *191*(47, 48), 5485–5498. https://doi.org/10.1016/S0045-7825(02)00464-4.

Xie, Y. M., & Steven, G. P., (1993). Technical note: a simple approach to structural optimization. *Compurers & Structures*, *49*(5), 885–896.

Yoon, Y., Sun, X., Huang, J. K., Hou, G., Rechowicz, K., & Mckenzie, F. D., (2013). Designing natural-tooth-shaped dental implants based on soft-kill option optimization. *Computer-Aided Design and Applications*, *10*(1), 59–72. https://doi.org/10.3722/cadaps.2013.59-72.

Young, V., Querin, O. M., Steven, G. P., & Xie, Y. M., (1999). 3D and multiple load case bi-directional evolutionary structural optimization (BESO). *Structural Optimization*, *18*(2, 3), 183–192. https://doi.org/10.1007/bf01195993.

Zhu, J., Zhou, H., Wang, C., Zhou, L., Yuan, S., & Zhang, W., (2021). A review of topology optimization for additive manufacturing: Status and challenges. *Chinese Journal of Aeronautics*, *34*(1), 91–110. https://doi.org/10.1016/j.cja.2020.09.020.

Ziervogel, G., Cartwright, A., Tas, A., Adejuwon, J., Zermoglio, F., Shale, M., & Smith, B., (2008). Climate change and adaptation in African agriculture. *Training*, *4179*(2015), 53. https://doi.org/10.1002/cnm.

CHAPTER 4

A Review on Hybrid Friction Stir Incremental Sheet Forming of Tailored Blanks

SUHAS CHAVHAN and PARNIKA SHRIVASTAVA

Department of Mechanical Engineering, National Institute of Technology, Hamirpur, Himachal Pradesh, India

ABSTRACT

In contrast to traditional sheet metal forming processes, which require specialized punches and dies, incremental sheet metal forming (ISF) is a die-less process. ISF facilitates localized deformation and is, therefore, proven as a cost-effective and flexible approach for the formation of customized and complex parts. However, ISF, when performed on sheets, localized stretching promotes localized sheet thinning, which leads to failure when the forming requirements are exceeded. In addition, parts formed by ISF suffer from geometrical inaccuracy due to springback. The issues pertaining to the process can be eliminated by joining ISF with a solid-state welding process, i.e., friction stir welding (FSW) technique. The hybrid friction stirs incremental sheet forming (FSISF) technique employs incremental sheet forming (ISF) of tailored blanks that are manufactured by FSW. FSW produces tailed blanks by local frictional heating and, therefore, creates a soft localized zone, which in turn enhances the formability of the same. The hybrid process is still under exploration, and most of the research have been focused on the influence of process parameter on the

Advances in Manufacturing and Processing of Materials: Characterization and Applications.
Narendra Kumar, PhD, Mohit Tyagi, PhD, Dilbagh Panchal, PhD & Ravi Pratap Singh, PhD (Eds.)

outcomes of the process. Preliminary studies have proven the efficiency of the hybrid FSISF technique however, the reported results are unstructured. Therefore, this chapter provides a critical review of the process and reports the recent work done, and concludes the reported results. In addition, the further scope for the research related to FSISF is also mentioned for the efficient manufacturing of sheet metal components.

4.1 INTRODUCTION

Incremental sheet forming (ISF) is a modification to conventional forming where sheets are incrementally formed into desired shapes using a forming tool in a CNC machine (Aerens et al., 2010; Jeswiet et al., 2005; Rubino et al., 2020). The process is a die-less forming operation and therefore, this new forming method is especially suitable for forming customized and complex shapes in small batches (Araghi et al., 2011; Hussain et al., 2020; Jeswiet et al., 2005). In ISF, there is no need for specialized tooling and thus enhanced flexibility, lowers lead time, and the smaller initial cost is proven as a significant advantage over the conventional forming operations (Andrade-Campos et al., 2020; Araghi et al., 2009).

ISF process has gained attention in manufacturing amongst automobiles, the aerospace industry, and the medical sector. ISF has numerous merits over conventional forming operations; however, when performed on sheets, localized stretching promotes localized sheet thinning, which originates failure when the forming requirements (wall angle and/or forming depth) are exceeded. In addition, due to localized deformation elastic-plastic zone builds up residual stresses which leads to spring back and results in a variation of the desired shape (Araghi et al., 2009, 2011; Jiang et al., 2017).

Instead of making developments to ISF, an interesting alternative to cope with the disadvantages of ISF has been introduced by combining it with other processes. Laser-assisted ISF is established to manufacture parts using materials such as titanium or magnesium alloys (Araghi et al., 2009; Li et al., 2017). Another interesting development, i.e., hybrid friction stir incremental sheet forming (FSISF) by combining ISF and friction stir welding (FSW) allows manufacturing of the sheet metals parts at enhanced formability. This development has been motivated not only by constraints of ISF but also by constraints of the formability of the material.

FSW is a solid-state welding technique that is used to join similar or dissimilar metals using a rotating tool (Cai et al., 2020; Jesus et al., 2018; Krishna et al., 2014). This technique was developed initially for welding metals and alloys which is difficult to join using traditional fusion welding techniques (Papadopoulos et al., 2013; Threadgill et al., 2009; Wang et al., 2004). Generally, Al and its alloys are difficult to weld as it suffers cracks and porosity after fusion welding susceptible to hydrogen damage (Araghi et al., 2009). Therefore, it is reported not to use shielding gas which is usually toxic in nature (Hasan et al., 2015; Krishna et al., 2014). The process reports advantages such as lesser sheet material distortion, lack of melting-related defects, higher joint strength, and lower hydrogen contents in the joints.

In hybrid FSISF, first, a tailored blank is manufactured using FSW by rotating tools on sheets to be joined. Thus, frictional heating is generated in FSW and a soft localized zone is produced near the weld in sheets. Further, a hemispherical shaped forming tool is utilized to deform incrementally the tailored blank in complex shape. A soft localized zone in welding helps to reduce spring back and improves plastic deformation capabilities. Hybrid FSISF results in the reduction of the forming forces and further enhances the formability of tailored blanks.

This chapter provides a critical review of the hybrid process reports the recent work done, and concludes the reported results. In addition, the further scope for the research related to FSISF is also highlighted for the efficient die-less manufacturing of the sheet metal components.

4.2 LIMITATION OF SPIF PROCESS

Although various advantages are associated with the process however, certain limitations do not allow the process to reach the shop floor. The major limitations are pointed out as follows:

1. Maji et al. (2019); Pasteur & Koch (1941); Silva et al. (2009); Skjoedt et al. (2010) highlighted that localized stretching promotes localized sheet thinning which further results in failure as forming requirements are outdone. Thus, a substantial amount of material flow occurs from the periphery of forming sheet which is not formed using localized deformation.

2. Araghi et al. (2011); Gohil & Modi (2021); John et al. (2021); Pasteur & Koch (1941) reported that due to the localized deformation, an elastic-plastic region will occur as forming tool proceeds incrementally. This induces residual stress, which further causes variation in target shape due to spring back. They have mentioned that the effect of residual stresses can be reduced by thermal treatment before forming operations.
3. Ahmed et al. (2017, 2021); Davarpanah et al. (2015) illustrated that in ISF, the processing speed cannot be increased significantly since it has an impact on the surface roughness of the part, further affecting the formability of sheets.

Because of these above restrictions, complex parts require committed forming approaches that reduce sheet stretching and also reduce the spring back effect in the formed part.

4.3 CHARACTERISTICS OF FRICTION STIR WELDING (FSW)

Friction stir welding (FSW) is a solid-state welding technique that is used to join similar or dissimilar metals using a rotating tool. This joining process was first used to join Al alloys,. Later it was expanded to join copper and titanium alloys, and it is now commonly used to join dissimilar materials. The results reported have been categorized as per the above-mentioned criteria. These are discussed in detail in subsections.

4.3.1 FSW OF SIMILAR ALUMINUM (AL) ALLOY

Due to its high strength-to-weight ratio, excellent corrosion resistance, and high ductility, Al and its alloys are widely employed in many manufacturing industries such as aerospace, marine, automobile, and food applications. Heidarzadeh et al. (2012) and Liu et al. (2012) carried out experiments that comprised process parameters as rotational speed and studied the effect on tensile strength and hardness after performing FSW on an Al alloy of grades 6,061 and 2,219 alloys. It was determined that as welding speed increases, tensile strength improves, but feed rate has a negative impact in turn. Alongside this, Costa et al. (2012) studied the fatigue behaviors of friction stir welded AA6082 and found that some serious defects (porosity, inclusion) affect the fatigue strength of the material to join.

Fahimpour et al. (2012) studied the corrosion resistance of AA6061. It was first welded using the FSW process, later with the gas tungsten arc welding (GTAW) process. Tafel polarization curve was used to find corrosion resistance, and it was found that the FSW joint has greater corrosion resistance than GTAW. Corrosion arises at a quicker rate at joint rather than when compared with the base metal.

Xu et al. (2012) investigated lap joint on AA6061-T6 joined using FSW and fusion-welded joint. A comparison analysis was carried out to determine the joint's fatigue strength and it was found that fatigue strength was lower than fusion welded joint, this was because of hooking defect. Upadhyay & Reynolds (2014) analyzed the effect of weld configuration and tool shape on mechanical and metallurgical properties of FSW on 25 mm thick AA6061, respectively. It can be concluded that the nugget zone shows a higher temperature throughout the thickness. In addition, the hardness and tensile tests were performed, which showed that stronger weld joints occur in the backplate due to low diffusivity.

Verma et al. (2017) investigated the temperature distribution while performing FSW on AA6082. The temperature distribution was measured using eight thermocouples, four of which were positioned on both the advancing and retarding sides of the centerline. Nur et al. (2017) performed an experimental investigation to study the effect of welding parameters on mechanical properties of welded components of AA5052 using FSW. The analysis was done to evaluate the tensile and bending strength of welded components at constant tool rotation. The maximum value of bending and tensile strength is 222 MPa and 422 MPa obtained at a different feed rate of 50 mm/min 208 mm/min, respectively. They have also mentioned that the shoulder diameter has a lower impact on the mechanical property of a welded component.

Amini et al. (2015) performed an experimental investigation to study the impact of tool geometry, i.e., the relative position of a pin to tool shoulder position on the welded joint which was prepared on the milling machine using FSW on AA5083-O. Four types of tools were used to perform FSW, i.e., tool C (tool pin is concentric with tool shoulder), tool O (tool pin is offset with tool shoulder), tool H (half tool pin), tool A (arched tool pin). Later results showed that with the increase in tool rotation speed, a significant amount of tool temperature increases. When welding speed is increased, welding forces can be reduced by offsetting the tool pin to the tool shoulder.

4.3.2 FSW OF DISSIMILAR MATERIAL

Ahmed et al. (2017) performed FSW on AA7075-T6 and AA5083-H111. These joints were welded with changing traverse rates and steady tool rotation. After the microstructural analysis using the electron backscatter diffraction (EBSD) technique, it was concluded that AA7075 and AA5083 have different grain structures in the base material and are recrystallized into fine grains after FSW. It was also reported that increase in welding speed resulted in reduction of grain size as observed in both AA7075 and AA5083. The joint's tensile strength ranged from 245 MPa to 267 MPa. The fracture occurred in a mixed-mode (ductile and brittle) and produced features of grain boundary cleavage after fracture.

Paoletti et al. (2015) performed Friction spot stir welding on Polycarbonate sheets. An analysis was done to evaluate welding forces and tool temperature. Also, the impact of the tool rotation and preheating time was studied. The joint strength was calculated by the shear test on a single lap joint. It was found that maximum strength found at the force of 1470 N with joint strength of approximately 60 MPa. The experiment showed that shear phenomena occurred at 224 N, which was lesser than joint strength.

Ramachandran et al. (2015) performed friction stir butt welding on Al alloy and HSLA steel. The impact of tool tilt angle and tool rotation speed was studied on mechanical properties. Optical microscopy was performed to study the microstructure, which showed that the intermetallic layer was formed at the interface. The range in which sound joint was performed was very narrow. Results reported that when the tool rotated at 450 rev/min with a welding speed of 45 mm/min and axial force of 7000 N, joint strength was 94% of UTS of base material.

Chu et al. (2016) joined 1.8 mm thick AA 2198-T8 plate with lithium plate using FSW spot configuration. The tilt angle was varied with which joining took place to analyze the impact on the joint's mechanical and microstructural properties. The results concluded that up to 45° tilt angles due to the flow of material from the nugget zone to the surrounding base material resulted in higher tensile strength. As soon as it passed this angle, there was a slight decrement in tensile strength.

Huang et al. (2016) studied material flow behavior using the marker insert technique while joining the sheets of AA6062-T6 with Copper foil using FSW. On the advancing side of joining, material flow crossed each other in a downward and upward direction. Tilt angle provided during weld formation caused excessive material flow, which was due to the effect of tool shape and selected welding parameters.

Ji et al. (2017) created intermetallic compounds (IMCs) layer by joining AA6061-T6 and magnesium alloy AZ31B using FSW which was associated with ultrasonic vibration. The vibration will not only increase interface length but also improved material flow. Due to mechanical interlocking, the tensile test reported a higher value of mechanical strength when compared to the results obtained without ultrasonic vibration. Ultrasonic vibration provided microstructure features having dimples with a mixed-mode fracture.

Shanavas et al. (2018) studied the underwater FSW (UFSW) on AA5052, in which heat conduction along the weld line was controlled to enhance the properties of the joint. A comparison of mechanical properties was done by performing UFSW and simple FSW. The effect of tool rotation speed and welding speed on ultimate tensile strength was investigated. Results reported that below 500 RPM, simple FSW is preferred over UFSW, but as soon as it passed this tool rotation speed UFSW is preferred. The heat-affected zone was not found in UFSW in the microstructural study. The study confirmed the feasibility of UFSW as it has resulted in more joint efficiency. Bakhtiari Argesi et al. (2018) investigated friction stir welded butt joints between AA5754 and pure copper of thickness 4 mm each. To have better material flow, a 1 mm offset was provided. The uniaxial test and optical microscopy technique were used to analyze the microstructure and mechanical properties of the resulting joints. The result showed that the lamellar layer was found in the stir zone. This activated dynamic recrystallization further created smaller grains than the base material. The maximum tensile strength resulted was 219 MPa at optimum process parameters valued as 1,000 RPM tool rotation speed and 100 mm/min transverse speed, which was 84% of the tensile strength of the Al alloy.

Safeen & Spena (2019) claimed some issues after performing FSW on Al alloy with steel sheets. The impact of various process parameters on the joint was also discussed. They have highlighted that the most important factor to choose the right process parameter; otherwise, defects will occur, leading to an increase in the stress concentration factor and a reduction in joint strength. Tool rotation speed and welding speed were the primary determinants for heat generation. The welding parameter created enough heat that was required to plasticize metal so that intermixing of material resulted in a sound joint.

Su et al. (2020) studied the importance of profile of tool pins while performing FSW on Al alloy. In addition, an investigation of thermo-mechanical, material flow, and welding forces was performed over the tool using the Computational Fluid Dynamics model. For better comprehension, a few cross-sections of the tool pin profile are also shown in Figure 4.1.

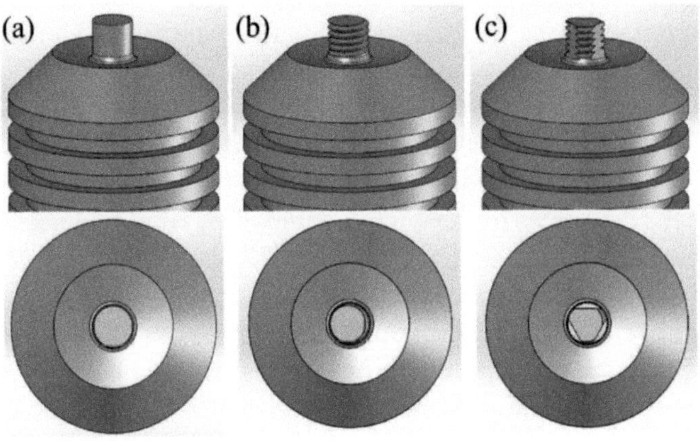

FIGURE 4.1 Cross-section of various tool pin profile.

Source: Reprinted with permission from Zhou et al., 2018. Copyright © 2018 The Society of Manufacturing Engineers. Published by Elsevier Ltd.

The obtained result of transverse force and microstructure were compared with and without flat features. The simulation model further was optimized by giving different torque components on flat features and studied tool wear. The result showed that enhancing the flat surface had further increased heat generation, but the impact of the profile on the tool pin on the torque produced was less than 5%. Study showed that enhancement in the flat surface not only reduces the transverse forces but also helps in achieving higher tool torque. The result revealed that T60 and T90 profile generated effective irrespective of high or low tool rotational speed.

4.4 REVIEW ON HYBRID FSISF PROCESS

The subsections discuss the hybrid FSISF process and report the results categorized as per workpiece materials joined by the FSW process.

4.4.1 FSISF ON SIMILAR MATERIAL

Various studies have been reported related to FSISF on similar material tailored blank. However, most of the work is concentrated on the process parameter analysis. The results are discussed in subsections accordingly.

4.4.1.1 IMPACT OF PROCESS PARAMETER ON FORMING FORCES

Li et al. (2017) and Liu (2017) studied process feasibility on the AA7075-O sheet which was done after performing ISF on tailored blanks produced by FSW process. Results highlighted that due to the increase in temperature, mechanical properties were greatly affected. However, when speed increases due to frictional heating, dynamic recrystallization is not activated, which further increases forming forces. When speed was above 3,000 RPM, frictional heating was increased and therefore microhardness and tensile strength increased by 50% and 28%, respectively. Thuillier et al. (2019) studied the mechanical behavior of incremental forming of friction stir welded blanks of AA6082-T6. Several tests were conducted on the formed part to analyze the impact of welding parameters on mechanical behavior. The result showed that the post-welding total strain obtained was lesser, and tensile strength was lower than that of yield strength.

4.4.1.2 IMPACT OF PROCESS PARAMETER ON FORMABILITY

Ambrogio et al. (2006); and Jeswiet et al. (2005) performed an experimental investigation of incremental forming on a tailored sheet of AA6181-T4 (1.2 mm) manufactured using the FSW technique. They compared the formability obtained post-ISF with the one obtained by conventional forming and found that improvement of maximum wall angle by 8°. They have attributed the improvement to the local brittleness phenomenon developing in FSW.

Li et al. (2017) and Liu (2017) studied process feasibility by performing ISF on the FSW-tailored blanks of AA7075-O. Experimental results showed that due to local frictional heating, formability increases with an increase in tool rotation speed, which occurs because of a positive reduction in friction.

Rubino et al. (2020) combined welding and forming techniques to produce customized components. A butt-welded joint was produced using the FSW technique and TIG welding technique. Strain analysis was performed after the completion of SPIF. The forming limit diagram is evaluated using SCAN PROBE and MATLAB software which stated that advancing the welding speed affected the welding joints, which negatively affected the blank formability. FSW welded blanks exhibited higher mechanical properties when compared to TIG welded joints. The obtained formability limit curve (FLC) for various Al grades is shown in Figure 4.2.

Otsu et al. (2014) investigated the formability of AA2017 welded sheet by performing SPIF post-welding. These sheets were shaped into a pyramid-shaped frustum. Results were compared with conventional forming, which showed that performing FSW and then ISF resulted in 69% more total strain, which was due to the reduction of dislocation by grain refinement.

Ebrahimzadeh et al. (2018) attempted to investigate the formability of FSW sheets of AA5083 sheets using SPIF. An experimental study was also performed at the joints, which were manufactured with 73% welding strength efficiency by parameter optimization. In SPIF and two-point incremental forming (TPIF), the comparison of formability in terms of forming height was made by varying the different wall angles.

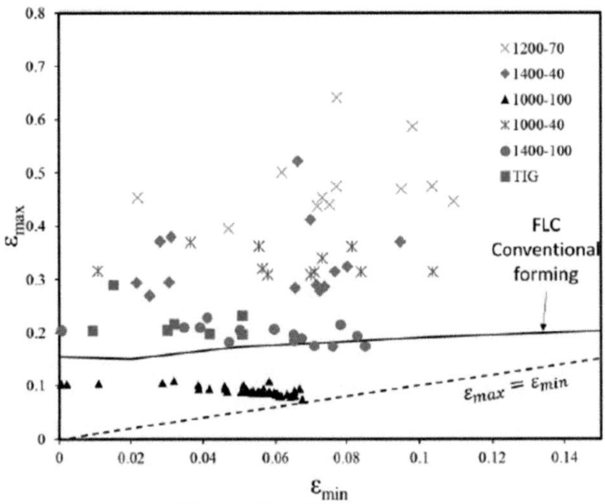

FIGURE 4.2 FLC for welded Al graded components.

Source: Reprinted with permission from Rubino et al., 2020. Copyright © 2020 The Author(s). Published by Elsevier B.V.

Figure 4.2 shows the formability limit curves of tailor welded blanks in ISF when compared to forming curves obtained for traditional forming process. It can be seen that the formability limit reached in incremental forming is higher. The plot of major and minor strain of FSWed joint

(Figure 4.3) shows increased minor strain which is distributed non uniformly though the welded region. This is due to loss of strength and predicts a lower formability in the welded region.

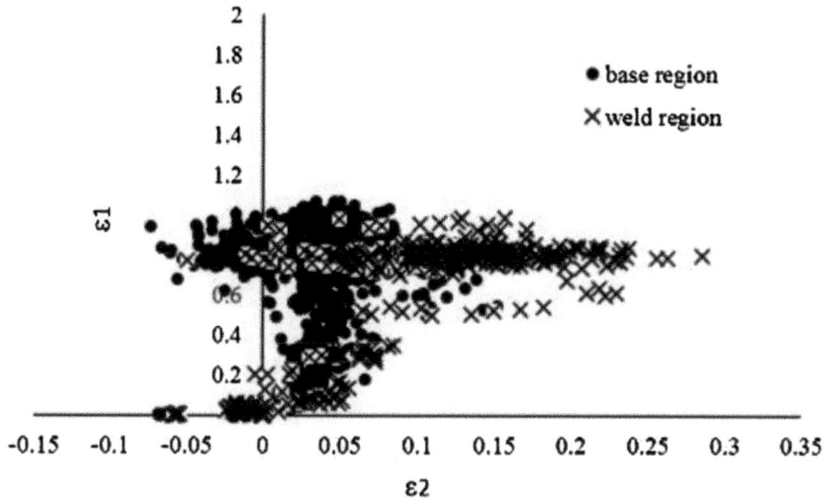

FIGURE 4.3 Variation of major and minor strain in SPIF of FSWed blank.

Source: Adapted from Tayebi et al., 2019

Otsu (2016) experimented the impact of tool rotation on light materials like Al, titanium and magnesium. Firstly, the sheet was joined using FSW then formed into a frustum of a pyramid. Result revealed that multi-stage forming improved the formability of this material, which further improved the forming accuracy also.

Andrade-Campos et al. (2020); and Carlone et al. (2021) performed FSW on the AA6082-T6 sheet and then SPIF. The forming was done until rupture occurred. Total strain and tensile stress were analyzed, which concluded that welded material had resulted in lower total strain and lower stresses than the yield stress. The strain changed drastically along the length had approximately zero value outside the welded area.

Maji et al. (2019) experimented the effect of friction stir incremental forming on formability. Aluminum alloy AA5083 sheet of 1 mm thickness was used as a raw material. Using FSW, these sheets were joined, and bending tests were conducted. The result showed that improvement in the feed rate of the tool caused decrement in bend angle. Later, SPIF was performed, and the Forming Limit Curve was evaluated, which showed that the formability of the parent as well as the welded sheet was more than the conventional forming. The comparison of forming limit curves for conventional forming sheet, parent sheet, FSISF sheet is shown in Figure 4.4.

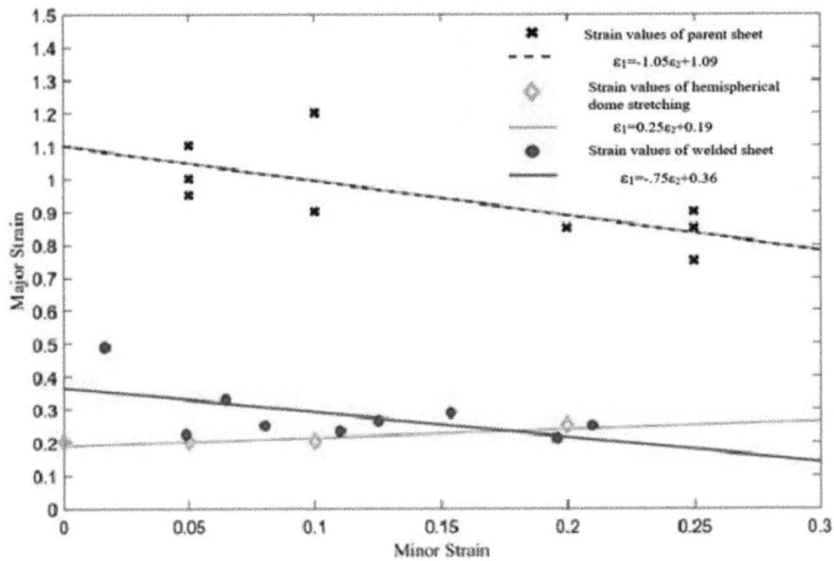

FIGURE 4.4 Comparison of forming limit curve.

Source: Reprinted with permission from Maji et al., 2019. https://creativecommons.org/licenses/by/3.0/

Grün et al. (2018) analyzed the formability of titanium alloy after performing the friction stir incremental forming by varying the tool rotation, step downsizes, and feed rates. Tool rotation decided the amount of localized frictional heating. The optimum tool rotation (of 2,000 RPM) generated the adequate thermal heating and enhanced formability and reduced not only tool wear but also forming forces.

4.4.1.3 IMPACT OF PROCESS PARAMETER ON SURFACE ROUGHNESS

Li et al. (2017); and Liu (2017) studied process feasibility by performing ISF on the FS welded AA7075-O. They have reported that surface roughness increased as tool rotation speed increased. They have concluded that this was due to the small, induced vibrations. This caused fish scale structures when speed increased from up to 2,000 RPM. Further, an increase in speed slightly reduced roughness.

Otsu et al. (2014) performed SPIF on AA2017 tailored sheets. Microstructural study showed that equiaxed grains were uniformly distributed due to elimination of dislocations. Some of the porosities were observed near the weld line as grain separation occurred.

4.4.1.4 IMPACT OF PROCESS PARAMETER ON SPRING BACK

Ebrahimzadeh et al. (2018) investigated the formability of tailored AA5083 using SPIF. Using surface response methods (RSM), an experimental investigation was also conducted to determine the influence of SPIF on dimensional accuracy and minimum thickness. Result revealed that tool rotation speed and step-down affected spring back more. The increase in tool rotation speed had less impact on spring back up to a certain speed, but further increase in speed caused variation in dimensional accuracy. RSM showed that with enhancement in step down thickness was decreased with increment in spring back.

Li et al. (2020) performed an experimental investigation on FSISF of Al alloy and steel sheets by synchronous bonding (SB). The impact of various parameters like stacking sequence and wall angle were investigated on forming and bonding quality. They have highlighted that manufacturing of laminated sheet metal components can be done using FSISF and SB. The optimized parameters in their study were 40° wall angles, the rotational speed of 3,000 RPM. Wall angle and step down were the main parameter that decided the failure modes.

4.4.2 FSISF ON DISSIMILAR MATERIAL

Similarly, a few research have been done targeting the FSIF on dissimilar tailored blanks. The reported results and findings are summarized in subsections.

4.4.2.1 IMPACT OF PROCESS PARAMETER ON FORMABILITY

Parente et al. (2016) investigated the formability of tailored blank manufactured by joining AA6061 and AA5182 using the FSW method.

When compared to parent materials, the results demonstrated that a major strain occurred along the weld line thus, lowering the formability requirements.

Tayebi et al. (2019) studied the formability of ISF on FSW welded dissimilar material such as AA6061 and AA5083. The investigation was done to analyze tensile test, microhardness, and microstructural behavior. The result showed that forming maximum wall angles are 60° and 62.5° respectively. Dissimilar Al alloy welded by FSW showed a reduction in formability.

Silva et al. (2009) presented forming of Al-tailored welded blanks (TWB). The industrial mechanical properties of joints were obtained by performing FSW on dissimilar AA1050-H111 alloys. SPIF was used to form tailored blanks into the desired shape. The comparison was done with conventional forming in a similar condition which pertains that FSW joints showed a higher depth forming that can be due to the reduction in local brittleness. The mechanical properties of TWB were compared with that of base material. The minimum global efficiency of tensile strength (GETS) and global efficiency of bending (GETS) were reported as 78% and 92%, respectively.

4.4.2.2 IMPACT OF PROCESS PARAMETER ON SURFACE ROUGHNESS

Tayebi et al. (2019) investigated the formability of ISF on FSW welded AA6061 and AA5083 dissimilar materials. The tensile, microhardness, and microstructural behavior were investigated. The result showed that grain size was reduced by 3 times when sheets were manufactured by hybrid process rather than when manufactured by ISF.

4.4.3 DEVELOPMENT OF FORMING TOOL

Novel designed tools were developed for forming tailored sheet for the concave and convex shape of parts. Jiang et al. (2017) proposed a new penetrating forming tool that was used to form welded sheets into concave-convex shapes. The commercially pure Al sheet was used as a workpiece component, which was formed into a truncated cone shape. Also, the effect of tool rotation direction was analyzed in SPIF. Result revealed that a grove-like defect occurred due to material flow towards retarding side from the advancing side.

4.5 CONCLUSION

SPIF is a flexible technology and adaptation of traditional forming in which sheets are incrementally imparted into complex geometries using a forming tool which moves in a specified pattern. However, in SPIF, there are a few limitations related to formability and geometrical accuracy. The said limitation can be minimized by hybrid friction stir incremental sheet forming (FSISF). FSW results in high-quality solid-state welding without distortion, shrinkage and filler metal or shielding gas requirements. As a result, FSW emerged as a viable alternative to fusion welding. It can be concluded that hybrid of FSW and ISF, i.e., FSISF leads to major advantages over conventional forming as far as tailored blanks are produced by joining of two similar alloy sheets. A soft localized zone in welding helps to reduce spring back and improves plastic deformation capabilities. Hybrid FSISF results in the reduction of the forming forces and further, improves the formability of tailored blanks. It is worth mentioning that dissimilar joining by FSW is still not commonly used due to inferior mechanical qualities and the creation of substantial amounts of Inter Metallic Compounds (IMCs). In dissimilar Cu–Al FSW systems, flaws such as fragmental defects, voids, pores, and cracks are frequent. These flaws are caused by incorrect process settings, which results in distinct IMCs and low mechanical characteristics. These IMCs also increase the hardness of the joint area, making it brittle, which is a driving factor for brittle fracture and low elongation. The formability of these joints is reduced after ISF, and they may also fracture along the weld line.

KEYWORDS

- **aluminum**
- **formability**
- **forming tool**
- **friction stir welding**
- **intermetallic compounds**
- **surface roughness**

REFERENCES

Aerens, R., Eyckens, P., Van, B. A., & Duflou, J. R., (2010). Force prediction for single point incremental forming is deduced from experimental and FEM observations. *International Journal of Advanced Manufacturing Technology, 46*(9–12), 969–982. https://doi.org/10.1007/s00170-009-2160-2.

Ahmed, M. M. Z., Ataya, S., El-Sayed, S. M. M., Ammar, H. R., & Ahmed, E., (2017). Friction stir welding of similar and dissimilar AA7075 and AA5083. *Journal of Materials Processing Technology, 242*, 77–91. https://doi.org/10.1016/j.jmatprotec.2016.11.024.

Ahmed, M. M. Z., Ataya, S., El-Sayed, S. M. M., Mahdy, A. M. A., Alsaleh, N. A., & Ahmed, E., (2021). Heat input and mechanical properties investigation of friction stir welded aa5083/aa5754 and aa5083/aa7020. *Metals, 11*(1), 1–20. https://doi.org/10.3390/met11010068.

Ambrogio, G., Fratini, L., & Micari, F., (2006). Incremental forming of friction stir welded tailored sheets. *Proceedings of 8th Biennial ASME Conference on Engineering Systems Design and Analysis; ESDA 2006, 2006*, 1–6. https://doi.org/10.1115/esda2006-95375.

Amini, S., Amiri, M. R., & Barani, A., (2015). Investigation of the effect of tool geometry on friction stir welding of 5083-O aluminum alloy. *International Journal of Advanced Manufacturing Technology, 76*(1–4), 255–261. https://doi.org/10.1007/s00170-014-6277-6.

Andrade-Campos, A., Thuillier, S., Martins, J., Carlone, P., Tucci, F., Valente, R., Paulo, R. M. F., & Alves De, S. R. J., (2020). Integrated design in welding and incremental forming: Material model calibration for friction stir welded blanks. *Procedia Manufacturing, 47*, 429–434. https://doi.org/10.1016/j.promfg.2020.04.327.

Araghi, B. T., Göttmann, A., Bambach, M., Hirt, G., Bergweiler, G., Diettrich, J., Steiners, M., & Saeed-Akbari, A., (2011). Review on the development of a hybrid incremental sheet forming system for small batch sizes and individualized production. *Production Engineering, 5*(4), 393–404. https://doi.org/10.1007/s11740-011-0325-y.

Araghi, B. T., Manco, G. L., Bambach, M., & Hirt, G., (2009). Investigation into a new hybrid forming process: Incremental sheet forming combined with stretch forming. *CIRP Annals – Manufacturing Technology, 58*(1), 225–228. https://doi.org/10.1016/j.cirp.2009.03.101.

Bakhtiari, A. F., Shamsipur, A., & Mirsalehi, S. E., (2018). Dissimilar joining of pure copper to aluminum alloy via friction stir welding. *Acta Metallurgica Sinica (English Letters), 31*(11), 1183–1196. https://doi.org/10.1007/s40195-018-0741-5.

Cai, S., Wu, R., Wang, Z., Li, M., & Chen, J., (2020). Numerical simulation of friction stir-assisted incremental forming with synchronous bonding of heterogeneous sheet metals. *International Journal of Advanced Manufacturing Technology, 106*(7, 8), 2747–2763. https://doi.org/10.1007/s00170-019-04792-x.

Carlone, P., Thuillier, S., Andrade-Campos, A., De Sousa, R. J. A., & Valente, R., (2021). Incremental forming of friction-stir welded aluminum blanks: An integrated approach. *International Journal of Material Forming, 14*(5), 1121–1137. https://doi.org/10.1007/s12289-021-01628-6.

Chu, Q., Yang, X. W., Li, W., & Li, Y., (2016). Microstructure and mechanical behavior of pinless friction stir spot welded AA2198 joints. *Science and Technology of Welding and Joining*, *21*(3), 164–170. https://doi.org/10.1179/1362171815Y.0000000078.

Costa, J. D., Ferreira, J. A. M., Borrego, L. P., & Abreu, L. P., (2012). Fatigue behavior of AA6082 friction stirs welds under variable loadings. *International Journal of Fatigue*, *37*, 8–16. https://doi.org/10.1016/j.ijfatigue.2011.10.001.

Davarpanah, M. A., Mirkouei, A., Yu, X., Malhotra, R., & Pilla, S., (2015). Effects of incremental depth and tool rotation on failure modes and microstructural properties in single point incremental forming of polymers. *Journal of Materials Processing Technology*, *222*, 287–300. https://doi.org/10.1016/j.jmatprotec.2015.03.014.

Ebrahimzadeh, P., Baseri, H., & Mirnia, M. J., (2018). Formability of aluminum 5083 friction stir welded blanks in the two-point incremental forming process. *Proceedings of the Institution of Mechanical Engineers, Part E: Journal of Process Mechanical Engineering*, *232*(3), 267–280. https://doi.org/10.1177/0954408917692370.

Fahimpour, V., Sadrnezhaad, S. K., & Karimzadeh, F., (2012). Corrosion behavior of aluminum 6061 alloys joined by friction stir welding and gas tungsten arc welding methods. *Materials and Design*, *39*, 329–333. https://doi.org/10.1016/j.matdes.2012.02.043.

Gohil, A., & Modi, B., (2021). Review of the effect of process parameters on performance measures in the incremental sheet forming process. *Proceedings of the Institution of Mechanical Engineers, Part B: Journal of Engineering Manufacture*, *235*(3), 303–332. https://doi.org/10.1177/0954405420961215.

Grün, P. A., Uheida, E. H., Lachmann, L., Dimitrov, D., & Oosthuizen, G. A., (2018). Formability of titanium alloy sheets by friction stir incremental forming. *International Journal of Advanced Manufacturing Technology*, *99*(5–8), 1993–2003. https://doi.org/10.1007/s00170-018-2541-5.

Hasan, A. F., Bennett, C. J., & Shipway, P. H., (2015). A numerical comparison of the flow behavior in friction stir welding (FSW) using unworn and worn tool geometries. *Materials and Design*, *87*, 1037–1046. https://doi.org/10.1016/j.matdes.2015.08.016.

Heidarzadeh, A., Khodaverdizadeh, H., Mahmoudi, A., & Nazari, E., (2012). Tensile behavior of friction stir welded AA 6061-T4 aluminum alloy joints. *Materials and Design*, *37*, 166–173. https://doi.org/10.1016/j.matdes.2011.12.022.

Huang, Y., Wang, Y., Wan, L., Liu, H., Shen, J., Dos, S. J. F., Zhou, L., & Feng, J., (2016). Material-flow behavior during friction-stir welding of 6082-T6 aluminum alloy. *International Journal of Advanced Manufacturing Technology*, *87*(1–4), 1115–1123. https://doi.org/10.1007/s00170-016-8603-7.

Hussain, G., Ilyas, M., Lemopi, I. B. B., & Khan, W. A., (2020). Mechanical properties and microstructure evolution in incremental forming of AA5754 and AA6061 aluminum alloys. *Transactions of Nonferrous Metals Society of China (English Edition)*, *30*(1), 51–64. https://doi.org/10.1016/S1003-6326(19)65179-4.

Jesus, J. S., Costa, J. M., Loureiro, A., & Ferreira, J. M., (2018). Assessment of friction stir welding aluminum T-joints. *Journal of Materials Processing Technology*, *255*(2017), 387–399. https://doi.org/10.1016/j.jmatprotec.2017.12.036.

Jeswiet, J., Micari, F., Hirt, G., Bramley, A., Duflou, J., & Allwood, J., (2005). Asymmetric single point incremental forming of sheet metal. *CIRP Annals – Manufacturing Technology*, *54*(2), 88–114. https://doi.org/10.1016/s0007-8506(07)60021-3.

Ji, S., Meng, X., Liu, Z., Huang, R., & Li, Z., (2017). Dissimilar friction stir welding of 6061 aluminum alloy and AZ31 magnesium alloy assisted with ultrasonic. *Materials Letters*, *201*, 173–176. https://doi.org/10.1016/j.matlet.2017.05.011.

Jiang, W., Miura, T., Otsu, M., Okada, M., Matsumoto, R., Yoshimura, H., & Muranaka, T., (2017). Development of friction stir incremental forming process using penetrating tool. *Procedia Engineering*, *207*, 789–794. https://doi.org/10.1016/j.proeng.2017.10.830.

John, J., Shanmuganatan, S. P., Kiran, M. B., Senthil, K. V. S., & Krishnamurthy, R., (2021). Friction stir processing combined with incremental forming effect on AA2014-T6. *Materials and Manufacturing Processes*, *36*(8), 950–966. https://doi.org/10.1080/10426914.2021.1885696.

Krishna, G. G., Reddy, P. R., & Hussain, M. M., (2014). Mechanical behavior of friction stir welding joints of aluminum alloy of AA6351 with AA6351 and AA6351 with AA5083. *International Journal of Engineering Trends and Technology*, *10*(4), 161–165. https://doi.org/10.14445/22315381/ijett-v10p231.

Li, M., Wu, R., Cai, S., Chang, Z., Wang, Z., & Chen, J., (2020). Experimental investigation on friction-stir-assisted incremental forming with synchronous bonding of aluminum alloy and steel sheets. *Journal of Materials Engineering and Performance*, *29*(2), 750–759. https://doi.org/10.1007/s11665-020-04600-8.

Li, Y., Chen, X., Liu, Z., Sun, J., Li, F., Li, J., & Zhao, G., (2017). A review of the recent development of the incremental sheet-forming process. *International Journal of Advanced Manufacturing Technology*, *92*(5–8), 2439–2462. https://doi.org/10.1007/s00170-017-0251-z.

Liu, H., Zhang, H., Pan, Q., & Yu, L., (2012). Effect of friction stir welding parameters on microstructural characteristics and mechanical properties of 2219-T6 aluminum alloy joints. *International Journal of Material Forming*, *5*(3), 235–241. https://doi.org/10.1007/s12289-011-1048-5.

Liu, Z., (2017). Friction stirs incremental forming of AA7075-O sheets: Investigation on process feasibility. *Procedia Engineering*, *207*, 783–788. https://doi.org/10.1016/j.proeng.2017.10.829.

Maji, K., Kumar, S., & Kumar, G., (2019a). Experimental study on the effects of incremental forming and friction stir welding on formability of AA5083 sheet. *Journal of Physics: Conference Series*, *1240*(1). https://doi.org/10.1088/1742-6596/1240/1/012090.

Maji, K., Kumar, S., & Kumar, G., (2019b). Experimental study on the effects of incremental forming and friction stir welding on formability of AA5083 sheet. *Journal of Physics: Conference Series*, *1240*(1). https://doi.org/10.1088/1742-6596/1240/1/012090.

Nur, R., Sultan, A. Z., & Suyuti, M. A., (2017). Mechanical properties on friction stir welding of aluminum alloy 5052. *ARPN Journal of Engineering and Applied Sciences*, *12*(15), 4445–4450.

Otsu, M., (2016). Excellent formability of light metals sheets by friction stir incremental forming. *Key Engineering Materials*, *716*, 3–10. https://doi.org/10.4028/www.scientific.net/KEM.716.3.

Otsu, M., Yasunaga, M., Matsuda, M., & Takashima, K., (2014). Friction stir incremental forming of A2017 aluminum sheets. *Procedia Engineering*, *81*, 2318–2323. https://doi.org/10.1016/j.proeng.2014.10.327.

Paoletti, A., Lambiase, F., & Di Ilio, A., (2015). Optimization of friction stir welding of thermoplastics. *Procedia CIRP*, *33*, 562–567. https://doi.org/10.1016/j.procir.2015.06.078.

Papadopoulos, M., Tavares, S., Pacchione, M., & Pantelakis, S., (2013). Mechanical behavior of AA 2024 friction stirs overlap welds. *International Journal of Structural Integrity*, *4*(1), 108–120. https://doi.org/10.1108/17579861311303663.

Parente, M., Safdarian, R., Santos, A. D., Loureiro, A., Vilaca, P., & Jorge, R. M. N., (2016). A study on the formability of aluminum tailor welded blanks produced by friction stir welding. *International Journal of Advanced Manufacturing Technology*, *83*(9–12), 2129–2141. https://doi.org/10.1007/s00170-015-7950-0.

Pasteur, L., & Koch, R., (1941). *1. Introduction*, *74*(1934), 535–546.

Ramachandran, K. K., Murugan, N., & Shashi, K. S., (2015). Friction stir welding of aluminum alloy AA5052 and HSLA steel. *Welding Journal*, *94*(9), 291s–300s.

Rubino, F., Esperto, V., Paulo, R. M. F., Tucci, F., & Carlone, P., (2020). Integrated manufacturing of AA6082 by friction stir welding and incremental forming: Strain analysis of deformed samples. *Procedia Manufacturing*, *47*(2019), 440–444. https://doi.org/10.1016/j.promfg.2020.04.331.

Safeen, M. W., & Spena, P. R., (2019). Main issues in quality of friction stir welding joints of aluminum alloy and steel sheets. *Metals*, *9*(5). https://doi.org/10.3390/met9050610.

Shanavas, S., Edwin, R. D. J., & Murugan, N., (2018). Weldability of marine grade AA 5052 aluminum alloy by underwater friction stir welding. *International Journal of Advanced Manufacturing Technology*, *95*(9–12), 4535–4546. https://doi.org/10.1007/s00170-017-1492-6.

Silva, M. B., Skjoedt, M., Vilaça, P., Bay, N., & Martins, P. A. F., (2009). Single point incremental forming of tailored blanks produced by friction stir welding. *Journal of Materials Processing Technology*, *209*(2), 811–820. https://doi.org/10.1016/j.jmatprotec.2008.02.057.

Silva, M. B., Skjoedt, M., Vilaçal, P., Bay, N., & Martins, P. A. F., (2009). Friction Stir welding of aluminum tailored blanks processed by single point incremental forming. In: *3rd International Conference on Integrity, Reliability, and Failure* (Vol. 1, pp. 599, 600).

Skjoedt, M., Silva, M. B., Martins, P. A. F., & Bay, N., (2010). Strategies and limits in multi-stage single-point incremental forming. *Journal of Strain Analysis for Engineering Design*, *45*(1), 33–44. https://doi.org/10.1243/03093247JSA574.

Su, H., Xue, L., & Wu, C., (2020). Optimizing the tool pin with three flats in friction stir welding of aluminum alloy. *International Journal of Advanced Manufacturing Technology*, *108*(3), 721–733. https://doi.org/10.1007/s00170-020-05479-4.

Tayebi, P., Fazli, A., Asadi, P., & Soltanpour, M., (2019). Formability analysis of dissimilar friction stir welded AA 6061 and AA 5083 blanks by SPIF process. *CIRP Journal of Manufacturing Science and Technology*, *25*, 50–68. https://doi.org/10.1016/j.cirpj.2019.02.002.

Threadgilll, P. L., Leonard, A. J., Shercliff, H. R., & Withers, P. J., (2009). Friction stir welding of aluminum alloys. *International Materials Reviews*, *54*(2), 49–93. https://doi.org/10.1179/174328009X411136.

Thuillier, S., Andrade-Campos, A., Carlone, P., Valente, R., & De Sousa, R. J. A., (2019). Integrated design in welding and incremental forming: Mechanical behavior of friction stir welded blanks. *AIP Conference Proceedings*, *2113*. https://doi.org/10.1063/1.5112605.

Upadhyay, P., & Reynolds, A., (2014). Effect of backing plate thermal property on friction stir welding of 25-mm-thick AA6061. *Metallurgical and Materials Transactions A: Physical Metallurgy and Materials Science*, *45*(4), 2091–2100. https://doi.org/10.1007/s11661-013-2121-0.

Verma, S., Meenu, & Misra, J. P., (2017). Study on temperature distribution during friction stir welding of 6082 aluminum alloy. *Materials Today: Proceedings*, *4*(2), 1350–1356. https://doi.org/10.1016/j.matpr.2017.01.156.

Wang, D., Liu, S., & (2004). Study of FSW of aluminum. *Journal of Materials Science*, *39*(5), 1689–1693. https://doi.org/10.1023/B:JMSC.0000016171.50107.e4.

Xu, X., Yang, X., Zhou, G., & Tong, J., (2012). Microstructures and fatigue properties of friction stir lap welds in aluminum alloy AA6061-T6. *Materials and Design*, *35*, 175–183. https://doi.org/10.1016/j.matdes.2011.09.

Zhou, L., Zhang, R. X., Li, G. H., Zhou, W. L., Huang, Y. X., & Song, X. G., (2018). Effect of pin profile on microstructure and mechanical properties of friction stir spot welded Al-Cu dissimilar metals. *Journal of Manufacturing Processes, 36*. https://doi.org/10.1016/j.jmapro.2018.09.01.

CHAPTER 5

3D Printing Parts: Mechanical Characterization, Applications, Challenges, and Techniques

PRADYUMN KUMAR ARYA, PANKAJ KUMAR, NEELESH KUMAR JAIN, GIRISH VERMA, and DAN SATHIARAJ

Department of Mechanical Engineering, Indian Institute of Technology, Indore, Madhya Pradesh, India

ABSTRACT

Additive manufacturing (AM), often known as 3D printing, has shown rapid growth over the last two decades and is being used in numerous applications. AM is one of the most continuously emerging manufacturing methods for creating prototypes and end users' products by layer-by-layer deposition. The mechanical and physical properties of a product formed using AM might be different from non-conventional processes. The mechanical performance of 3D-printed objects is influenced by materials, layer thickness, energy, scan speed, depositing layout, post-processing, and other process parameters. AM provides numerous advantages, including design flexibility, customization, and creating highly complicated structures. In addition, it also has limitations such as higher setup costs, low-scale production, inferior mechanical qualities, material limits, and defects. The AM played a significant role in the recent research in several sectors such as aerospace, biomedical, automotive, and construction. The materials such as filaments, wires, powders, paste, films, and

Advances in Manufacturing and Processing of Materials: Characterization and Applications.
Narendra Kumar, PhD, Mohit Tyagi, PhD, Dilbagh Panchal, PhD & Ravi Pratap Singh, PhD (Eds.)

liquids could be employed for 3D printing, while polymers are the most prevalent materials produced for printing. Numerous academic researchers employed ASTM and ISO mechanical test standards for measuring the strength of 3D printed objects. This chapter provides a comprehensive overview of mechanical properties, applications, process parameters, and future development trends in 3D printing.

5.1 INTRODUCTION

3D printing is a methodology for creating various objects and complicated geometries using 3D model data. This method generates three-dimensional objects by layering materials on top of one another. Throughout the last three decades, Additive Manufacturing (AM) has seen a remarkable transformation (Sing, Tey, Tan, Huang, & Yeong, 2020). According to the materials deposition forms, AM techniques can be categorized into three groups. The first one is sintering, which involves heating the materials without liquifying them to make complicated structures. The second method requires completely melting the materials, which can be achieved using an electron beam or a laser light. The third method is stereolithography (SLA), which uses a photopolymerization process (Kumar, Kumar, & Chohan, 2021; Yang, Fuh, Loh, & Wong, 2003). The time it takes for AM to build a product depends on the speeds, quality, and number of materials. The AM process is different from the conventional manufacturing process. The conventional manufacturing process consists of forming and subtractive processes in which the parts are manufactured by material removal or plastic deformation (Low et al., 2017). Therefore, subtractive manufacturing techniques waste many materials. AM allows for the use of optimal material and the creation of near-net-shaped parts. Conventional manufacturing often requires joining multiple elements or cutting unwanted material in order to build an object. In comparison to conventional manufacturing, AM provides for the layer-by-layer fabrication of a product (Keleş, Blevins, & Bowman, 2017). Conventional manufacturing remains preferred for convenient and easy products, while AM is increasingly being used to manufacture highly complicated items. As a result, AM does not entirely replace traditional manufacturing. So, AM is useful for making lightweight, more complicated designs that would be difficult or expensive to make with conventional manufacturing methods. AM has a number of benefits for prototype and manufacturing, as it eliminates the need for molds, machining, and processing (Beaman et al., 1997).

Though it might be an exaggeration to argue that AM will "transform the world," it is already making a good influence. It provides the fabrication of intricate designs with minimal wastage of materials compared to components that need machining (Ivanova, Williams, & Campbell, 2013). AM is needed to manufacture lightweight structures for the aerospace and automotive industries, saving fuel and reducing environmental impact. Metal and its alloys, ceramics, biological materials, and thermoplastics are common materials used in AM. AM technique is becoming a popular way to construct upgraded ceramics for tissue engineering and biomaterials such as dental and bone platforms (Vaezi, Seitz, & Yang, 2013). Currently, AM is used to create 3D prints from concrete and ceramics, free of cracks and pores. Alumina, zirconia, and bioactive glasses are certain ceramic materials that are used in building construction, aviation, and medical industries. AM is also helpful to manufacture 3D prints of polymer and polymer matrix composites (PMCs) for biomaterials, medical equipment, and orthopedic devices (Bhushan & Caspers, 2017). AM is beneficial for the fabrication of complicated geometries of metal and alloys with good physical and mechanical properties, which is used for aerospace and automotive industries. Ti alloys, Ni-based alloys, cobalt-based alloys, steel, and Al alloys are some of the most commonly used materials in AM (Bahuguna, Arya, & Patel, 2020; Sawant & Jain, 2018). Plastics are a popular AM material that may be utilized in FDM, SLS, and SLA processes. Plastics such as acrylonitrile butadiene styrene (ABS), polylactic acid, and polycarbonate are the most often used for components parts in automobiles, mobile phones, and appliances. AM is recently being used to deposit shaper memory alloys and high entropy alloys (HEAs) for a variety of industrial applications. The most commonly used materials in AM of various components are discussed in Figure 5.1(a). Figure 5.1(b) shows the AM application in various industries. AM is utilized to produce various products in an extensive range of industries such as aerospace, automobile, electronics, construction, biomedical, etc. AM is well suited to aerospace industries because of its weight-saving capabilities, reduced costs, reduced lead times, and ability to make complex structural components. Aerospace parts such as nozzles, combustion chambers, fittings, connectors, accessories, molds, and dies are manufactured and repaired using AM. AM is widely used in the automobile sector since it allows for fast prototyping while also reducing weight and price (Bahuguna, Arya, & Patel, 2020; Sawant & Jain, 2018; Arya, Jain, Murugesan, & Patel, 2021; Sawant & Jain, 2017). AM is widely used in the biomedical sector because it can easily manufacture complex structural components such as implants, tissues, and prosthetics.

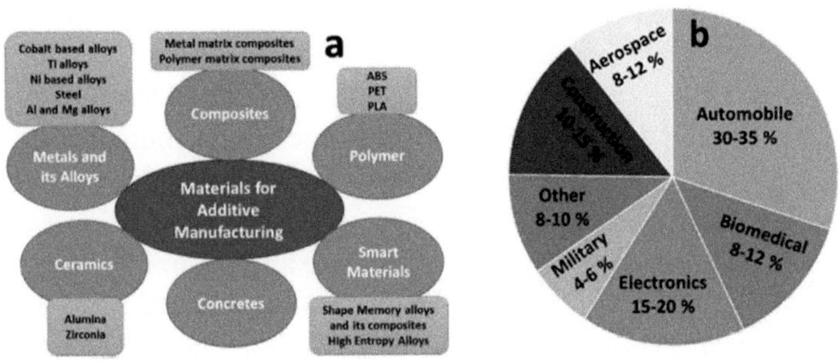

FIGURE 5.1 (a) Materials used for additive manufacturing; and (b) application of AM in different industries.

5.2 OVERVIEW OF ADDITIVE MANUFACTURING (AM)

The concept of 3D printing goes back to the 1940s when the first industrial 3D printing equipment was introduced in 1980 (Huang, Liu, Mokasdar, & Hou, 2013). AM originated in Japan at the beginning 1980s when Hidep Kodama introduced SLA. SLA is the oldest 3D printer which was invented in 1987. It creates 3D objects by solidifying thin layers of liquid polymer with UV light and a laser. Following that, lots of different techniques known as 3D printing were developed. FDM (fused deposition modeling) is the 2nd most common AM technology, introduced by S Scott Crump and Stratasys in 1987. After three years, a newer AM technology known as SLS (selective laser sintering) was developed, and the first commercialized SLS was released in 1990. The first bio-printer was invented at the end of the 20th century, and the first 3D implants were manufactured using biomaterials (Pereira, Kennedy, & Potgieter, 2019). Furthermore, after 10 years, the first 3D printers were introduced to the marketplace. Nowadays, full-scale printers with the ability to produce large 3D products for automobiles and civil constructions are commercially available.

Manufacturing, transportation, construction, academia, hobbies, space flight, energy, and defense all benefit from AM goods and applications nowadays (Kumar, Sawant, & Jain, 2021). In the biomedical, automobile, and aviation industries, AM has been widely used to manufacture finished products directly. If present trends persist, we will shortly reach a novel step in which AM becomes a manufacturing method as well as designing model. In AM techniques, a computerized data flow provides machine code to AM, which is accompanied by a physical workflow that converts raw materials into completed products (Gibson et al., 2006). The AM procedure is divided into five categories outlined below and illustrated in figure 2. The development of three-dimensional drawings using CAD/CAM software or surface modeling tools is the preliminary step of AM. The next phase is to convert the data to 3D printing specifications (Rosso et al., 2021). The most prevalent data format is STL, which is made up of a triangulated surface mesh that describes the entire geometry. The next phase is slicing, which is a set of tools that build machine code based on the system's properties, variables, and paths. The significant parameters of various techniques have in common are layer width, height, processing speed, temperature, and fill percentage. Printing is the technical procedure of materializing things utilizing 3D printers, and it has different features depending on the technique used (Williams et al., 2016). The last phase is post-processing, which, dependent on the technical features, more or less involvement may be required to get the desired finish and qualities of the printed parts (Figure 5.2).

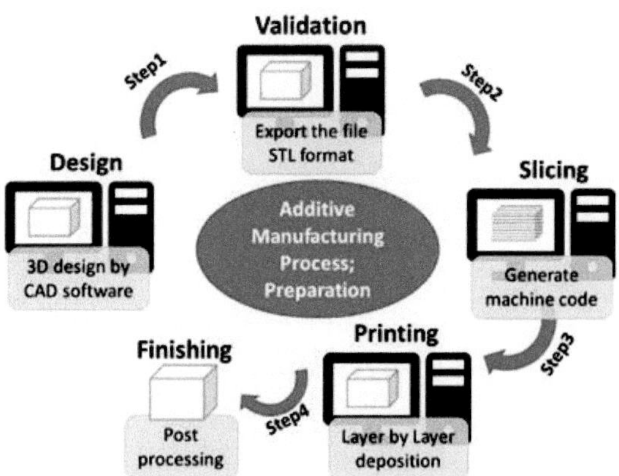

FIGURE 5.2 Additive manufacturing (AM) workflow.

5.3 ADDITIVE MANUFACTURING (AM) TECHNIQUES OF 3D PRINTED PARTS

AM technique is grouped into the following categories based on the starting materials and procedure, as shown in Figure 5.3. FDM is the most common AM technology that primarily uses polymer material. In recent times, progressive organizations are expected to use a variety of AM techniques (Kumar & Jain, 2021).

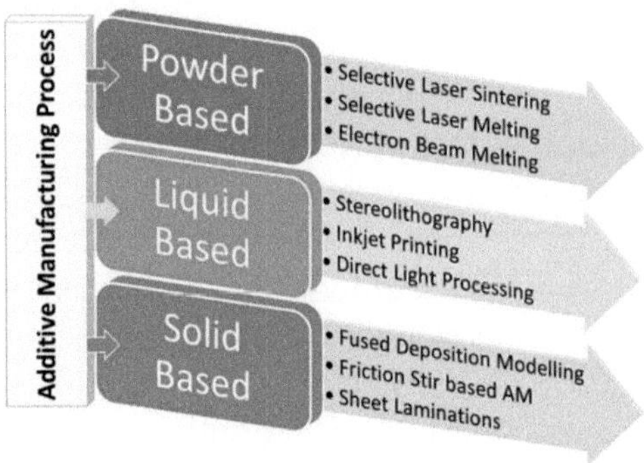

FIGURE 5.3 Classification of additive manufacturing.

Several AM processes for polymers and materials are available, such as SLM, SLS, EBM, LBM, LOM, and IJP (Kruth, Leu, & Nakagawa, 1998). On the other hand, Each AM technique has its own set of benefits and drawbacks. The essential aspects for choosing the best AM techniques for a given application are costs, duration, materials, qualities, etc. A detailed overview of all AM processes is presented in Table 5.1. Table 5.1 provides the benefits, limitations, applications, and materials used in different AM technologies.

TABLE 5.1 Description of Various Additive Manufacturing Technologies

AM Methods	Application	Advantages	Limitations	Materials Used
SLA (stereolithography)	Biomedical and automobile parts	Good quality and best resolution	Limited use of materials and slow process	Resin or monomers, i.e., epoxy or acrylic-based.

TABLE 5.1 *(Continued)*

AM Methods	Application	Advantages	Limitations	Materials Used
FDM (fused deposition modeling)	Composites materials, parts, and toys.	Low cost and fast process.	Limited use of materials and low mechanical properties.	Thermoplastic, polymer matrix composites.
Powder-based fusion (SLM, SLS)	Aerospace, biomedical, and automobile parts.	Best quality, high resolution.	Slow process and expensive.	Metal and alloys, ceramics, polymers.
LOM (laminated object manufacturing)	Foundry industry, paper, and electronics industry.	Low cost, suitable for large objects.	Poor surface quality and limitation in complicated shapes.	Ceramics, polymer composite, paper, metal roll.
Inkjet printing	Large structures and buildings.	High-speed printing.	Coarse structure, layer-by-layer finish.	Ceramics, concrete, and soil.
Friction stirs additive manufacturing.	Automobile parts and coating application.	High deposition rate, no solidification problem.	Limitation for high-temperature materials.	Metals and alloys

5.3.1 FDM

FDM (Fusion deposition modeling) was first developed by 'Scott Crump' in the 1800s and commercialized in the 1990s by a US corporation. The FDM process is based on the principles of heat energy and surface chemistry. A continuous filament of a thermoplastic polymer is intended to produce 3D components layer by layer, as shown in Figure 5.4. The filament is preheated to a semi-liquid condition at the nozzle before being injected onto substrate over already produced layers (Vyavahare, Teraiya, Panghal, & Kumar, 2019). This method is often known as extrusion type AM, and it involves the usage of thinner thermoplastic polymeric material to build the necessary part. The filament materials are available in wire coil form must be heated but not melted before being extruded and deposited on a surface.

The filament material is melted with each other while manufacturing and then cooled at ambient temperature afterward. The key process factors that influence the mechanical properties of deposited materials are layer thickness, the width of filaments, inclination, and spacing (Alafaghani, Qattawi, Alrawi, & Guzman, 2017). Inter-layer deformation had been found to be the principal cause of lower mechanical properties. The fast speed, flexibility, low expense, and capacity to make customizable objects are the best features of the FDM process. Moreover, this FDM technique has several disadvantages, such as mechanical properties qualities, low surface finish, higher production time, layer-by-layer form, etc.

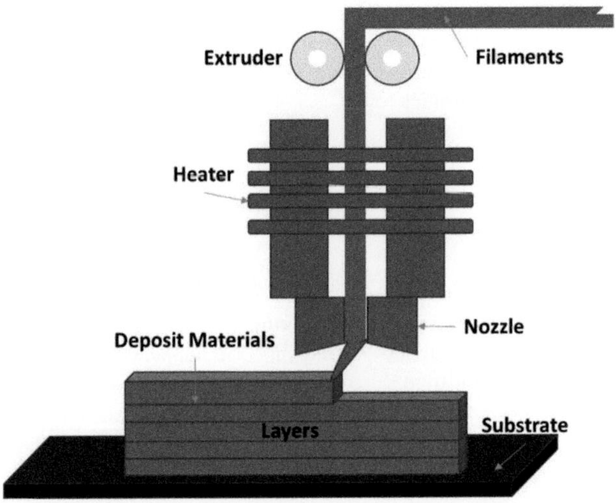

FIGURE 5.4 Schematic diagram of FDM process.

5.3.2 SLA

SLA is among the oldest known faster prototyping processes, commonly produced in 1986. SLA is a laser-based 3D printing technique that is based on photopolymerization.

The schematic block diagram of SLA is depicted in Figure 5.5. Ultraviolet light would be focused on a pool of photopolymer resin, and the UV light would sketch a programmed pattern or structure on the surface of the photopolymer vat using CAM/CAD software. Because photopolymers are UV sensitive, they can be solidifying and produce one layer of the required 3D shape (Huang, Qin, & Wang, 2020). The build pattern then lowers down one layer, and a sweeper re-coats the tank's surface with resin. This procedure is continued for every layout layer till the 3D product is finished. The final component was washed with a solution to remove the wet resin from the substrates. SLA manufactures 3D-printed products with the best accuracy, precise resolution, and perfect surface quality. Moreover, it is a slower process, costly, and has a restricted choice of printed materials. In addition, the reaction kinetics and the curing procedure are complicated (Yang, Li, & Zhao, 2019). The major aspects regulating the thickness of every layer are the light source's power and exposure. SLA can be required to develop complicated nanocomposites in a 3D printing. 3D printing objects may require a post-process operation like heat or photo-curing to get the appropriate mechanical properties. To manufacture ceramic polyester composites, a dispersing of ceramic fragments in monomers is used.

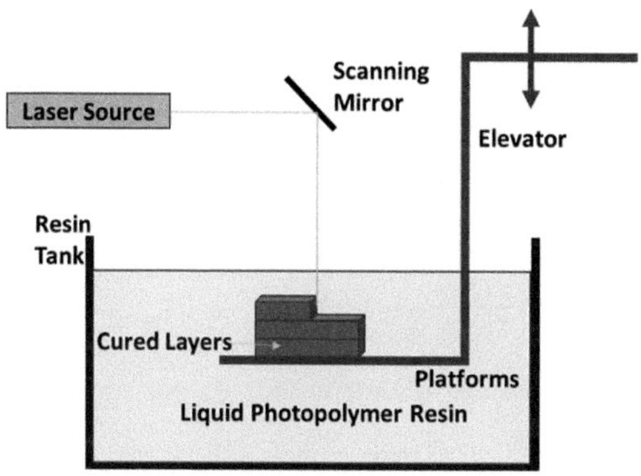

FIGURE 5.5 Schematic diagram of SLA process.

5.3.3 *IJP*

IJP (inkjet printing) is one of the essential non-contact approaches for 3D printing of metals, alloys, composites, and ceramic. Solids-cape Inc. invented this IJP technique, which starts with wax-based supporting materials and thermoplastic-based building material (Singh, Haverinen, Dhagat, & Jabbour, 2010). These thermoplastic materials get melted within two overheated chambers. Such melted materials are injected into the IJP unit, which sprays droplets to the appropriate position, forming a single layer of materials. As a result, a similar procedure is repeated for every successive layer. The wax supporting materials are therefore melted by removing the final part. Figure 5.6 shows a general layout of the IJP process. The IJP process is most commonly used for producing more complex and sophisticated ceramics objects, such as scaffolding for tissue and biomedical applications (Derby, 2010). IJP is cost-effective, quick, customizable, and capable of producing fine printing and complex structural design. The coarser quality, lesser durability, poor layers adhesion, and inability to print huge volumes are some of the key drawbacks of this method.

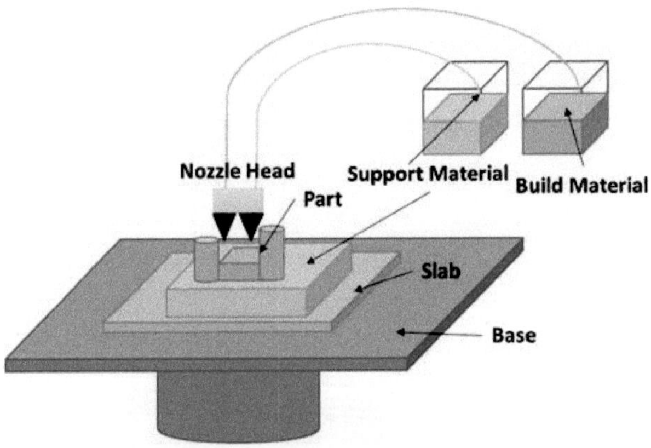

FIGURE 5.6 Schematic diagram of IJP process.

5.3.4 SLM

SLM (selective laser melting) is among the best extensively utilized powder-based methods in the AM industry. SLM works by melting and fusing metal particles with laser light, depositing a small powder layer above a substrates sheet or on a previously deposited layer. Laser energy, laser scan rate, hatching length, hatching overlapping, hatching design are all essential process parameters in the SLM process (Kruth, Mercelis, Van, Froyen, & Rombouts, 2005). In order to create a defect-free object with excellent mechanical properties, certain processing parameters must be optimized. The entire procedure is carried out inside a sealed tank that is generally filled with argon gas. The inclusion of an argon atmosphere in the container reduces oxygen contamination during the procedure. In the SLM process, base plate heating is commonly used to slow the cooling rates during production to avoid probable cracks during solidification. SLM is viewed as one of the best adaptable AM techniques because that can print a large range of metals, such as Ti, Ni, Co, Fe, and Al-based alloys and their composites. Furthermore, due to the rapid cooling rates observed throughout this technique, SLM is able to produce amorphous materials (Hassanin, 2021). The benefits of this method include the ability to deposit a variety of materials, greater functionality, minimal cost, and the manufacture of near-net-shaped parts. Slow process, exact size constraints, high power consumption, high initial expenses, difficulty to handle powder, and higher surface roughness are some of the drawbacks of the SLM method. Figure 5.7 shows a general layout of the SLM process.

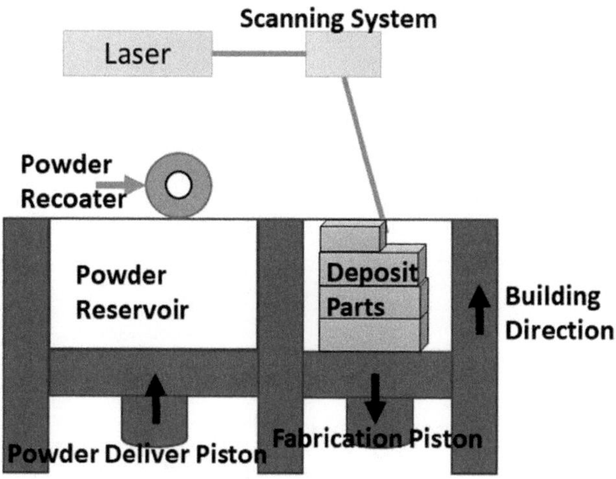

FIGURE 5.7 Schematic diagram of SLM process.

5.3.5 FSAM

Friction stir additive manufacturing (FSAM) is a fast-growing AM approach that employs the principle of solid-state friction stir welding (FSW)/processing to create parts layer-by-layer. White first used and copyrighted FSAM in 2002, while Thomas et al. proposed an equivalent approach in 2005, and in 2006, FSAM was commercialized. FSAM is comprised of a hollowed circular tool that delivers powder or wires direct to the nugget region. Frictional heat is generated at the contact of the tool surface and the base sheet by the fast rotating of the FSW tool. The rotating tool generated friction heat, which heated and plastically deformed the feeding materials, which are deposited layer by layer on the surface. The FSAM method has a number of advantages, including solid-state existence, wrought microstructural characteristics, materials feeding versatility, and capacity to process nearly all types of metals and alloys (Srivastava, Rathee, Maheshwari, Noor, & Kundra, 2019). The melting point problem can be minimized because the technique is non-melting, and the peak temperature ranges from 50 to 80% of the melting point. As a result, FSAM can manufacture a product with little porosity, impurities, particles segregation, interior voids and hot crack dispersion. Figure 5.8 shows a general layout of the FSAM process.

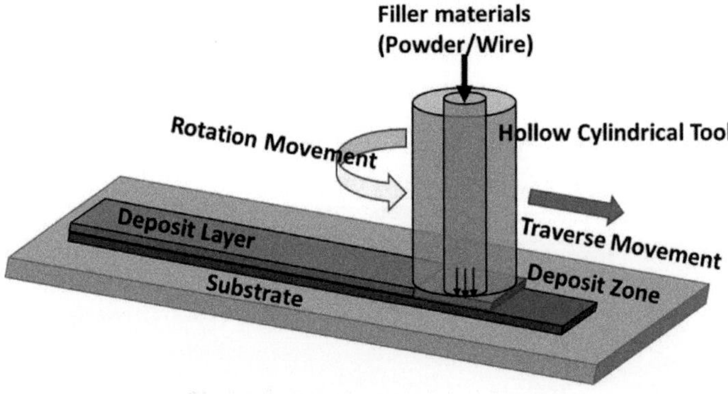

FIGURE 5.8 Schematic diagram of FSAM process.

5.4 TRENDING APPLICATIONS

5.4.1 AEROSPACE INDUSTRY

AM techniques are one of the first to be implemented in the aerospace industry, and according to Wohlers' reports, AM today accounts for 18–20% of the overall market (Kianian, 2017). AM techniques have been used in all sections of the aerospace industry, including corporate aircraft, defense applications, and missile platforms. AM technologies are appropriate for aircraft parts because they efficiently manufacture complicated geometries, difficult-to-machine materials, maintenance, and integrated parts. AM can be used to fabricate or repair both metal and non-metal components for aerospace industries. Non-metal such as plastics, composite materials, and ceramics parts are fabricated by following AM methods such as SLA, laminate object manufacturing, and FDM. NASA used FDM to print 70 parts for the Mars mission, while Bell Helicopter used it to build polycarbonate wiring tubes for its V-22 Osprey. The Aeronautics Research Organization of NASA is working to develop a non-metal gasoline turbine engine through AM. Metal such as Ti alloys, Ni alloys, and high resistance superalloys parts are manufactured using the DED AM technique (Sacco & Moon, 2019). Arconic is manufacturing titanium fuselage and engine pylon parts and NASA satellite vents, using powder bed fusion AM technology. GE Aviation has used powder bed fusion AM techniques to fabricate its next-generation of aircraft engine parts. The brackets for the Airbus A350 XWB developed by AM are 30% lightweight and consume 75% less time to fabricate. General Electric Aeronautics has used DED AM techniques to produce gasoline nozzles that are 25% powerful and lightweight than the conventional nozzle. The aerospace industry can reduce up to 65% on the lead time, saving up to 40% on non-recurring costs, and reduce up to 30% on weight by employing AM technique.

5.4.2 BIOMEDICAL INDUSTRY

Metal AM techniques are particularly suited to the manufacturing of finished components in the dental and surgical sectors. According to a previous Wohlers' research, the AM market is expected to increase from $6 billion in 2016 to $24 billion by 2021. Nowadays, the biomedical industry contributes 11% of the global AM sector (Kianian, 2017). AM has been employed as a flexible and cost-effective approach for fabricating complex geometric shapes in the medical sector. Furthermore, AM technique has been used to manufacture products varying from implants to heart tissue, hip, and knee joint repairs or replacements. AM is commonly used to produce three-dimensional reconstructions of human bone defects, head models, porous titanium implants, and titanium implants suited to the skull designs (Vyas & Udyawar, 2019). Tissues such as cardiac muscle, bones, tracheal prosthesis, cartilaginous structures, vascular grafts, and multi-layered skin can be manufactured and transplanted via bioprinting. Bio-printing is classified into three primary technologies: laser-based AM, IJP AM, and extrusion type AM. Different orthopedic devices, dental surgical equipment, and guides are manufactured using different AM techniques. AM technology is ideal for biomedical parts because it can manufacture customized layouts, highly complicated components, small-scale manufacturing quantities, and the availability of novel materials. Lima Corporation is already working on titanium-based alloy orthopedic implants using Metal PBF and EBM processes. 3D printing is used to manufacture facial, cranial, mandibular, and orthopedic implants (Zadpoor & Malda, 2017). Nexxt Spine is producing additively manufactured porous titanium-based alloy spinal implants. Egan Dental Laboratory, Johnson & Johnson, Osseus Fusion Technologies, Endocon GmbH, and OMX Solutions are now just a few organizations that have employed AM to make biomedical parts.

5.4.3 AUTOMOTIVE INDUSTRY

For fabricating automotive parts, AM has provided the following advantages such as highly adjustable, efficient, and reliable design, as well as lightweight, stronger, and safer objects, faster customization, and shorter lead time and costs. As a result, various automotive companies have started to adopt metal AM technology. Formula Senior Germany in 2012 utilized DMLS AM techniques to develop a lightweight weighted column that was 30% lighter than the previous cast component (Vafadar, Guzzomi, Rassau, & Hayward, 2021). HP and EOS, a BMW Company member, utilize a number of AM techniques and utilized the HP Dual Jets Fusion metals AM to produce a window guiding rail for the i8 Roadster. These AM techniques presently have been used to produce 100 rails in 24 hours. A powder-based AM technique is used to make BMW engine parts, such as the cylinder head. Bugatti, a Spanish vehicle manufacturer, has used AM techniques to produce a number of metal parts, including a Ti_6Al_4V brake caliper and an $AlSi_{10}Mg$ caliper bracket. Audi, a German automotive industry, has teamed up with SLM technology center to produce distinct items, spare components, and Audi W12 engine other parts. Honda Automobile Group has been using metal AM to make the body shells portions of its customizable electric vehicle. Germany Airbus company is fabricating lightweight structures of scalmalloy (an alloy of Sc, Al, and Mg) through metal AM.

5.4.4 OTHER INDUSTRY

In recent years, metal AM has been used to fabricate innovative electronics, electricals, and communications parts like resistors, circuit boards, antennas, microwaves, splitters, etc. Electronic, electrical, and communication components are manufactured from metals, semiconductors, and polymers and are fabricated using AM techniques such as SLA, SLM, and FDM. Metal AM is applied in the oil and gas sector because it reduces lead times, improves production efficiency, and allows final components production (Espera, Dizon, Chen, & Advincula, 2019). GE company uses metal AM techniques to fabricate complicated geometries in control valve components like hollow structures, curvature geometries, and meshes, among other things. Siemens manufactures various components for the oil and gas sector like gas turbine blades, fuel nozzles, gas turbine burners, heat shielding impellers, etc., via AM techniques (Sireesha, Lee, Kiran, Jagadeesh, Kee, & Ramakrishna, 2018). The use of metal AM techniques for construction applications is currently limited because of long printing periods, small print volumes, high initial costs, and environmental concerns. Mold and tool manufacturers are rapidly adopting Metal AM for the production of mold inserts, jigs, fixtures, and gauges, etc. (Le Néel, Mognol, & Hascoët, 2018). The railway sector has begun to use metal AM techniques to manufacture lightweight equipment and perform maintenance.

5.5 MAIN CHALLENGES FOR FORMATION OF PARTS

AM techniques are associated with several advantages, such as the ability to manufacture complex structures, design flexibility, ease of use, and customized product (Wang, Jiang, Zhou, Gou, & Hui, 2017). However, AM techniques have various drawbacks and challenges which need to be investigated for advanced technological development. Low manufacturing productivity, high tooling cost, limitation on the part size, poor dimensional accuracy and surface finish, anisotropic mechanical properties, layer misalignment, and limitations in the choice of metallic materials are the biggest problems that need to be resolved in the field of AM techniques (Zareiyan & Khoshnevis, 2017). The following challenges and difficulties associated with AM methods are listed in subsections.

5.5.1 *EVALUATION OF VOID*

The major drawback of AM techniques is the formation of the void between the consequent layers of deposited metallic materials. This issue arises as a consequence of insufficient bonding among successive layers, resulting in lower mechanical properties of deposited samples (Zareiyan & Khoshnevis, 2017). FDM, EBM, and SLM-based AM processes, for example, cause voids to emerge in successive layers, resulting in anisotropic mechanical properties of deposited samples (Arya, Sathiaraj, Jain, Jayaprakash, & Saxena, 2022). The quantity of porosity induced by void generation depends on the category of AM process employed and the material utilized. Paul et al. (2018) examined the impact of nozzle design on void development between succeeding layers in order to reduce the effect of void generation. The key findings of this research revealed that rectangle nozzles outperformed cylinder nozzles. The high porous structure of metallic components is not always a defect and can be utilized in biomedical applications such as the porous structure of scaffolds in tissue engineering.

5.5.2 *STAIR-STEPPING APPEARANCE*

Another of the most significant issues with AM manufactured components is the stair-stepping or stairway impact. Internally, such kind of defect is minor; yet, it has a substantial impact on the performance of externally created surfaces. The appearance of AM products may not be an essential aspect in the application of tissue engineering. External appearance is crucial in the automobile and aerospace industries, as flat surfaces are favored over stair-stepping look (Zocca, Colombo, Gomes, & Günster, 2015). Post-processing techniques like sand sintering can help to minimize this effect, but these techniques increase the processing time and production cost. Khoshnevis (2017) utilized a trowel-like apparatus to reduce the staircase appearance. The number of layers is expressed by the thickness and height of the component. As a result, reducing the number of layers of parts may reduce the difficulty of the staircase's structure.

5.5.3 ANISOTROPIC MECHANICAL PROPERTIES

Anisotropic mechanical properties and microstructure are other significant challenges associated with AM manufactured components (Carroll, Palmer, & Beese, 2015). AM-manufactured parts often result in various mechanical properties and microstructures. The components manufactured using FDM result in better mechanical strength in the x and y direction compared to the z-direction, which is also a build direction (Zocca, Colombo, Gomes, & Günster, 2015). In metallic metals and alloys, the components are manufactured by the fusion of heat such as SLM, SLS, or EBM; the additional successive layers reheat the boundaries of earlier layers, which results in anisotropic mechanical behavior as well as distinct microstructure due to large thermal gradient (Cooke, Anne, Burguete, Johns, & Vanard, 2011).

5.5.4 DIVERGENT FROM DESIGN TO EXECUTION

Computer added design (CAD) software is the primary tool to design the components to be manufactured through AM techniques is (CAD) software. The manufactured components can consist of a variety of defects that were not predicted because of the limitations of AM processes. A CAD model is a set of boundaries and solid part geometry which imitates a 3D model using tessellation techniques. However, errors and flaws are common when converting a CAD model into a 3D printed object, especially in curved regions (Cooke, Anne, Burguete, Johns, & Vanard, 2011). Extremely precise tessellation may be able to solve this issue to some level, but the computed processing and manufacturing will be intricate as well as time-consuming. Therefore, post-processing of AM manufactured components to eliminate these defects are occasionally considered. The powder-bed AM approach benefits using unbound powders as support that may be easily removed using air pressure. In the situation of part manufacture with FDM, external support is provided by counter crafting, which is hard to eliminate. Moreover, process parameters (varied according to different AM techniques) and material properties can significantly affect the exterior and mechanical properties of the 3D manufactured components (Quan, Suhr, Yu, et al., 2018).

5.5.5 SMALL VOLUME

The corporation that uses the AM techniques is encountering the challenge of small build volume of parts. It is understood as one of the main disadvantages of AM techniques. Commonly, the large parts are scaled or cut down to subcomponents, which add up time as well as energy. Moreover, the scaling down of the components in most cases is not practical and valuable. After scaling down, the assembly of subcomponents possesses reduced mechanical strength if adhesives are employed in the case of mechanical fasteners (Quan, Suhr, Yu, et al., 2018). Till now, AM techniques have been unsuccessful for large-scale manufacturing applications (Coniglio, Sivarupan, & El Mansori, 2018).

5.5.6 COMPLIANCE WITH FOOD AND DRUG ADMINISTRATION (FDA) SAFETY STANDARDS

The surgical implants, orthopedic implants, and medical devices manufactured using AM techniques must comply with the guidelines defined by the local authority of the Food and Drug Administration (FDA). Since AM techniques are relatively new to the medical manufacturing sector, the FDA is still working on the formulation of rules and regulations on AM techniques in the medical industry. Furthermore, the current guidelines are complicated and tricky to execute; as a result, corporations are uncertain about using AM in the healthcare industry.

5.6 MECHANICAL PROPERTIES OF 3D PRINTED PARTS

AM techniques are a rapidly growing trend in the manufacturing industries sector. It is very pleasing due to the ease of implementation of design as well as eliminating the need for fasteners and assemblies. The mechanical properties of components manufactured through AM techniques vary with the change in conventional manufacturing processes like casting, forging, extrusion, etc. (Ivanova, Williams, &

Campbell, 2013). These mechanical properties are evaluated with the universal testing machine (UTM) to confirm that the AM manufactured components meet the desired standards. Sometimes AM manufactured components have localized stress concentrations due to which components can fail prematurely. It may also be possible that due to fixtures of UTM equipment, stress concentration can occur to components during testing. The testing direction is essential to determine and record the results, as AM components tend to be anisotropic. In several cases, the test must be completed in various directions to establish the correct direction of the material during production. For this instance, a strain measurement device is crucial to determine the effect of applied force on components precisely. The various tests for determining mechanical properties associated with the AM manufactured components are discussed in subsections.

5.6.1 TENSILE TEST

The purpose of a tensile test on a metallic material is to investigate its performance under axial tensile loading. This type of test is performed by fixturing the test specimen into the experimental apparatus and then applying an applied load to the test specimen by splitting the crosshead of the testing equipment. The strain rate is controlled by varying the crosshead speed. The obtained results are used to evaluate yield strength, tensile strength, and the young's modulus. ASTM E8 is a standard test procedure for conducting a tensile test of metallic materials at room temperature and elevated temperature (Godec, Cano, Holzer, & Gonzalez-Gutierrez, 2020). Tensile testing of specimen's entails increasing the load until it reaches the failure point. The specimen is steadily held through the top as well as bottom grips attached to the UTM machine. The grips are moved apart at a constant rate to pull and stretch the test specimen during the tensile testing. The applied load and displacement are constantly monitored until failure. The final length of the test specimen is measured by putting them together, and this measure is compared to the original length to determine elongation. The failed test samples were further investigated through microscopy to study the type of fracture and point of fracture.

5.6.2 COMPRESSION TEST

The compression test is performed to evaluate the material's behavior, such as proportional as well as the elastic limit, yield point, compression strength, and yield strength under an applied compression load. This test is performed by loading the specimen between the two flat plates, and then compression load is applied to the test specimen by moving the crosshead at a given speed together. The test specimen is compressed throughout the compression test, and results are recorded in terms of deformation against applied load. The failed test samples were further investigated through microscopy to study the fracture type and point of fracture. ASTM E9 is a standard test procedure for conducting metallic material compression tests at normal room temperature (Goh, Yap, Tan, Sing, Goh, & Yeong, 2020).

5.6.3 CREEP TEST

Creep is defined as the service failure by a mechanism at a high temperature. A creep test is performed using a tensile test specimen where constant stress is applied at a constant test temperature, which is surrounded by the thermostatically controlled furnace. The temperature is being controlled by a thermocouple which is attached to the gauge length of the test specimen.

A slow failure mechanism occurs in a metallic material exposed to an applied load of prolonged-time below its elastic limit. The metallic materials start to increase in length to the direction of applied stress. The deformation of materials occurs at ambient temperature, and the rate of deformation increases as the temperature increases. In metallic materials, creep failure occurs at the grain boundary to provide intergranular fracture (Kuo, Horikawa, & Kakehi, 2017). Figure 5.9 represents the schematic diagram of the creep test. ASTM E139 standard is a guideline to perform creep tests of metallic materials.

5.6.4 FATIGUE TEST

A fatigue test is conducted for AM components to evaluate the material response, stress concentrations, and reliability under repeated stresses (Fischer & Schöppner, 2017). A structural application-based fatigue test is performed to investigate various structural designs' fatigue life, manufacturing processes, and stress concentrations. The reliability of the AM components was examined through actual service type fatigue testing (Shanmugam et al., 2021). Fatigue is defined as structural damage of components because of cyclic loading, whose strength is less than the ultimate tensile strength of components. Fatigue test equipment is classified based on the methods of various load applied, such as axial, rotational, or torsional load. In axial loads, uniform stress or strain is applied to the cross-section of the components in the tension or compression mode of the applied load. In rotational loading, the specimen is revolved at a constant frequency against applied load. In every rotational cycle, the mode of applied stresses changes from tension to compression or vice versa. Torsion fatigue test was conducted at each end of the test specimen, either clamped or twisted to a specified value through stress or strain. Fatigue failure in metallic materials occurs in the form of mechanical failure where crack initiation starts at some points and propagates through different grain boundaries or grains (Ziemian, Ziemian, & Haile, 2016).

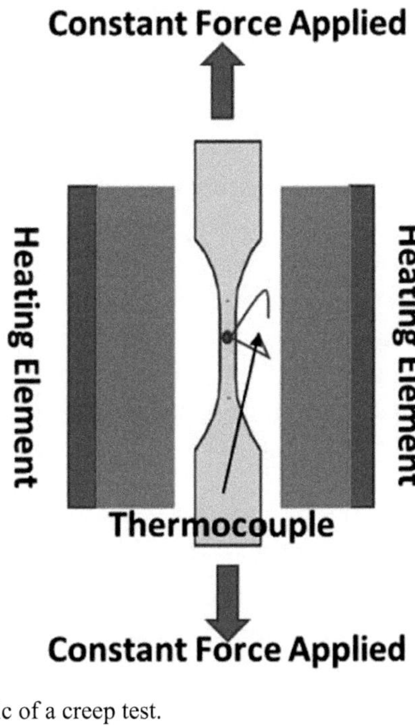

FIGURE 5.9 Schematic of a creep test.

5.6.5 *FRACTURE TOUGHNESS TEST*

The main objective of the fracture toughness test is to evaluate the resistance of a material in terms of the required load to cause a ductile or brittle fracture. The results of fracture toughness are expressed in terms of KIC or critical J. ASTM E1820 standards are used for the measurement of fracture toughness (Yu et al., 2020). Measurement of fracture toughness consists of the following steps:

- Notch formation on a test specimen as per standard dimension.
- Development of pre-crack fatigue by applying cyclic loading at normal road temperature.
- Connection of displacement measure gauge through the crack estuary.
- Maintaining a stable temperature of the test specimen.
- Applying monotonically increasing applied load while monitoring applied load as well as crack initiation.
- Calculation of the relevant fracture toughness parameters KIC.
- Validation of the measured results.

5.7 CURRENT AM STANDARD

International Organization for Standardization (ISO standard) and the American Society for Testing and Materials (ASTM standard) coordinate the development of AM techniques' standards. Till date, ISO and ASTM society developed standards related to AM terminology, individual AM processes, quality parameters, test methods for AM components, customer-supplier deals, and fundamental elements. In 2013, ISO and ASTM society defined a common objective to create a set of worldwide standards, including general standards that apply to most AM techniques, materials, and several applications (Huang, Leu, Mazumder, & Donmez, 2015). The various countries are making a great effort for developing AM standards, such as Germany (VDI FA 105), France (AFNOR UNM 920), Spain (AEN 116), Sweden (SIS 563), the UK (BSI AMT/8), and the US (SAE AMS-AM) developed standards related to AM techniques. AM standards support a general perception of the AM area and a shared dictionary from which to work for various AM techniques. These standards are crucial for expanding AM techniques associated with tools design and procedures. For instance, ISO/ASTM DIS 20195 "Guide for Design for Additive Manufacturing" (Rosen, 2014) is presently under development.

5.8 CONCLUSIONS

AM techniques have progressed significantly over the years, and their items have now been evaluated to replace components and materials produced using traditional processes. This chapter provided a comprehensive overview of various AM processes, including their benefits, limits, applications, and challenges encountered during AM procedures. The generally accepted ASTM and ISO mechanical testing standards have been described, numerous research institutions use that to assess the strength of AM components. AM has a several benefits, including design flexibility, customization, and the ability to create extremely complex structures, but it also has some drawbacks, including higher setup costs, limited scale production, material limitations, and defects. Applying such techniques in the medical, aviation, automotive, electronics, and other industries has accelerated research into materials, components feature, and sustainability. The success of each 3D printed product is decided by the materials, technique, and parameters selections. As a result, future researchers will find this analysis highly beneficial to selecting the best method and material for a given application.

KEYWORDS

- **3D printed parts**
- **3D printing**
- **additive manufacturing**
- **aerospace**
- **anisotropic mechanical properties**
- **automotive industry**
- **fatigue test**
- **mechanical properties**
- **tensile test**

REFERENCES

Alafaghani, A., Qattawi, A., Alrawi, B., & Guzman, A., (2017). Experimental optimization of fused deposition modeling processing parameters: A design-for-manufacturing approach. *Procedia Manufacturing, 10*, 791–803. https://doi.org/10.1016/j.promfg.2017.07.079.

Arya, P. K., Jain, N. K., Murugesan, J., & Patel, V. K., (2021). Developments in friction stir welding of aluminum to magnesium alloy. *Journal of Adhesion Science and Technology, 36*(13), 1365–1402. https://doi.org/10.1080/01694243.2021.1975614.

Arya, P. K., Sathiaraj, D., Jain, N. K., Jayaprakash, M., & Saxena, K. K., (2022). *Microstructure, Mechanical and Corrosion Behavior of Friction Stir Welding of AA6061 Al Alloy and AZ31B Mg Alloy, 119*, 413. https://doi.org/10.1051/metal/2022057.

Bahuguna, S., Arya, P. K., & Patel, V. K., (2020). Mechanical and abrasive wear properties of friction stir welded joints of aluminum alloy AA6061-T6 with/without nickel coating. *Strojnícky Časopis-Journal of Mechanical Engineering, 70*(2), 21–36. https://doi.org/10.2478/scjme-2020-0017.

Beaman, J. J., Barlow, J. W., Bourell, D. L., Crawford, R. H., Marcus, H. L., & McAlea, K. P., (1997). Introduction. In: Beaman, J. J., Barlow, J. W., Bourell, D. L., Crawford, R. H., Marcus, H. L., & McAlea, K. P., (eds.), *Solid Freeform Fabrication: A New Direction in Manufacturing: With Research and Applications in Thermal Laser Processing* (pp. 1–21). Springer US. https://doi.org/10.1007/978-1-4615-6327-3_1.

Bhushan, B., & Caspers, M., (2017). An overview of additive manufacturing (3D printing) for microfabrication. *Microsyst. Technol., 23*(4), 1117–1124. https://doi.org/10.1007/s00542-017-3342-8.

Carroll, B. E., Palmer, T. A., & Beese, A. M., (2015). Anisotropic tensile behavior of Ti–6Al–4V components fabricated with directed energy deposition additive manufacturing. *Acta Materialia, 87*, 309–320. https://doi.org/10.1016/j.actamat.2014.12.054.

Coniglio, N., Sivarupan, T., & El Mansori, M., (2018). Investigation of process parameter effect on anisotropic properties of 3D printed sand molds. *Int. J. Adv. Manuf. Technol., 94*(5), 2175–2185. https://doi.org/10.1007/s00170-017-0861-5.

Cooke, W., Anne, T. R., Burguete, R., Johns, D., & Vanard, G., (2011). Anisotropy, homogeneity and ageing in an SLS polymer. *Rapid Prototyping Journal, 17*(4), 269–279. https://doi.org/10.1108/13552541111138397.

Derby, B., (2010). Inkjet printing of functional and structural materials: Fluid property requirements, feature stability, and resolution. *Annual Review of Materials Research, 40*(1), 395–414. https://doi.org/10.1146/annurev-matsci-070909-104502.

Espera, A. H., Dizon, J. R. C., Chen, Q., & Advincula, R. C., (2019). 3D-printing and advanced manufacturing for electronics. *Prog Addit Manuf., 4*(3), 245–267. https://doi.org/10.1007/s40964-019-00077-7.

Fischer, M., & Schöppner, V., (2017). Fatigue behavior of FDM Parts manufactured with Ultem 9085. *JOM., 69*(3), 563–568. https://doi.org/10.1007/s11837-016-2197-2.

Gibson, I., Cheung, L. K., Chow, S. P., et al., (2006). The use of rapid prototyping to assist medical applications. *Rapid Prototyping Journal, 12*(1), 53–58. https://doi.org/10.1108/13552540610637273.

Godec, D., Cano, S., Holzer, C., & Gonzalez-Gutierrez, J., (2020). Optimization of the 3D printing parameters for tensile properties of specimens produced by fused filament fabrication of 17-4PH stainless steel. *Materials*, *13*(3), 774. https://doi.org/10.3390/ma13030774.

Goh, G. D., Yap, Y. L., Tan, H. K. J., Sing, S. L., Goh, G. L., & Yeong, W. Y., (2020). Process–structure–properties in polymer additive manufacturing via material extrusion: A review. *Critical Reviews in Solid State and Materials Sciences*, *45*(2), 113–133. https://doi.org/10.1080/10408436.2018.1549977.

Hassanin, H., (2021). Powder-based laser hybrid additive manufacturing of metals: A review. *The International Journal of Advanced Manufacturing Technology*. Published online. https://repository.canterbury.ac.uk/item/8x3wy/powder-based-laser-hybrid-additivemanufacturing-of-metals-a-review (accessed on 11 September 2023).

Huang, J., Qin, Q., & Wang, J., (2020). A review of stereolithography: Processes and systems. *Processes*, *8*(9), 1138. https://doi.org/10.3390/pr8091138.

Huang, S. H., Liu, P., Mokasdar, A., & Hou, L., (2013). Additive manufacturing and its societal impact: A literature review. *Int. J. Adv. Manuf. Technol.*, *67*(5), 1191–1203. https://doi.org/10.1007/s00170-012-4558-5.

Huang, Y., Leu, M. C., Mazumder, J., & Donmez, A., (2015). Additive manufacturing: Current state, future potential, gaps and needs, and recommendations. *Journal of Manufacturing Science and Engineering*, *137*(1). https://doi.org/10.1115/1.4028725.

Ivanova, O., Williams, C., & Campbell, T., (2013). Additive manufacturing (AM) and nanotechnology: Promises and challenges. *Rapid Prototyping Journal*, *19*(5), 353–364. https://doi.org/10.1108/RPJ-12-2011-0127.

Keleş, Ö., Blevins, C. W., & Bowman, K. J., (2017). Effect of build orientation on the mechanical reliability of 3D printed ABS. *Rapid Prototyping Journal*, *23*(2), 320–328. https://doi.org/10.1108/RPJ-09-2015-0122.

Kianian, B., (2017). *Wohlers Report 2017: 3D Printing and Additive Manufacturing State of the Industry, Annual Worldwide Progress Report: Chapters Titles: The Middle East, and Other Countries*. Wohlers Associates, Inc. http://lup.lub.lu.se/record/ac880c85-f9ef-477b-b5db-3a7d2d575ea5 (accessed on 11 September 2023).

Kruth, J. P., Leu, M. C., & Nakagawa, T., (1998). Progress in additive manufacturing and rapid prototyping. *CIRP Annals.*, *47*(2), 525–540. https://doi.org/10.1016/S0007-8506(07)63240-5.

Kruth, J., Mercelis, P., Van, V. J., Froyen, L., & Rombouts, M., (2005). Binding mechanisms in selective laser sintering and selective laser melting. *Rapid Prototyping Journal*, *11*(1), 26–36. https://doi.org/10.1108/13552540510573365.

Kumar, P., & Jain, N. K., (2021). Finite element analysis of femoral prosthesis using Ti-6Al-4 V alloy and TiNbZrTaFe high entropy alloy. *Materials Today: Proceedings*, *44*, 1195–1201. https://doi.org/10.1016/j.matpr.2020.11.239.

Kumar, P., Sawant, M. S., & Jain, N. K., (2021). Optimization of process parameters in micro-plasma transferred arc deposition process for cobalt-based alloy. *Materials Today: Proceedings*, *44*, 1681–1686. https://doi.org/10.1016/j.matpr.2020.11.840.

Kumar, R., Kumar, M., & Chohan, J. S., (2021). Material-specific properties and applications of additive manufacturing techniques: A comprehensive review. *Bull Mater Sci.*, *44*(3), 181. https://doi.org/10.1007/s12034-021-02364-y.

Kuo, Y. L., Horikawa, S., & Kakehi, K., (2017). Effects of build direction and heat treatment on creep properties of Ni-base superalloy built up by additive manufacturing. *Scripta Materialia.*, *129*, 74–78. https://doi.org/10.1016/j.scriptamat.2016.10.035.

Le Néel, T. A., Mognol, P., & Hascoët, J. Y., (2018). A review on additive manufacturing of sand molds by binder jetting and selective laser sintering. *Rapid Prototyping Journal*, *24*(8), 1325–1336. https://doi.org/10.1108/RPJ-10-2016-0161.

Low, Z. X., Chua, Y. T., Ray, B. M., Mattia, D., Metcalfe, I. S., & Patterson, D. A., (2017). Perspective on 3D printing of separation membranes and comparison to related unconventional fabrication techniques. *Journal of Membrane Science*, *523*, 596–613. https://doi.org/10.1016/j.memsci.2016.10.006.

Paul, S. C., Tay, Y. W. D., Panda, B., & Tan, M. J., (2018). Fresh and hardened properties of 3D printable cementitious materials for building and construction. *Archives of Civil and Mechanical Engineering*, *18*(1), 311–319. https://doi.org/10.1016/j.acme.2017.02.008.

Pereira, T., Kennedy, J. V., & Potgieter, J., (2019). A comparison of traditional manufacturing vs additive manufacturing, the best method for the job. *Procedia Manufacturing*, *30*, 11–18. https://doi.org/10.1016/j.promfg.2019.02.003.

Quan, Z., Suhr, J., Yu, J., et al., (2018). Printing direction dependence of mechanical behavior of additively manufactured 3D preforms and composites. *Composite Structures*, *184*, 917–923. https://doi.org/10.1016/j.compstruct.2017.10.055.

Rosen, D. W., (2014). Research supporting principles for design for additive manufacturing. *Virtual and Physical Prototyping*, *9*(4), 225–232. https://doi.org/10.1080/17452759.2014.951530.

Rosso, S., Uriati, F., Grigolato, L., Meneghello, R., Concheri, G., & Savio, G., (2021). An optimization workflow in design for additive manufacturing. *Applied Sciences*, *11*(6), 2572. https://doi.org/10.3390/app11062572.

Sacco, E., & Moon, S. K., (2019). Additive manufacturing for space: Status and promises. *Int. J. Adv. Manuf. Technol.*, *105*(10), 4123–4146. https://doi.org/10.1007/s00170-019-03786-z.

Sawant, M. S., & Jain, N. K., (2017). Characteristics of single-track and multi-track depositions of stellite by micro-plasma transferred arc powder deposition process. *J. of Materi. Eng. and Perform*, *26*(8), 4029–4039. https://doi.org/10.1007/s11665-017-2828-y.

Sawant, M. S., & Jain, N. K., (2018). Investigations on additive manufacturing of Ti–6Al–4V by microplasma transferred arc powder deposition process. *Journal of Manufacturing Science and Engineering*, *140*(8). https://doi.org/10.1115/1.4040324.

Shanmugam, V., Das, O., Babu, K., et al., (2021). Fatigue behavior of FDM-3D printed polymers, polymeric composites and architected cellular materials. *International Journal of Fatigue*, *143*, 106007. https://doi.org/10.1016/j.ijfatigue.2020.106007.

Sing, S. L., Tey, C. F., Tan, J. H. K., Huang, S., & Yeong, W. Y., (2020). 3D printing of metals in rapid prototyping of biomaterials: Techniques in additive manufacturing. In: Narayan, R., (ed.), *Rapid Prototyping of Biomaterials* (2nd edn., pp. 17–40). Woodhead Publishing Series in Biomaterials. Woodhead Publishing. https://doi.org/10.1016/B978-0-08-102663-2.00002-2.

Singh, M., Haverinen, H. M., Dhagat, P., & Jabbour, G. E., (2010). Inkjet printing process and its applications. *Advanced Materials*, *22*(6), 673–685. https://doi.org/10.1002/adma.200901141.

Sireesha, M., Lee, J., Kiran, A. S. K., Jagadeesh, B. V. T., Kee, B. B., & Ramakrishna, S., (2018). A review on additive manufacturing and its way into the oil and gas industry. *RSC Advances*, *8*(40), 22460–22468. https://doi.org/10.1039/C8RA03194K.

Srivastava, M., Rathee, S., Maheshwari, S., Noor, S. A., & Kundra, T. K., (2019). A review on recent progress in solid state friction based metal additive manufacturing: Friction stir additive techniques. *Critical Reviews in Solid State and Materials Sciences*, *44*(5), 345–377. https://doi.org/10.1080/10408436.2018.1490250.

Vaezi, M., Seitz, H., & Yang, S., (2013). A review of 3D micro-additive manufacturing technologies. *Int. J. Adv. Manuf. Technol.*, *67*(5–8), 1721–1754. https://doi.org/10.1007/s00170-012-4605-2.

Vafadar, A., Guzzomi, F., Rassau, A., & Hayward, K., (2021). Advances in metal additive manufacturing: A review of common processes, industrial applications, and current challenges. *Applied Sciences*, *11*(3), 1213. https://doi.org/10.3390/app11031213.

Vyas, D., & Udyawar, D., (2019). A review on current state of art of bioprinting. In: Kumar, L. J., Pandey, P. M., & Wimpenny, D. I., (eds.), *3D Printing and Additive Manufacturing Technologies* (pp. 195–201). Springer. https://doi.org/10.1007/978-981-13-0305-0_17.

Vyavahare, S., Teraiya, S., Panghal, D., & Kumar, S., (2019). Fused deposition modeling: A review. *Rapid Prototyping Journal*, *26*(1), 176–201. https://doi.org/10.1108/RPJ-04-2019-0106.

Wang, X., Jiang, M., Zhou, Z., Gou, J., & Hui, D., (2017). 3D printing of polymer matrix composites: A review and prospective. *Composites Part B: Engineering*, *110*, 442–458. https://doi.org/10.1016/j.compositesb.2016.11.034.

Williams, S. W., Martina, F., Addison, A. C., Ding, J., Pardal, G., & Colegrove, P., (2016). Wire + arc additive manufacturing. *Materials Science and Technology*, *32*(7), 641–647. https://doi.org/10.1179/1743284715Y.0000000073.

Yang, Y., Fuh, J. Y. H., Loh, H. T., & Wong, Y. S., (2003). A volumetric difference-based adaptive slicing and deposition method for layered manufacturing. *Journal of Manufacturing Science and Engineering*, *125*(3), 586–594. https://doi.org/10.1115/1.1581887.

Yang, Y., Li, L., & Zhao, J., (2019). Mechanical property modeling of photosensitive liquid resin in stereolithography additive manufacturing: Bridging degree of cure with tensile strength and hardness. *Materials & Design*, *162*, 418–428. https://doi.org/10.1016/j.matdes.2018.12.009.

Yu, X., Lin, X., Liu, F., et al., (2020). Influence of post-heat-treatment on the microstructure and fracture toughness properties of Inconel 718 fabricated with laser directed energy deposition additive manufacturing. *Materials Science and Engineering: A*, *798*, 140092. https://doi.org/10.1016/j.msea.2020.140092.

Zadpoor, A. A., & Malda, J., (2017). Additive manufacturing of biomaterials, tissues, and organs. *Ann. Biomed. Eng*, *45*(1), 1–11. https://doi.org/10.1007/s10439-016-1719-y.

Zareiyan, B., & Khoshnevis, B., (2017). Interlayer adhesion and strength of structures in contour crafting – effects of aggregate size, extrusion rate, and layer thickness. *Automation in Construction*, *81*, 112–121. https://doi.org/10.1016/j.autcon.2017.06.013.

Ziemian, C. W., Ziemian, R. D., & Haile, K. V., (2016). Characterization of stiffness degradation caused by fatigue damage of additive manufactured parts. *Materials & Design*, *109*, 209–218. https://doi.org/10.1016/j.matdes.2016.07.080.

Zocca, A., Colombo, P., Gomes, C. M., & Günster, J., (2015). Additive manufacturing of ceramics: Issues, potentialities, and opportunities. *Journal of the American Ceramic Society*, *98*(7), 1983–2001. https://doi.org/10.1111/jace.13700.

CHAPTER 6

3D and 4D Printing Technologies and their Applications: Current Status and Future Perspectives

HUSSAM BIN MEHARE, MOHAMMED ZAINUL AREFEEN, ISRAR AHMAD, and FAISAL TALIB

Department of Mechanical Engineering, Zakir Husain College of Engineering and Technology, Aligarh Muslim University, Aligarh, Uttar Pradesh, India

ABSTRACT

Additive manufacturing (AM) is taking giant leaps in the manufacturing industry, and day by day there is multiple research taking place on its effectiveness to be the manufacturing process of the future as it does have multiple benefits of waste reduction, production of more complex and intricate designs and geometries. In order to be the manufacturing method of the future, it must be economically viable for all sorts of production so that it can replace the old and the traditional methods of subtractive manufacturing which have been in use for hundreds of years now. This study takes into account the techniques being used, namely 3D and 4D printing, along with their various added benefits, but taking into concern the job quality and the economic viability of this ever-growing popular method of manufacturing.

Advances in Manufacturing and Processing of Materials: Characterization and Applications.
Narendra Kumar, PhD, Mohit Tyagi, PhD, Dilbagh Panchal, PhD & Ravi Pratap Singh, PhD (Eds.)

6.1 INTRODUCTION

As we move further and further ahead in time and in order to keep up with the pace of development, it has become quite necessary for us to keep introducing new methods of manufacturing which can provide us with better end products and are as efficient as possible. Manufacturing is the process of producing finished goods from raw materials, in which the raw materials undergo various processes like design, selection of raw materials, and the order of process applied on the raw materials (Serope, n.d.; Rao, n.d.). The complete process of manufacturing involves both human labor and the help of computers/robots according to the needs of the process being carried out on the job (Mehrabi, Ulsoy, & Koren, 2000). Furthermore, manufacturing has been broadly classified into two, namely Subtractive Manufacturing and AM.

The term subtractive manufacturing refers to the old and common processes which have been in use for so many years by now, they consist of processes like designing, cutting, casting, etc. (González-Henríquez, Sarabia-Vallejos, & Rodriguez-Hernandez, n.d.). The main reason why this technique got so popular (Steve & Richard, n.d.) in industries is that it is quite easy to implement in industries, and it gives out excellent job results.

For example, CNC Machining is quite a commonly used subtractive manufacturing method, and the jobs obtained have excellent finish and it does not take much time and effort either (Van, Henri, & Guillaume, n.d.; Zhu, Dhokia, Nassehi, & Newman, n.d.). But there are a few shortcomings in this method of manufacturing, and they in themselves have such a massive impact that they can't be ignored. Subtractive manufacturing cannot be used to produce jobs that have complex designs, instead, they are used to produce simple jobs with a high-quality finish and in a time-efficient manner. Also, the subtractive manufacturing industry has a humongous impact on nature as every small piece of waste adds up to an already growing waste that just keeps on affecting the environment (Stephen, Zicheng, Vimal, & Alborz, n.d.).

In subtractive manufacturing, there are various levels of the materials which are removed by different processes thus adding up to the net waste and hence further degrading the environment and the machines which are used are quite complex to use (Marion, Daniel, Adam, & Franziska, n.d.). That is why in recent times there has been a shift towards AM and there has been research taking place (both for the academic and industrial processes) on this method for the past 40 years (Satish, Nancharaiah, & Subba, n.d.).

6.2 ADDITIVE MANUFACTURING (AM)

Additive manufacturing (AM) is also known as 3D printing or as rapid prototyping was introduced to the world in the year 1984 is a layer-based manufacturing technique developed by Charles Hull (Wei et al., n.d.; Bethany, Jayda, Sarah, Chengpeng, & Dana, 2014). According to ASTM, AM is defined as a method of combining materials to fabricate objects from the 3D model surface by surface. The biomedical industry along with the aerospace industry, musical instruments industry, and automotive industry are the ones that have reaped benefits from the advent of 3D printing (SPIE-Professional, n.d.; Amit & Sumita, n.d.). The main reason why 3D printing was introduced was for providing faster and easier prototyping. In comparison with traditional injection-molded prototypes, it is more efficient and produces better quality end products. It covers a host of processes and technologies that offer a full spectrum of capabilities for the production of parts and products in different materials (Vayre, Vignat, & Villeneuve, n.d.). It has also further reduced the waste of materials to almost zero as only the exact amount of materials needed are being utilized (Henri, Hossein, Eric, Matthieu, & Inigo, n.d.). Figure 6.1 shows the difference in the amount of waste created by both methods. Also, AM methods are quite economically viable as well (Weller, Kleer, & Piller, 2015).

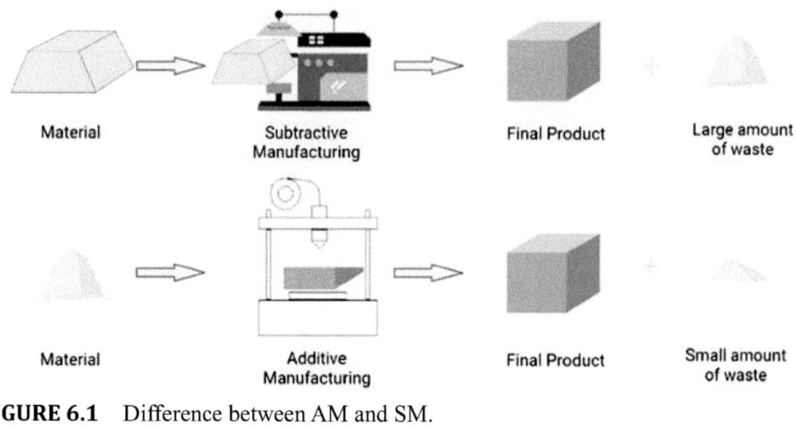

FIGURE 6.1 Difference between AM and SM.

Source: Adapted and modified from Nirish and Rajendra, 2020

Despite its ability to be complicated and bio-based, 3D printing is not yet ready to be adopted in large-scale manufacturing because the relative quality and tolerance of the job produced are poor. Because of this major disadvantage, the focus had turned towards improvising the current technique, and thus 4D printing was introduced to us. Since being released to the world in 2013, by Tibbits at TED Talk, it has got the world fascinated and left dreaming of its complete potential (Stephen, Zicheng, Vimal, & Alborz, n.d.). 4D printing is an upgrade from the current 3D printing method, which itself is yet to be developed and deployed completely, and adds an extra dimension to 3D printing. 4D printing can be said to be the progression in 3D printing and originally 4D printing was defined as 3D printing plus time thereby adding an extra dimension to it in which the 3D material being developed can change its properties namely structure, size, and the functions with respect to time. The material thus developed is usually dependent on time, the printer and its properties can be predicted. The material in some cases is also capable of repairing itself on its own and provides multiple functions (Xiao et al., 2018; Tibbits, 2013). These qualities of a 4D printed object make it an extremely viable option for the development of certain machine parts, human organs, etc. (Figure 6.2).

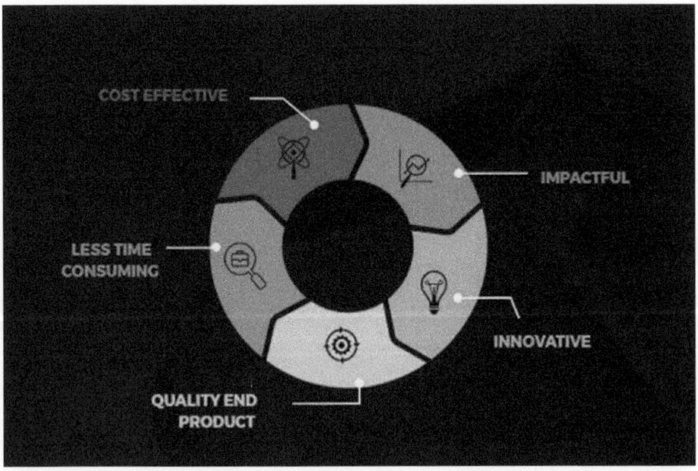

FIGURE 6.2 Benefits of additive manufacturing.

In this work of ours, we will be trying to review the two most commonly known methods of AM, i.e., 3D and 4D printing while taking into account the method of each, most viable materials which can be used along with the advantages and disadvantages of those methods.

6.3 3D PRINTING

3D printing, as mentioned above, is an AM method in which the materials are laid down, one layer at a time, to form a 3-dimensional object, as shown in Figure 6.3. As mentioned earlier above, 3D printing is a revolutionary technique that is evolving more and more as per our needs. From the construction sector, where it's been used to fabricate complex parts, to the biomedical industry, where it's used for the fabrication of various organs of the human body to replace the damaged ones having almost similar capabilities to that of the real ones (Schubert, Van, & Donoso, 2014), it's definitely improving our lives by making everything much easier.

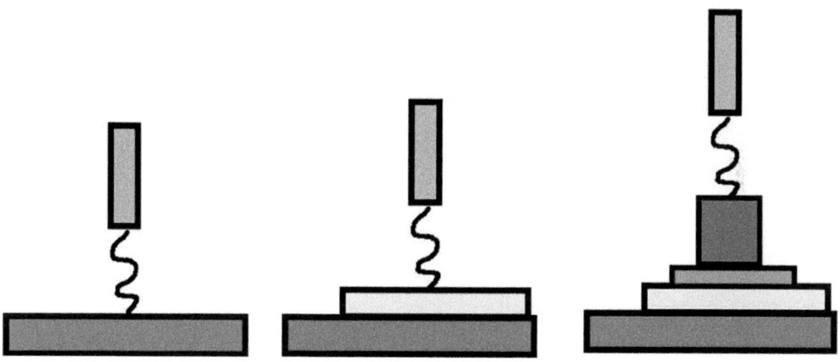

FIGURE 6.3 Schematic representation of 3D printing.

Also, 3D printing can be done using various sorts of materials from metals like Al, to ceramics, to powders, alloys, and even bio-compatible materials (Randolph, 2018). The materials are then placed at top of each other, one layer at a time according to the shape and design desired and this process continues till the desired shape of the job has been obtained (Campbell, Williams, Ivanova, & Garrett, 2011). Hence this makes this process quite cost-effective, accommodating towards the product requirements, and easy to use (Peterson, Larsen, Ganter, Storti, & Boydston, 2014). 3D printing is synonymous with terms like Computer Automated Manufacturing, Solid Freeform Technology, etc.

6.3.1 *TECHNIQUES OF 3D PRINTING*

In order to print the job using the 3D Printing method, a 3-Dimensional design needs to be made, and in order to do so, a digital designing software (some examples of these software are: Google Sketch Up, Mimics, Onshape, AutoDesk, AutoCAD & Solidworks) (Roopavath & Kalaskar, 2017; Patwardhan, 2017) and 3-Dimensional digital scanners are needed. Once the design is ready, it is converted to a STL (standard tessellation language) file format which is supported by nearly all the different techniques of 3D printing. The .STL file format consists of a register of triangular slants which include all the information about the design of the 3D structure to be created (Gross, Erkal, Lockwood, Chen, & Spence, 2014).

Usually, this procedure of turning the 3D design into .STL format is self-regulating as it is present on most of the 3D printing systems. Software such as Materialize are used to sort out any errors during the conversion of the file as the .STL file is unable to hold details such as material type, surface, colors, etc. Due to these drawbacks, file types like AMF and 3MF have been brought in to trounce these issues (Gibson, Rosen, & Stucker, 2010). After this step, software like Cura or PrusaSlicer are used to slice the .STL file into 2D parallel, cross-sectional parts to a kind of special file called G file and then the foundation of the 3D structure of formed by moving the print head of the printer in the x-y plane (Horvath & Cameron, 2020). As the layers keep getting completed, the print head prints in the z-direction and continues depending upon the structure of the job being developed. 3D printing itself is a very vast and expansive topic, hence ASTM divided 3D printing into seven major separate categories, by the ASTM Standard F2972 (ASTM F2792-12a, 2012), i.e., binder jetting, directed energy deposition (DED), material extrusion (fused filament fabrication), material jetting, powder bed fusion (direct metal laser melting or laser powder bed fusion), sheet lamination and vat photopolymerization. The factors which were considered while classifying these processes were namely manufacturing speed, strength and finish of the job, economic viability, and resolution. Out of all these factors, the speed of manufacturing and resolution were considered as the most important ones and hence these are the factors on which we evaluate the job (Jian-Yuan, Jia, & Chee, n.d.). Each of the following seven has been explained in detail below:

1. **Binder Jetting:** This technology was developed at the Massachusetts Institute of Technology in the year 1992 (Sachs, Cima, Williams, Brancazio, & Cornie, 1992) and then later in 1995 a company named Z Corporation gained the permit to use this technology and further later on many companies took over (Nguyen, 2002). In this process, an industrial pinhead is used to precisely deposit a layer of liquid that binds on a thin layer of powdered particles, and this process keeps repeating until the job is completed. The main advantages of this process are that it does not take much time and also there has been a lot of work done on this process thus making it a relatively well-known process and it's cheaper compared to the other methods as well. But the jobs created cannot be used for structural parts, and more processing can lead to further changes. The materials which are commonly used in this process are shown in Table 6.1, along with the characteristics and the uses (Ligon, Liska, Stampfl, Gurr, & Mulhaupt, 2017; Costa, Duarte, & Bartolo, 2017; Zhang et al., 2018).

TABLE 6.1 Materials Viable for Binder Jetting

Materials	Characteristics	Uses
Ceramics	High hardness, elastic module and melting point, resistance to chemicals, wear and corrosion.	Dentist implants, toilets, electronic devices
Stainless steel	High ductility, resistance towards corrosion and heat.	Kitchen sinks, cutlery, medical equipment.
H13 tool steel	High tensile strength and melting point.	Extrusion dies, hot-cold work applications.

2. **Direct Energy Deposition (DED):** Developed at the Sandia National Laboratories at the beginning of the 1990s (Syed et al., 2017), now this method is usually used in the repair of existing jobs or the addition of new materials in an already constructed structure. This method is fairly more difficult to execute compared to the other techniques of 3D printing available. Although this method works well enough on non-metals but is usually used on metals and their alloys (Mudge & Wald, 2007; Liu et al., 2017). The material being used is liquefied using a source that provides a focused energy beam like arc beam and laser beams and gets

deposited on the surface in the form of a wire or powder and it hardens once it lands on the surface (Rybicki & Grant, 2017; Horn & Harrysson, 2012). The material gets deposited in fine lines, and this gives it an added benefit for repairing cracks or fault lines (Pinkerton, Wang, & Li, 2015), especially in the aerospace industry (Lang, 2017). Also, the parts made using this process are cost-effective and take much less time to fabricate. Apart from all these benefits, it's still very difficult to produce ceramics jobs of greater complexity. Also, even though the process might be cost-effective to a great extent, the setup cost of the complete unit as a whole is quite high. Some of the commonly used materials, along with their properties and uses, are mentioned in Table 6.2.

TABLE 6.2 Materials Viable for Direct Energy Deposition

Materials	Characteristics	Uses
Niobium	Malleable and ductile, high melting point.	Jet engines, girders for oil rings.
Stainless steel 300 series	Resistance towards corrosion and heat, easy to maintain.	Mining and petroleum industries, medical equipment.
Aluminum alloy 2319	High structural strength, ductile, heat treatable.	Structures that need high strength, aircrafts.

3. **Powder Bed Fusion:** In this method, focused heat energy which can be a laser or an electron beam and is used to integrate the components by melting as they get deposited and forming a three-dimensional structure (International Standard Reference for Principles and Terminology of AM ISO/ASTM 52900:2015(en), 2015; Goodridge & Ziegelmeier, n.d.). It consists of the following techniques; Electron beam melting (EBM), direct metal laser sintering (DMLS), selective laser sintering (SLS), and selective heat sintering (SHS) (https://www.lboro.ac.uk/research/amrg/about/the7categoriesofadditivemanufacturing/powderbedfusion/). In the EBM process electrons are used to melt the components and the temperatures usually range from around 700°C to around 1,000°C and the complete process takes place under a vacuum which makes it a perfect method to be used on highly excited elements. In the DMLS method, a laser beam is used and each and

every single surface gets scanned by the laser, which integrates the powder to the layer, one at a time until the structure is completed (Khaing, Fuh, & Lu, 2001). The jobs made using this process are dense in nature and can substitute the jobs made by the traditional manufacturing methods (O'Neill, Chris, & Hon, 1999).

In the SLS process, the 3-dimensional structure is created by solidifying melted powders by a laser beam one layer at a time and one of its major advantages over other processes is that almost any type of material can be used (Kruth, Leu, & Nakagawa, n.d.). In the SHS method. The material used is a powder and laser beams are used to fuse the powder and then spread out on a minutely thin layer using a roller and each layer is scanned by a laser and this process continues layer by layer till the structure is obtained (Deckard, 1989; Bai, Goodridge, Yuan, Zhou, Chua, & Wei, 2015). The materials which can be used in the Powder Bed Fusion method are mentioned in Table 6.3 (Sudarmadji, Tan, Leong, Chua, & Loh, 2011; Liao, Chang, Jiang, Chen, & Lee, 2011; Kanczler et al., 2009). But one major drawback of this method is that it's quite costly and the speed of printing is slow.

TABLE 6.3 Materials Viable for Powder Bed Fusion

Materials	Characteristics	Uses
Titanium alloys	High strength and biocompatibility, low density, fatigue resistance.	Aerospace sector, medical implants, lightweight structures.
Copper ($CuCr_1Zr$)	Favorable thermal, electrical, and biological characteristics.	Electronic devices and electrodes.
Inconel (624)	Maintains properties at high temperatures.	Rocket engine nozzles and high-temperature gas turbine.

4. **Material Extrusion:** This, also known as fused filament fabrication, consists of liquefying the material being used and then forcing it through a nozzle in a pre-set part. As the liquefied material solidifies, it hardens and bonds well with the neighboring material present (Valkenaers, Vogeler, Ferraris, Voet, & Kruth, 2013; Sun, Rizvi, Bellehumeur, & Gu, 2008). This process makes it easier to use this method to develop complex parts with ease and still, there are many steps being taken to improve the end

result's job quality and finish (Hopkinson, Hagur, & Dickens, 2006; Vasudevarao, Natarajan, Razdan, & Henderson, 2000). Also, the equipment involved in this process are affordable, easy to use, feedstock variety but also on the other hand, the print speed and job quality aren't of the best quality. In this process, the job is created by liquefying the material first and then it's released in a specific predefined pattern onto a base, it cools down once it leaves the nozzles, solidifies, and joins the material together but it should be made sure that a proper amount of pressure should be exerted while forcing the fused material out of the nozzle (Jiaqi, Qian, Hao, Bin, Lei, & Jikai, n.d.). The materials being used in this method are usually thermoplastics like ABS, polycarbonate (PC), etc. (Shaqour et al., 2021). Table 6.4 contains the list of the materials viable for this process.

TABLE 6.4 Materials Viable for Material Extrusion

Materials	Characteristics	Uses
Nylon 12	Resistant to fatigue and chemicals.	Antenna covers, automobiles and aerospace.
Polyphenyl sulfone	Heat resistance, good mechanical strength.	Sterilizable medical equipment, automotives.
Polylactic acid	Heat resistance and good mechanical strength.	Environmental friendly industries.

5. **Material Jetting:** This also known as PolyJet Printing or Multijet modeling is a 3D Printing Technique which uses Photo polymer-based inkjet printing (IJP) where the complete job or structure gets built one layer at a time (Gibson, Rosen, Stucker, & Khorasani, 2021; Ahmad, n.d.). Photopolymer droplets are dropped on the surface via multiple inkjet print heads, exposed to UV rays. Structures made from this process have a really good surface finish and the job is extremely accurate as per its original design (Stephen, George, & Stefan, 2013; Vaezi, et al., 2013; Report, 2016). This process is used for the fabrication of biomaterials and living cells (Bhattacharjee, Urrios, Kang, & Folch, 2016). The materials used for this kind of process along with their applications and properties are mentioned in Table 6.5 (Königshofer, Stoiber, Unger, Grasl, & Moscato, n.d.).

TABLE 6.5 Materials Viable for Material Jetting

Materials	Characteristics	Uses
VeroWhitePlus	Rigid and high dimensional accuracy.	Medical equipment, electronic housing.
Biomaterials	Transparent, high dimensional stability, rigid.	Ideal for applications requiring short term [<24 hours] and long term [>30 days] skin contact.
FullCure720	Transparent and smooth surface finish.	Medical applications, eyewear

6. **Vat Photopolymerization:** It's a process quite similar to Material Jetting Procedure which we had discussed earlier above. A photopolymer, which is a light-curable resin, is cured with UV light and starts off the polymerization reaction to form bonds or cross-links thus forming a solidified resin. This process further has further subdivisions based on how the polymer is cured, when cured using a laser, the process is known as stereolithography (SLA), when cured using a projector the process is known as Digital light processing (DLP), and when cured using light emitting diodes and oxygen the process is CDLP or CLIP (ASTM International, 2015; Desimone, Alexander, Nikita, Edward, & Smulski, 2014; Tumbleston et al., 2015). Amongst all these processes, SLA was the first one to be developed in the late 1980s by Chuck Hull (Liaw & Guvendiren, 2017). The jobs developed using Vat Photopolymerization do have excellent accuracy and job finish. Table 6.6 contains the list of the suitable materials used and their applications for this process (https://www.carbon3d.com/news/riddell-partners-with-carbon-to-produce-next-gen-football-helmet/; https://www.8sole.com/; https://www.konstrukter.cz/derek-luther-zahajil-jsem-evoluci-aditivni-vyroby-v-adidasu/).

TABLE 6.6 Materials Viable for Vat Photopolymerization

Materials	Characteristics	Uses
Propylene fumarate	Biocompatible and biodegradable copolyester.	Bone tissue engineering
Polyethylene oxide	Swell-ability in water, hydrophilic.	Tissue engineering, pharmaceutical industries.
Elastomeric polyurethane [EPU 40]	High load resistance, does not absorb water, UV, ozone resistance.	Footwear, automobile parts, blood bags, roller blade wheels, football helmet liners.

7. **Sheet Lamination:** In this process of manufacturing, the feedstock used is usually sheets of metal, plastic, or fused paper and ASTM defines it as the process where sheets are either fused together or on top of each other to produce a 3D object. Using a concentrated source of energy, the sheets are fused together to form the structure (Obikawa, n.d.). The sheets also get laminated one at a time before a 3D structure gets created (Chia & Wu, 2015). This procedure is extremely very easy and even the leftover materials can be used over and over again but again there is a lot of wastage being produced as the end product (Gebhardt, 2012). This process can be used to create very intricately designed parts with very minimal cost of manufacturing and even in a lesser amount of time (Vikayavenkataraman, Jerry, & Wen, 2017). The materials used for this process along with the characteristics and uses are mentioned in Table 6.7 (Wong & Hernandez, 2012; Nikhil, n.d.).

TABLE 6.7 Materials Viable for Sheet Lamination

Materials	Characteristics	Uses
Carbon fiber	Low weight, high stiffness and tensile strength.	Formula 1 cars, sport equipment.
Thermoplastics	Recyclable, remolding capabilities, chemical resistance.	Drinking bottles, bags, sport equipment.
Zirconia	Traction and compression resistance.	Pump seals, oxygen sensors, medical prostheses.

6.3.2 APPLICATIONS OF 3D PRINTING

3D printing is continuously being developed to fabricate human body parts (Miller et al., 2012; Jeong et al., 2012; Leukers et al., 2005; Wohlers Associates, n.d.) and the main reason why this method is being used is because once it reaches its complete potential it will definitely be a game-changer in the medical industry and could potentially save and improve the lives of millions of people all around the globe but as of now the complete 3D industry is worth around \$700 million but only around 2% of this is used for the medical industry buts it's projected to jump to around \$ 1.9 billion in the next upcoming 10 years [138]. It is further being used for Surgical Planning, medical research, medical education and training (Paiva, Amorim, Bezerra, & Masini, 2007; D'Urso et al., 1999; Suzuki et al., 2004). Apart from the medical industry as mentioned above, the aerospace sector, the automobile sector, and various others can benefit even more once 3D printing is developed completely to its complete potential. Table 6.8 briefly sums up all seven methods of 3D printing along with the print quality, accuracy, and how financially viable that particular process is.

TABLE 6.8 Brief Summary of Methods of 3D Printing

Method	Materials	Cost	Accuracy
Binder jetting	Stainless steel, ceramics	Cheap	Bad
Direct energy deposition	Niobium, stainless steel [300 series]	Average	Excellent
Powder bed fusion	Titanium alloys	Costly	Average
Material extrusion	Polyphenylsulfone, Nylon 12	Cheap	Bad
Material jetting	VeroWhitePlus, FullCure720	Average	Excellent
Vat photopolymerization	Propylene fumarate, polyethylene oxide	Average	Excellent
Sheet lamination	Carbon fiber, zirconia	Cheap	Bad

3D printing is also being used in the food industry to develop food materials, especially for athletes which need highly nutritious food and removing the unnecessary materials presented and by using 3D Printing various materials are deposited one by one, and layer by layer (Lili, Yuanyuan, Ke, & Yang, 2018; Dankar et al., 2018; Liu, Zhang, Bhandari, & Wang, 2017). 3D printing is also increasingly used to print out full-scale buildings and many such buildings have already come up in many countries (Koslow, 2016; Sakin & Kiroglu, 2017) and the main reason why 3D printing is used for the printing of complete buildings is because this process is quite eco-friendly and does not leave behind much waste. Whenever 3D printing enters the Fashion industry, it will drastically bring down costs of fabrication and make the complete process much easier than it is right now (Richardot, 2018). 3D printing is fast making its presence known even in the electronic market and is being used to fabricate the more complex components (Jeongwoo et al., 2018; Chuan, Hong, Mahdi, Wahid, & Nay, 2018). 3D printing has its advantages, but it also has its fair share of difficulties and disadvantages. The main issue that pops up is that the quality of the surface finish is not very good, and it depends from one method to another. Also, surface porosity is another huge issue that can lead to structural failure over a long period of time. One more drawing concern is that the precision is very less compared to the other technologies we have developed already. In short, as it's still in the development stage, there is still a massive scope for improvements in various sectors.

6.4 4D PRINTING

4D printing is an upgrade to the current 3D printing as there is an introduction of a new dimension to the current 3D printing, which is Time, and it has been proven that the structures fabricated using 4D printing respond to their surroundings like light, pH level, temperatures, and moisture presence. This further allows the structures to repair themselves or assemble on their own once exposed to factors like light, pH, magnetic field, temperature, and moisture (Hassan, n.d.). The change in the structure after the material is printed using 3D is the main characteristic of 4D printing. The 4D Printed materials depend on time but they are quite foreseeable (Momeni, Hassani, Liu, & Ni, 2017). Hydrogels and shape memory polymers are the two most commonly used materials used for 4D printing but there are other materials also which are being used which will be discussed further in subsections (Hong et al., n.d.; Wehner et al., 2016; Osada & Matsuda, 1995; Ionov, 2011; Behl & Lendlein, 2007).

6.4.1 TECHNIQUE OF 4D PRINTING

A brief explanation of the technique of 4D Printing is shown in the flowchart in Figure 6.4. The basic building blocks of 4D printing usually tend to include a 3D Printer, Smart Materials, an Interlinkage Technique, Mathematical Modeling System, and Stimuli.

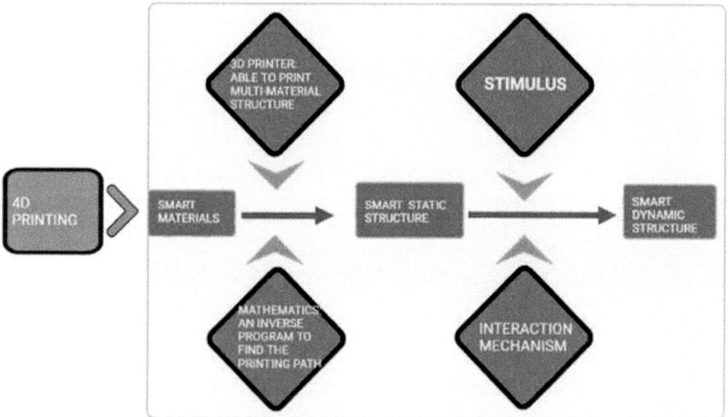

FIGURE 6.4 4D printing process.

All these are explained in further detail below:

1. **3D Printer:** A single 4D structure is fabricated of multiple layers of materials that are integrated together into a single structure. The various materials properties have to be examined carefully before it's integrated into the 4D printed structure, and this job of integrating and developing it into a single structure is done by a 3D printer (Raviv et al., 2014).
2. **Stimuli:** The main properties of the 4D printed objects comes because of the material's response to stimuli and its necessary to set off the shape or property changing characteristics so that the desired job is produced. The most commonly used stimuli responses are moisture or water, a mix of heat and light energy, and a mix of heat and water/moisture (Gladman n.d.; Jamal et al., 2013; Yu, Dunn, & Qi, 2015; Zhang, Zhang, & Hu, 2016). The selection and the choice of the stimuli to be applied depends on structure to structure and is usually dependent on the properties expected from the structure and where it is going to be implemented.
3. **Smart Materials:** Selection of smart materials is one of the most important steps in this complete process. Smart materials are basically any sort of material that shows or provides any sort of response when exposed to stimuli. The main characteristics to be considered for Smart Materials are their shape memory, decision

making, their response, adaptability to various conditions, etc. (Khoo et al., 2015; Sun et al., 2012).

4. **Interlinkage Technique/Interaction Mechanism:** In certain sort of instances, we are not able to get the required shape of the 4D printed structure just by just bringing it to the exposure of stimuli but in needs to be bought in contact in certain steps, under a particular series or arrangement keeping time in consideration. And this is where this process comes into usage.
5. **Mathematical Modeling:** This process is quite necessary for the process of 4D printing so that the viability of the materials being used along with their dispensation is monitored and the shape and design of the structure need to be calculated and designed beforehand in conceptual designs. This step glues together the other four parts of a 4D printing process.

6.4.2 CLASSIFICATION OF THE MATERIALS USED

The materials used in this process are broadly classified on the basis of their response to stimuli, and we have discussed some of them in brief detail in Figure 6.5.

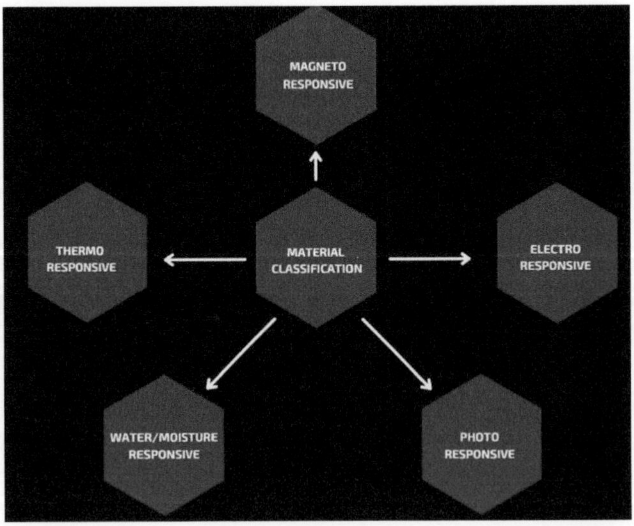

FIGURE 6.5 Types of material used in 4D-printing.

1. **Magneto-Responsive:** Smart Materials that respond when exposed to a magnetic field as their stimuli. They are usually bioinspired materials and can be further divided into MR fluids, MR elastomers, and MR gels according to the physical state in the absence of magnetic (Tong, Dong, & Qi, 2017; Wan, Xiong, & Zhang, 2019; Xu, Gong, & Xuan, 2013). These kinds of materials are still in the dormant state and are still to be utilized to their full ability, but the main shortcoming of this material is that the size of the structure being printed is constricted.
2. **Thermo-Responsive:** SME (shape memory effect) or SCE (shape change effect) are the main driving factors behind these kinds of materials. Materials that are dependent on SME are commonly known as SMM (shape memory materials) (Hager, Bode, Weber, & Schubert, 2015). Shape memory polymers, shape memory gels, shape memory alloys, shape memory hybrids, shape memory ceramics (SMC) are some of the commonly used materials (Ge et al., 2015; Choong, Maleksaeedi, Eng, Wei, & Su, 2017).
3. **Water/Moisture-Responsive:** Materials that respond to the presence of water or moisture are known as water/moisture responsive materials; they are also called Hydro-Responsive Materials. These are the most commonly used materials as they have a wide range of usage compared to the other sorts of materials being used. Hydrogels are the most commonly used Water-Responsive material, as their hydrophilic nature makes them an excellent choice to be used in a variety of environmental settings. Also, they are biocompatible and, in some cases, they have a tendency to swell up to more than 200% of their original size (Zhang et al., 2017). But the processing of this material is quite slow and takes up a considerable amount of time to dry out.
4. **Photo-Responsive:** These are the materials that show a response towards the presence of a light source and these materials are quite different from the other ones mentioned in this section as light acts as indirect stimuli and the material needs to absorb light in order to bring about the changed we need, these materials have various advantages like better rapid switching, accurate focusing and sustainable properties (Jochum & Theato, 2013; Yang, Yuan, Zhang, & Scherman, 2014).

5. **Electro-Responsive:** In these kinds of materials the process is similar to that of the Photo-Responsive Materials. Similar to light, even an electric current can be used as a stimulus to bring about the changes needed. These materials have been used to develop muscles comprising a mixture of silicone elastomer and ethanol. By fluctuating the external environment conditions, i.e., when current is passed through, phase change takes place, thus lengthening the complete structure (Okuzaki, Kuwabara, Funasaka, & Saido, 2013; Miriyev, Stack, & Lipson, 2017).

6.4.3 APPLICATIONS OF 4D PRINTING

Most of the robots or machinery being developed all across the globe consists of robots made up of firm materials, and it leads to complications and issues in certain cases when dealing with organic life forms. Hence there has been a shift of attention towards Soft Robotics where soft materials, especially designed elastomers, were being used to replace the earlier used firm materials. This even helped provide gentle contact with the objects which were brought into contact with (Shian, Bertoldi, & Clarke, 2015; McCoul, Rosset, Schlatter, & Shea, 2017; López-Valdeolivas et al., 2018).

The medical industry is where 4D printing, like 3D printing, can have a massive footprint. Due to 4D printing, it is now possible to tailor-make tablets and drugs for a specific person by controlling the quantity of each drug in it (Khaled, Burley, Alexander, Yang, & Roberts, 2015). Furthermore, technologies like TDD (targeted drug delivery) have improved manifolds by the introduction of 4D printing. The drugs get delivered to a particular site depending upon their response to the stimuli and its environmental conditions (Azam, Laflin, Jamal, Fernandes, & Gracias, 2011; Malachowski et al., 2014). There has been development taking place in the fabrication of new and better splints, which are medical devices used to hold together parts (Miao et al., 2016). Also, stents which are small, tiny tubes that are placed inside blood vessels to ease the flow are being developed using 4D printing in a much better way (Bodaghi, Damanpack, & Liao, 2016). Also, 4D printing is being used to develop containers using SMP materials that can sustain plastic deformation thus making them much stronger than the ones used traditionally (Ge, Qi, & Dunn, 2013; Kuribayashi-Shigetomi, Onoe, Takeuchi, & Han, 2012). Multifunctional materials, by controlling the nature of the materials being used, are also in advanced stages of research

(Wu, Zhao, Kuang, Hamel, Fang, & Qi, 2018). In the Aerospace industry as well 4D Printing is being sought after at a rapid pace and materials like SMA and SMP are being incorporated into various different kinds of technologies (Ishizawa, Imagawa, Minami, Hayashi, & Miwa, 2003; Hinkle, Lin, & Kling, 2014; Koehler, Lyman, Summers, & Danforth, 2017; Yu, Sun, Liu, Zhang, Liu, & Leng, 2009; Meo, Marulo, Guida, & Russo, 2013). Some other applications do also include structures which can assemble on their own without any need for human or computer help. As some 4D materials can repair themselves, this property is also being exploited to create self-healing pipes (Campbell, Tibbits, & Garrett, 2014).

6.5 CONCLUSION

AM definitely is the future of the manufacturing sector mainly due to its advantages regarding the fabrication time and the amount of waste generated when compared to the current manufacturing techniques being employed all round the world. Due to these reasons, both 3D printing and 4D Printing are being developed intensively and are growing at an increasingly rapid pace. AM has more than enough proven itself in many conditions to be a more viable option for the future and it definitely is the need of the future as it may hold the key to impact our future in a more positive manner. 3D printing can be considered as a base for 4D Printing and hence it's important to improvise on 3D printing first. Because only if the main drawbacks of 3D printing are conquered, we can work towards developing and applying 4D printing in many more fields. Although, as mentioned, 4D Printing is still taking its first steps, and once incorporated and developed completely, the sky will be the limit.

KEYWORDS

- **3D printing**
- **4D printing**
- **additive manufacturing**
- **rapid prototyping**
- **robots**
- **subtractive manufacturing**
- **targeted drug delivery**

REFERENCES

8sole. Available online: https://www.8sole.com/ (accessed on 11 September 2023).

Al-Ali, A. B., Griffin, M. F., & Butler, P. E., (2015). Three-dimensional printing surgical applications. *Eplasty, 15*, 352–367.

Alshahrani, H. A., (2021). Review of 4D printing materials and reinforced composites: Behaviors, applications and challenges. *Journal of Science: Advanced Materials and Devices, 6*(2), 167–185.

Amit, B., & Sumita, B., (2019). *Additive Manufacturing* (2nd edn.).

ASTM F2792-12a, (2012). *Standard Terminology for Additive Manufacturing Technologies*. ASTM International. West Conshohocken, PA.

ASTM International, (2015). *ISO/ASTM 52900:2015 Additive Manufacturing—General Principles—Terminology* (p. 5). ASTM: West Conshohocken, PA, USA.

Azam, A., Laflin, K. E., Jamal, M., Fernandes, R., & Gracias, D. H., (2011). Self-folding micropatterned polymeric containers. *Biomed. Microdevices, 13*(1), 51–58.

Bai, J., Goodridge, R. D., Yuan, S., Zhou, K., Chua, C. K., & Wei, J., (2015). Thermal influence of CNT on the polyamide 12 nanocomposites for selective laser sintering. *Molecules, 20*, 19041–19050.

Behl, M., & Lendlein, A., (2007). Shape-memory polymers. *Mater. Today, 10*, 20–28.

Bhattacharjee, N., Urrios, A., Kang, S., & Folch, A., (2016). The upcoming 3D-printing revolution in microfluidics. *Lab a Chip., 16*, 1720–1742.

Bodaghi, M., Damanpack, A. R., & Liao, W. H., (2016). Self-expanding/shrinking structures by 4D printing. *Smart Mater. Struct., 25*(10), 1–15.

Campbell, T. A., Tibbits, S., & Garrett, B., (2014). The programmable world. *Sci. Am., 311*, 60–65.

Campbell, T., Williams, C., Ivanova, O., & Garrett, B., (2011). *Could 3D printing Change the World, Technologies, Potential, and Implications of Additive Manufacturing*. Washington, DC: Atlantic Council.

Chia, H. N., & Wu, B. M., (2015). Recent advances in 3D printing of biomaterials. *J. Biol. Eng., 9*, 4.

Choong, Y. Y. C., Maleksaeedi, S., Eng, H., Wei, J., & Su, P. C., (2017). 4D printing of high-performance shape memory polymer using stereolithography. *Mater. Des., 126*, 219–225.

Chuan, Y. F., Hong, N. L., Mahdi, M. A., Wahid, M. H., & Nay, M. H., (2018). Three-dimensional printed electrode and its novel applications in electronic devices. *Scientific Report, 1*, 1–11.

Costa, E. C. E., Duarte, J. P., & Bartolo, P., (2017). A review of additive manufacturing for ceramic production. *Rapid Prototype. J., 23*(5), 954–963.

D'Urso, P., Barker, T., Earwaker, W., Bruce, L., Atkinson, R., Lanigan, M., Arvier, J., & Effeney, D., (1999). Stereolithographic biomodelling in cranio-maxillofacial surgery: A prospective trial. *J. Craniomaxillofac. Surg., 27*, 30–37.

Dankar, I., Pujola, M., Omar, F. E., Sepulcre, F., & Haddarah, A., (2018). Impact of mechanical and microstructural properties of potato puree-food additive complexes on extrusion-based 3D printing. *Food and Bioprocess Technology, 1*, 1–11.

Deckard, C., (1989). *Method and Apparatus for Producing Parts by Selective Sintering*. US Patent 5597589.

Derek Luther: Zahájil Jsem Evoluci Aditivní Výroby v Adidasu. Available online: https://www.konstrukter.cz/derek-luther-zahajil-jsem-evoluci-aditivni-vyroby-v-adidasu/ (accessed on 11 September 2023).

Desimone, J., Alexander, M., Nikita, E., Edward, E., & Smulski, T., (2014). *Continuous Liquid Interphase Printing*. U.S. Patent PCT/US2014/015506.

Gao, W., Yunbo, Z., Devarajan, R., Karthik, R., Yong, C., Christopher, B. W., Charlie, C. L. W., Yung, C. S., Song, Z., & Pablo, D. Z., (2015). The status, challenges, and future of additive manufacturing in engineering. *Computer-Aided Design, 69*, 65–89.

Ge, Q., Qi, H. J., & Dunn, M. L., (2013). Active materials by four-dimension printing. *Appl. Phys. Lett., 103*(13), 131901.

Ge, Q., Sakhaei, A. H., Lee, H., Dunn, C. K., Fang, N. X., & Dunn, M. L., (2016). Multimaterial 4D printing with tailorable shape memory polymers. *Sci. Rep., 6*(1), 31110.

Gebhardt, A., (2012). *Understanding Additive Manufacturing*. Hanser, Munich, Germany.

Gibson, I., Rosen, D., & Stucker, B., (2010). *Additive Manufacturing Technologies, 3D Printing, Rapid Prototyping, and Direct Digital Manufacturing*. New York Heidelberg Dordrecht London: Springer.

Gibson, I., Rosen, D., Stucker, B., & Khorasani, M., (2021). Material jetting. In: *Additive Manufacturing Technologies*. Springer, Cham. https://doi.org/10.1007/978-3-030-56127-7_7.

Gladman, A. S., Matsumoto, E. A., Nuzzo, R. G., Mahadevan, L., & Lewis, J. A., (2016). Biomimetic 4D printing. *Nature Materials*.

González-Henríquez, C. M., Sarabia-Vallejos, M. A., & Rodriguez-Hernandez, J., (2019). *Polymers for Additive Manufacturing and 4D-Printing: Materials, Methodologies, and Biomedical Applications*.

Goodridge, R., & Ziegelmeier, S., (2017). Powder bed fusion of polymers. In: *Laser Additive Manufacturing* (pp. 181–204). Woodhead Publishing.

Gross, B. C., Jayda L. E., Sarah, Y. L., Chengpeng, C., & Dana, M. S., (2014). *Evaluation of 3D Printing and its Potential Impact on Biotechnology and the Chemical Sciences*, 3240–3253.

Hager, M. D., Bode, S., Weber, C., & Schubert, U. S., (2015). Progress in polymer science shape memory polymers: Past, present, and future developments. *Prog. Polym. Sci., 49, 50*, 3–33.

Hinkle, J., Lin, J., & Kling, D., (2014). Design and materials study on secondary structures in deployable planetary and space habitats. In: *52nd AIAA/ASME/ASCE/AHS/ASC Struct. Struct. Dyn. Mater. Conf. 19th AIAA/ASME/AHS Adapt. Struct. Conf. 13t* (pp. 1–10, 2024).

Hong, S., Dalton, S., Hon, F. C., Shaoting, L., Gabriel, P. L., Farshid, G., Kam, W. L., & Xuanhe, Z., (2015). 3D printing of highly stretchable and tough hydrogels into complex, cellularized structures. *Advanced Materials, 27*, 4035–4040.

Hopkinson, N., Hagur, R. J. M., & Dickens, P. H., (2006). *Rapid Manufacturing: An Industrial Revolution for the Digital Age*. John Wiley, Chichester.

Horn, T. J., & Harrysson, O. L., (2012). Overview of current additive manufacturing technologies and selected applications. *Sci. Prog., 95*, 255–282.

Horvath, J., & Cameron, R., (2020). 3D printer workflow and software. In: *Mastering 3D Printing*. Apress, Berkeley, CA.

https://link.springer.com/chapter/10.1007/978-1-4842-5842-2_3 (accessed on 11 September 2023).

https://www.lboro.ac.uk/research/amrg/about/the7categoriesofadditivemanufacturing/powderbedfusion/ (accessed on 11 September 2023).

Huang, J., Chen, Q., Jiang, H., Zou, B., Li, L., Liu, J., & Yu, H., (2020). A survey of design methods for material extrusion polymer 3D printing. *Virtual and Physical Prototyping, 15*(2), 148–162.

International Standard Reference for Principles and Terminology of AM ISO/ASTM 52900:2015(en), (2015). *Additive Manufacturing d-General Principles and Terminology*. Geneva, Switzerland: International Organization for Standardization (ISO).

Ionov, L., (2011). Soft micro-origami: Self-folding polymer films. *Soft Matter., 7*, 6786–6791.

Ishizawa, J., Imagawa, K., Minami, S., Hayashi, S., & Miwa, N., (2003). Research on application of shape memory polymers to space inflatable systems. In: *Proc. 7th Int. Symp. Artif. Intell. Robot. Autom. Sp. I-SAIRAS*. NARA, Japan.

Jamal, M., Kadam, S. S., Xiao, R., Jivan, F., Onn, T. M., Fernandes, R., Nguyen, T. D., & Gracias, D. H., (2013). Bio- origami hydrogel scaffolds composed of photocrosslinked PEG bilayers. *Advanced Healthcare Materials, 2*, 1142–1150.

Jeong, J. E., Chan, V., Cha, C., et al., (2012). "Living" microvascular stamp for patterning of functional neovessels; orchestrated control of matrix property and geometry. *Adv. Mater., 24*, 58–63.

Jeongwoo, L., Ho-Chan, K., Jae-Won, C., & In, H. L. A., (2018). A review on 3D printed smart devices for 4D printing. *International Journal of Precision Engineering and Manufacturing-Green Technology, 4*(3), 373–383.

Jochum, F. D., & Theato, P., (2013). Temperature-and light-responsive smart polymer materials. *Chem. Soc. Rev., 42*(17), 7468–7483.

Kanczler, J. M., Mirmalek-Sani, S. H., Hanley, N. A., Ivanov, A. L., Barry, J. J. A., Upton, C., Shakesheff, K. M., et al., (2009). Biocompatibility and osteogenic potential of human fetal femur-derived cells on surface selective laser sintered scaffolds. *Acta Biomater., 5*(6), 2063–2071.

Khaing, M. W., Fuh, J. Y. H., & Lu, L., (2001). Direct metal laser sintering for rapid tooling: Processing and characterization of EOS parts. *Journal of Materials Processing Technology, 113*(1–3), 269–272. doi: 10.1016/S0924-0136(01)00584-2.

Khaled, S. A., Burley, J. C., Alexander, M. R., Yang, J., & Roberts, C. J., (2015). 3D printing of tablets containing multiple drugs with defined release profiles. *Int. J. Pharm., 494*(2).

Khoo, Z. X., Teoh, J. E. M., Liu, Y., Chua, C. K., Yang, S., An, J., Leong, K. F., & Yeong, W. Y., (2015). 3D printing of smart materials: A review on recent progress in 4D printing. *Virtual and Physical Prototyping, 10,* 103–122.

Koehler, F. B., Lyman, W. D., Summers, M. H., & Danforth, J. C., (2017). *Spacecraft with Shape Memory Polymer Deployment Mechanism*. US14/921,545, 2017. https://patents.google.com/patent/US20170113817A1/en (accessed on 11 September 2023).

Königshofer, M., Martin, S., Ewald, U., Christian, G., & Francesco, M., (2021). Mechanical and dimensional investigation of additive manufactured multimaterial parts. *Frontiers in Physics*, *9*, 635736.

Koslow, T., (2016). *DUS Architects Constructs 3D Printed "Urban Cabin."* In Amsterdam. Retrieved from 3D print.com.

Kruth, J. P., Leu, M. C., & Nakagawa, T., (1998). Progress in additive manufacturing and rapid prototyping. *Cirp. Annals.*, *47*(2), 525–540.

Kuang, X., Devin, J. R., Jiangtao, W., Craig, M. H., Zhen, D., Tiejun, W., Martin, L. D., & Hang, J. Q., (2019). Advances in 4D printing: Materials and applications. *Advanced Functional Materials, 29*(2), 1805290.

Kuribayashi-Shigetomi, K., Onoe, H., Takeuchi, S., & Han, A., (2012). Cell origami: Self-folding of three-dimensional cell-laden microstructures driven by cell traction force. *PLoS One, 7*(12), 1–11.

Lang, M., (2017). *An Overview of Laser Metal Deposition*. A publication of the Fabricators & Manufacturers Association.

Lee, J. Y., An, J., & Chua, C. K., (2017). Fundamentals and applications of 3D printing for novel materials. *Applied Materials Today, 7*, 120–133.

Leukers, B., Gulkan, H., Irsen, S. H., et al., (2005). Hydroxyapatite scaffold for bone tissue engineering made by 3D printing. *J. Mater. Sci. Makter. Med., 16*, 1121–1124.

Liao, H. T., Chang, K. H., Jiang, Y., Chen, J. P., & Lee, M. Y., (2011). Fabrication of tissue-engineered PCL scaffold by selective laser-sintered machine for osteogenesis of adipose-derived stem cells. *Virtual Phys. Prototyping, 6*(1), 57–60.

Liaw, C. Y., & Guvendiren, M., (2017). Current and emerging applications of 3D printing in medicine. *Biofabrication., 9*, 024102.

Ligon, S. C., Liska, R., Stampfl, J., Gurr, M., & Mulhaupt, R., (2017). Polymers for 3D printing and customized additive manufacturing. *Chem. Rev., 117*(15), 10212–10290.

Lili, L., Yuanyuan, M., Ke, C., & Yang, Z., (2018). 3D printing complex egg white protein objects: Properties and optimization. *Food and Bioprocess Technology, 1*, 1–11.

Liu, R., et al., (2017). *13–Aerospace Applications of Laser Additive Manufacturing A2—Brandt, Milan, Laser Additive Manufacturing* (pp. 351371). Woodhead Publishing.

Liu, Z., Zhang, M., Bhandari, B., & Wang, Y., (2017). 3D printing: Printing precision and application in the food sector. *Trends in Food Science & Technology, 2*(1), 1–36.

López-Valdeolivas, M., Liu, D., Broer, D. J., & Sánchez-Somolinos, C., (2018). 4D printed actuators with soft-robotic functions. *Macromol. Rapid Commun., 39*(5).

Malachowski, K., Breger, J., Kwag, H. R., Wang, M. O., Fisher, J. P., Selaru, F. M., & Gracias, D. H., (2014). Stimuli-responsive theragrippers for chemomechanical controlled release. *Angew. Chemie – Int. Ed., 53*(31), 8045–8049.

McCoul, D., Rosset, S., Schlatter, S., & Shea, H., (2017). Inkjet 3D printing of UV and thermal cure silicone elastomers for dielectric elastomer actuators. *Smart Mater. Struct. 26*(12), 125022.

Mehrabi, M. G., Ulsoy, A. G., & Koren, Y., (2000). Reconfigurable manufacturing systems: Key to future manufacturing. *Journal of Intelligent Manufacturing, 11*, 403–419.

Meo, M., Marulo, F., Guida, M., & Russo, S., (2013). Shape memory alloy hybrid composites for improved impact properties for aeronautical applications. *Compos. Struct., 95*, 756–766.

Merklein, M., Junker, D., Schaub, A., & Neubauer, F., (2016). Hybrid additive manufacturing technologies – an analysis regarding potentials and applications. *Physics Procedia, 83*, 549–559.

Miao, S., Zhu, W., Castro, N. J., Nowicki, M., Zhou, X., Cui, H., Fisher, J. P., & Zhang, L. G., (2016). 4D printing smart biomedical scaffolds with novel soybean oil epoxidized acrylate. *Sci. Rep., 6*, 1–10.

Miller, J. S., Stevens, K. R., Yang, M. T., et al., (2012). Rapid casting of patterned vascular networks for perfusable engineered three-dimensional tissues. *Nat. Mater., 11*, 768–774.

Miriyev, A., Stack, K., & Lipson, H., (2017). Soft material for soft actuators. *Nat. Commun., 8*(1), 1–8.

Momeni, F., Hassani, N. S. M. M., Liu, X., & Ni, J., (2017). A review of 4D printing. *Mater. Dès., 12*, 42–79.

Mudge, R. P., & Wald, N. R., (2007). Laser engineered net shaping advances additive manufacturing and repair. *Weld. J., 86*(1), 4448.

Newman, S. T., Zhu, Z., Dhokia, V., & Shokrani, A., (2015). Process planning for additive and subtractive manufacturing technologies. *CIRP Annals*, *64*(1), 467–470.

Nguyen, N. T., (2002). *MEMS-Micropumps: A Review* (Vol. 124, p. 384). American Society of Mechanical Engineers, New York, N.Y.

Nikhil, A. *3D Printing Processes – Sheet Lamination*. https://www.engineersgarage.com/articles/3d-printing-processes-sheet-lamination (accessed on 11 September 2023).

Nirish, M., & Rajendra, R. (2020). Suitability of metal additive manufacturing processes for part topology optimization–A comparative study. Materials Today: Proceedings, 27, 1601-1607.

O'Neill, W., Chris, J. S., & Hon, K. K. B., (1999). Investigation on multi-layer direct metal laser sintering of 316L stainless steel powder beds. *Article in CIRP Annals – Manufacturing Technology*.

Obikawa, T., Yoshino, M., & Shinozuka, J., (1999). Sheet steel lamination for rapid manufacturing. *J. Mater. Process. Technol., 89, 90*, 171–176.

Okuzaki, H., Kuwabara, T., Funasaka, K., & Saido, T., (2013). Humidity-sensitive polypyrrole films for electro-active polymer actuators. *Adv. Funct. Mater., 23*(36), 4400–4407.

Osada, Y., & Matsuda, A., (1995). Shape-memory in hydrogels. *Nature, 376*, 219.

Paiva, W., Amorim, R., Bezerra, D., & Masini, M., (2007). Application of the stereolithography technique in complex spine surgery. *Arq. Neuropsiquiatr., 65*, 443–445.

Paris, H., Mokhtarian, H., Coatanéa, E., Museau, M., & Ituarte, I. F., (2016). Comparative environmental impacts of additive and subtractive manufacturing technologies. *CIRP Annals, 65*(1), 29–32.

Patwardhan, A., (2017). How 3D printing will change the future of borrowing, lending, and spending? *Handb. Blockchain Digital Finance Inclusion, 2*, 493–520. Elsevier.

Peterson, G. I., Larsen, M. B., Ganter, M. A., Storti, D. W., & Boydston, A. J., (2014). 3D-printed mechanochromic materials. *ACS Appl. Mater. Interfaces, 7*, 577–583.

Pinkerton, A. J., Wang, W., & Li, L., (2015). (Vol. 222, No. 7, pp. 827–836).

Prakash, K. S., Nancharaih, T., & Rao, V. S., (2018). Additive manufacturing techniques in manufacturing: An overview. *Materials Today: Proceedings, 5*(2), 3873–3882.

Randolph, S. A., (2018). *3D Printing: What Are the Hazards?* Workplace Health & Safety; 2018:2165079917750408.

Rao, P. N., (2019). *Manufacturing Technology* (Vol. 1, 4th edn.).

Raviv, D., Zhao, W., McKnelly, C., Papadopoulou, A., Kadambi, A., Shi, B., Hirsch, S., et al., (2014). Active printed materials for complex self-evolving deformations. *Scientific Reports, 4*.

Report, W. S., (2016). *Additive Manufacturing and 3D Printing State of the Industry Annual Worldwide Progress Report*, Wohler's Associates, Inc.

Richardot, A., (2018). *3D Printed Fashion: Why is Additive Manufacturing Interesting for Fashion?* Sculpteo. Available: https://www.sculpteo.com/blog/2018/01/24/3d-printed-fashion-why-is-additive-manufacturing-interesting-for-fashion/

Rideell Partners with Carbon to Produce First-Ever 3D Printed Football Helmet Liner. Available online: https://www.carbon3d.com/news/riddell-partners-with-carbon-to-produce-next-gen-football-helmet/ (accessed on 11 September 2023).

Roopavath, U. K., & Kalaskar, D. M., (2017). Introduction to 3D printing in medicine. *3D Print Med.* (pp. 1–20). Elsevier.

Rybicki, F. J., & Grant, G. T., (2017). *3D Printing in Medicine: A Practical Guide for Medical Professionals.* Springer.

Sachs, E., Cima, M., Williams, P., Brancazio, D., & Cornie, J., (1992). Three-dimensional printing: Rapid tooling and prototypes directly from a CAD model. *J. Inst. Eng., 114*, 481–488.

Sakin, M., & Kiroglu, Y. C., (2017). 3D printing of building: Construction of the sustainable houses of the future by BIM. *Energy Procedia, 134*, 702–711.

Schubert, C., Van, L. M. C., & Donoso, L. A., (2014). Innovations in 3D printing: A 3D overview from optics to organs. *Br. J. Ophthalmol., 98*, 159–161.

Serope, K., (2002). *Manufacturing Engineering Technology Textbook* (4th edn.).

Shaqour, B., Abuabiah, M., Abdel-Fattah, S., et al., (2021). Gaining a better understanding of the extrusion process in fused filament fabrication 3D printing: A review. *Int. J. Adv. Manuf. Technol., 114,* 1279–1291. https://doi.org/10.1007/s00170-021-06918-6.

Shian, S., Bertoldi, K., & Clarke, D. R., (2015). Dielectric elastomer-based 'grippers' for soft robotics. *Adv. Mater., 27*(43), 6814–6819.

SPIE-Professional. Chuck Hull: Pioneer in Stereolithography. http://spie.org/x91418.xm (accessed on 11 September 2023).

Stephen, J. F., George, B., & Stefan, S., (2013). Design and fabrication of a bat-inspired flapping flight platform using shape memory alloy muscles and joints. *Smart Mater. Struct., 22*(1), 014011.

Sudarmadji, N., Tan, J. Y., Leong, K. F., Chua, C. K., & Loh, Y. T., (2011). Investigation of the mechanical properties and porosity relationships in selective laser-sintered polyhedral for functionally graded scaffolds. *Acta Biomater., 7*(2), 530–537.

Sun, L., Huang, W. M., Ding, Z., Zhao, Y., Wang, C. C., Purnawali, H., & Tang, C., (2012). Stimulus-responsive shape-memory materials: A review. *Materials & Design, 33*, 577–640.

Sun, Q., Rizvi, G. M., Bellehumeur, C. T., & Gu, P., (2008). Effect of processing conditions on the bonding quality of FDM polymer filaments. *Rapid Prototyp. J., 14*, 72–80.

Suzuki, M., Ogawa, Y., Kawano, A., Hagiwara, A., Yamaguchi, H., & Ono, H., (2004). Rapid prototyping of temporal bone for surgical training and medical education. *Acta Otolaryngol., 124*, 400–402.

Syed, A. M. T., Elias, P. K., Amit, B., Susmita, B., Lisa, O., & Charitidis, C., (2017). Additive manufacturing: Scientific and technological challenges, market uptake and opportunities. *Materials Today, 1*, 1–16.

Tibbits, S., (2013). *The Emergence of "4D Printing.*" TED conference.

Tong, Y., Dong, X., & Qi, M., (2017). High performance magnetorheological fluids with flower-like cobalt particles. *Smart Materials and Structures*.

Tumbleston, J. R., Shirvanyants, D., Ermoshkin, N., Janusziewicz, R., Johnson, A. R., Kelly, D., & De Simone, J. M., (2015). Continuous liquid interface production of 3D objects. *Science, 347*, 1349–1352.

Upcraft, S., & Fletcher, R., (2003). The rapid prototyping technologies. *Assembly Automation, 23*(4), 318–330.

Vaezi, M., et al., (2013). Multiple material additive manufacturing—Part 1: A review. *Virtual Phys. Prototyping, 8*(1), 1950.

Valkenaers, H., Vogeler, F., Ferraris, E., Voet, A., & Kruth, J. P., (2013). A novel approach to additive manufacturing: Screw extrusion 3D-printing. In: Azcarate, S., & Dimov, S., (eds.), *Proceedings of the 10th International Conference on Multi-Material Micro Manufacture, San Sebastian, Spain* (pp. 235–238). Research Publishing: San Sebastian, Spain.

Vasudevarao, B., Natarajan, D. P., Razdan, A., & Henderson, M., (2000). Sensitivity of RP surface finish to process parameter variation. *Solid Freeform Fabrication Proceedings* (pp. 252–258). The University of Texas, Austin.

Vayre, B., Frédéric, V., & François, V., (2012). Designing for additive manufacturing. *Procedia CIrP3*, 632–637.

Vikayavenkataraman, S., Jerry, Y. H. F., & Wen, F. L., (2017). 3D printing and 3D bioprinting in pediatrics. *Bioengineering, 4*(63), 1–11.

Wan, Y., Xiong, Y., & Zhang, S., (2019). Temperature effect on viscoelastic properties of anisotropic magnetorheological elastomers under compression. *Smart Materials and Structures*.

Wehner, M., Truby, R. L., Fitzgerald, D. J., Mosadegh, B., Whitesides, G. M., Lewis, J. A., & Wood, R. J., (2016). An integrated design and fabrication strategy for entirely soft, autonomous robots. *Nature, 536*, 451–455.

Weller, C., Kleer, R., & Piller, F. T., (2015). Economic implications of 3D printing: Market structure models in light of additive manufacturing revisited. *Int. J. Prod. Econ., 164*, 43–56.

Wohlers Associates. *What Is 3D printing*? Wohlers Report 2013. https://cir.nii.ac.jp/crid/1130282269717519360 (accessed on 11 September 2023).

Wong, K. V., & Hernandez, A., (2012). A review of additive manufacturing. *ISRN Mechanical Engineering*.

Wu, J., Zhao, Z., Kuang, X., Hamel, C. M., Fang, D., & Qi, H. J., (2018). Reversible shape change structures by grayscale pattern 4D printing. *Multifunct. Mater., 1*(1).

Xu, Y., Gong, X., & Xuan, S., (2013). Soft magnetorheological polymer gels with controllable rheological properties. *Smart Materials and Structures*.

Yang, H., Yuan, B., Zhang, X., & Scherman, O. A., (2014). Supramolecular chemistry at interfaces: Host-guest interactions for fabricating multifunctional biointerfaces. *Acc. Chem. Res., 47*(7), 2106–2115.

Yu, K., Dunn, M. L., & Qi, H. J., (2015). Digital manufacture of shape-changing components. *Extreme Mechanics Letters, 4*, 9–17.

Yu, K., Sun, S., Liu, L., Zhang, Z., Liu, Y., & Leng, J., (2009). Novel deployable morphing wing based on SMP composite. In: *Second Int. Conf. Smart Mater. Nanotechnol. Eng.*

Zhang, Q., Zhang, K., & Hu, G., (2016). Smart three-dimensional lightweight structure triggered from a thin composite sheet via 3D printing technique. *Scientific Reports, 6.*

Zhang, Y. S., Yue, K., Aleman, J., Mollazadeh-Moghaddam, K., Bakht, S. M., Yang, J., Jia, W., Dell'Erba, V., Assawes, P., Shin, S. R., Dokmeci, M. R., Oklu, R., Khademhosseini, A., et al., (2017). 3D bioprinting for tissue and organ fabrication. *Ann. Biomed. Eng., 45*(1), 148–163.

Zhang, Y., et al., (2018). Additive manufacturing of metallic materials: A review. *J. Mater. Eng. Perform., 27*(1).

Zhu, Z., Dhokia, V. G., Nassehi, A., & Newman, S. T., (2013). *A Review of Hybrid Manufacturing Processes – State of the Art and Future Perspectives*, 596–615.

CHAPTER 7

Feature Optimization Methods: A Step Towards Improvement in Medical Image Processing

SHELLY GARG[1] and BALKRISHAN JINDAL[2]

[1]*Research Scholar, Department of Electronics and Communication, Punjabi University, Patiala, Punjab, India*

[2]*Assistant Professor, Computer Engineering Section, Punjabi University, Guru Kashi Campus, Talwandi Sabo, Punjab, India*

ABSTRACT

In the area of machine learning and pattern recognition, features are essential for forecasting, classification of the image, medical imaging, and a few other applications of features that are well-known. Health agencies are gathering vast amounts of data by virtue of the rapid growth of knowledge expenditures in medical data archives and healthcare procurement. These data repositories hold a wealth of knowledge that can help doctors to make better diagnoses and develop a better quality of patient responsibility. On the contrary, this expansion has made it more crucial to understand and use this information for a variety of purposes. Because of the presence of extraneous features in larger datasets, the findings of imaging data can become skewed. Feature selection allows for a reduction in the number of constituents in these large datasets. Using selection techniques, the insignificant features are eliminated, and a fragment of components is chosen that yields high characterization fidelity. A precise grouping model

Advances in Manufacturing and Processing of Materials: Characterization and Applications.
Narendra Kumar, PhD, Mohit Tyagi, PhD, Dilbagh Panchal, PhD & Ravi Pratap Singh, PhD (Eds.)

is generated by making the right agreement to asset a good virtue, which improves learning speed and forecast control. An analysis of feature choice strategies and attribute choice steps for medical field imaging is presented in this chapter. This survey aims to explain feature selection strategies in the medical domain, as well as their advantages and disadvantages, and to demonstrate their use in imaging data and data mining algorithms. The study presents the flaws in current multi-sourced data features and attributes selection methods. Furthermore, the influence of feature selection on the proper distribution of medical infections is discussed in this study. Finally, critical reviews and recommendations for the future are presented.

7.1 INTRODUCTION

To eliminate non-essential features, feature selection method is to be used and choosing a fragment of constituents that produces high characterization precision. Data mining algorithms or inductive machine learning are both concerned with determining an attribute's quality. When insignificant properties are included in the forecasting method, effective mining is not possible. In a research problem, such properties are expressed as features. The general feature selection process is depicted in Figure 7.1. A subset of features is created from the original feature set using a search approach. Already the subset gets created, all of these features must pass the enlargement criteria and then these are aggregated in the eventual feature set to continue the evaluation process. It is used in many applications of medical domains and also useful in a wide range of applications. It's helpful when there's a lot of trivial and insignificant data mixed with high-dimensional data. These methods have recently been used in a variety of regions like image retrieval, image recognition, bio-informatics data analysis, skin cancer, lungs cancer, brain cancer, and name a few more. Computations are speed up and time is saved when the right attributes are chosen. The importance of function selection in various medical fields, such as in image recognition. The need for these selection methods is more critical than before, even though decades ago, image processing researchers first recognized the need for knowing which features needed to be extracted from every pixel. Image analysis encompasses a broad range of applications and, as a result, various techniques (Bolón-Canedo, & Remeseiro, 2020). For image retrieval, feature selection on the base of two developmental computations, namely the Genetic method and also Binary Bat method are used to find out the optimal features analogous to estimate the calculation on ranking base for the reduction in dimensionality and also to refine the best features from a comprehensive feature set (Kushwaha & Welekar, 2016). As for Bio-informatic data analysis has a long history of sequence analysis. The content analysis examines a sequence's broad characteristics, such as its proclivity for coding for proteins or its ability to perform a specific biological role (Saeys, Inza, & Larrañaga, 2007). To optimize pattern recognition systems in use, feature selection is a must to detect skin cancer (Handels, Ross, Kreusch, Wolff, & Pöppl, 1999). The utility of feature selection methods is primarily actuated by an event that occurs when the classifiers are equipped with a constrained number of training samples. After a peak, the classification rate of the classifier reduces as the number of features is increased. Moreover, while developing a recognition system, the need for reducing the number of features is often motivated by computational considerations. Cellular breakdown in the lungs is the main disease executioner among the two people (Kancherla & Mukkamala, 2012; WHO, 2004). The indications of cellular breakdown in the lungs don't show up until the malignancy spreads to different zones, accordingly, prompting disease location of just 24% in beginning phases (Fontana, Sanderson et al., 1986). It needs an exact early recognition of cellular breakdown in the lungs, for expanding the endurance rate. Different techniques such as computed tomography (CT) scan, microarray information examination are utilized for the cellular breakdown in the lungs discovery (Nanda, McCory, Myers et al., 2000). CT scan is a promising technique used for the chest mass screening for the cellular breakdown in the lungs identification. Anyway, this strategy isn't suggested because it is exorbitant and the long-haul wellbeing of this technique isn't set up because of the danger of openness to radiation (Taxdal, Ward, & Figge, 1955). The utilization of microarray information for malignant growth is examined by Galeotti et al. (1984). Anyway, the utilization of microarray information is an expensive methodology and approaches. An expanding death rate among various age classes is brought about brain cancer (Kharrat & Neji, 2018). This sickness is showed by the development of unusual cells inside or over the brain (Gopi & Ramashri, 2016). There exist different kinds of brain cancer, like cancerous and non-cancerous. Quick revelation in brain cancer is of incredible significance and a significant test for additional examinations. Hence, image processing equipment are utilized in dissecting cancer and the cancer territory (Shah & Chauhan, 2016).

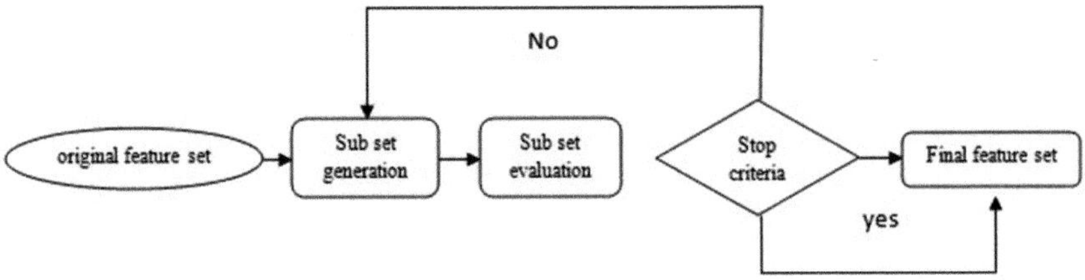

FIGURE 7.1 Irrelevant redundant feature detection.

To make a computer-aided detection (CAD) framework, different image handling strategies, for example, segmentation, feature extraction and feature selection, and classification are coordinated. Feature selection addresses a functioning examination space in pattern detection (Zhou, Gao, Wang, Yu, Wang, & Chi, 2017), machine learning, and also data mining (Bischl et al., 2016). Insignificant and repetitive features appeal to further hunt as to make design with less noticeable and rules essential for anticipating or characterization less clear, notwithstanding the high over fitting hazard. The choice of feature subsets requires deciding the desired feature to expand the precision of forecast or order. A significant point of this chapter is to decide an ideal element subset. Selecting features is typically founded on the boundaries of computational time, and the nature of the produced includes subset arrangements. Fast and accurate classification, utilizing the minimum number of features is regularly settled on. This can clearly be obtained through feature selection. A brief literature survey is described in Section 7.2. Various feature selection methods are investigated in Section 7.3. In Section 7.4, research challenges and future direction has been described. The conclusion is being discussed in the last section.

7.2 LITERATURE REVIEW

The uttermost fundamental step in image processing is the feature selection process. This feature selection algorithm is utilized to determine the most applicable feature from the element feature vector and assists with giving up the undesirable ascribe. Furthermore, this is essential for research territory in machine learning method as well as in and pattern recognition (Kharrat & Mahmoud, 2019). Several images are being taken for the process of recognition in clinical medical imaging (Majid et al., 2020). There are several pictures with high resolution and comprises of more ascribes which are utilized for research work (Wang, Tang, Sun, & Zhang, 2019). Because of this huge information, a more noteworthy dimensionality vector set is acquired in the feature extraction technique. These techniques are used to decrease the extensity along with lessen the expenses of evocation of data and comprehension of the model (Bolón-Canedo, Sánchez-Maroño, & Alonso-Betanzos, 2010). Genetic algorithm (GA) is a typical component choice strategy utilized these days. GA can give the specific or assessed best arrangements. This depends on the hypothesis of enlargement and the genetics of universal selection (Goldberg, 1989). A different aggregate hypergraph method (Shao et al., 2020) is presented for an arrangement of given Alzheimer's contamination. A cross-breed-directed feature selection technique (Othi, 2016) is suggested for the grouping of brain tumors from MRI examine. Besides, other distinctive methods are being created for the perception of experimental medical images (Yu et al., 2020). Recent procedures and operations are efficiently attainable to produce an enormous extent of information (Zhang, Govindaraj, Tang, Zhu, & Sun, 2019). It may be represented in design visuals or other format (Solorio-Fernández, Carrasco-Ochoa, & Martínez-Trinidad, 2020; Wang et al., 2019). The expansion in the extent of datasets forge it hard to understand and use the information for different pursue (Bommert, Sun, Bischl, Rahnenführer, & Lang, 2020). High-geometric data is gathered in online medical images databases to make it more impose (Habib, Aljarah, Faris, & Mirjalili, 2020). In Data Mining, observation is found from recorded information. Information is breaking down, and Information is figured out how to extricate information (Shaheen, Shahbaz, & Guergachi, 2013). With regards to information from the executives, revelation is made on information to discover significant examples. Knowledge observation is not essential in databases and demonstrated its scientific effectiveness for data inquiry (Shaheen & Khan, 2016). This feature selection is finished by setting a couple of standards that evaluate the worth of each contender quality (Batool et al., 2020). Numerous words are utilized for quality assurance like feature selection, feature reduction (Hussain et al., 2020). While, trait choice, is a numerical portrayal inside which a rule is determined against which quality is assessed (Alaoui & Elberrichi, 2020; Alirezanejad, Enayatifar, Motameni, & Nematzadeh, 2020). The sorts of information that are dug in clinical utilization for finishing up one of the previously mentioned measures incorporate microarray information (Selvaraj & Natarajan, 2011), information identified with heart issues (Pavithra & Jayalakshmi, 2019), clinical imaging information (Parmar, Barry, Hosny, Quackenbush, & Aerts, 2018), and others. The information utilized for examination in the clinical area is typically thick and scanty, for instance, the information might be as straightforward and complex pictures, clusters, X-beams, radiotherapy data, and immune histo-chemistry (López et al., 2018). The insufficiency of data increases the number of features and can cause oath of extent (Bellman, 1957). Another significant impediment to the utilization of selection methods on medical image is numerous of the calculations which are all firmly identified with one another (Bolón-Canedo, Sánchez-Maroño, & Alonso-Betanzos, 2013). As referenced before, feature selection methods are additionally utilized on pathological information for genome examination (Tadist, Najah, Nikolov, Mrabti, & Zahi, 2019). A productive analysis of selection technique is represented by Wang, Wang, & Chang (2016) employed in bioinformatics. This scientific categorization of selection techniques along several benefits based on filter method, wrapper method, implanted techniques is introduced by Saeys, Inza, & Larrañaga (2007). A few ensemble feature selectors are likewise evaluated by Khaire & Dhanalakshmi (2019). A bit of feature selection methods are utilized for the investigation of biomedical signal (Wang, Jing, & Niu, 2017). These biomedical signals are conspicuously examined in this survey with regards to electroencephalogram (EEG) beacon examination (Schroder, Bogdan, Hinterberger, & Birbaumer, 2003), heart cancer forecast (Too, Abdullah, Mohd, & Tee, 2019). Many groupings examination techniques endeavor to perceive short, pretty much-monitored signs to indulge protein procession. One more citable endeavor is identified with the use of any feature selection method on microarray information (Hira & Gillies, 2015). Microarray data sets are utilized for tumor differential analysis (Eisen, Spellman, Brown, & Botstein, 1998). These genetic techniques are likewise famous for the selection of features from microarray information (Fakoor, Ladhak, Nazi, & Huber, 2013). A survey of various feature selection methods utilized over medical care datasets is introduced by Manonmani & Balakrishnan (2020). Various feature selection methods are utilized on this given information for examining significant features only.

7.3 VARIOUS FEATURE SELECTION TECHNIQUES

Systems are assembled in three orders dependent on choice standards (Agnihotri, Verma, & Tripathi, 2017). Various Filters, Wrapper, and Embedded algorithms have been described with some advantages and disadvantages.

7.3.1 FILTER ALGORITHMS

These techniques need to assess the norm across which each attribute is computed and filter out insignificant aspects before the enlistment action begins. The way towards to choose trait cease when each trait is calculated in contrast to some allowance. Some attributes are chosen with the finest assessment criteria. These filter algorithms are portrayed in a way.

7.3.1.1 LVF ALGORITHM

LVF algorithm (Nandi, 2011) irregularly does an arbitrary pursuit with any assessment method. In this algorithm, consistency is utilized as an assessment measurement. With every emphasis, the best subset is chosen as a set apart. It can deal with small-scale noisy datasets. It has high time complexity. The selected attributes is normally larger in number for accomplishing always a steady arrangement (Chen, Xia, & Liu, 2014).

7.3.1.2 B&B ALGORITHM

This algorithm is exponential pursuit in a reverse way with a part of monotonic method for the assessment in clinical purposes (Li et al., 2020). It is an excellent search algorithm where a fixed value is characterized. Nodes with lesser fixed values are not investigated as a result of monotonicity supposition, which expresses that its sub-hubs won't create an ideal arrangement. The algorithm finds important ascribes by doing fanning of the dataset. This algorithm stops when the chance of expanding finish. It is highly expensive in nature.

7.3.1.3 RELIEF ALGORITHM

This algorithm has an irregular inquiry in the recommended pursuit space and appoints loads to the property that are close to the ideal worth (He, Zhou, Lin, & Zhu, 2019). Distance is assessment parameter in this calculation. It picks a property haphazardly and discovers its closest hit-and-miss. The closest hit is viewed as the nearest occurrence of that trait with the same class through the closest miss is the nearest occasion with an alternate class (Chen, Tadikamalla, Shang, & Song, 2020).

7.3.1.4 MRMR ALGORITHM

This algorithm chooses ascribes based on two assessment estimates reliance and pertinence (Bugata & Drotar, 2020). Importance implies firmly associated. Pertinence among credits is settled based on common data divided among property and class esteem. It selects those highlights just whose significance is greatest and reliance is least (Rana & Ahmed, 2020).

7.3.1.5 JOINT MUTUAL INFORMATION MAXIMIZATION ALGORITHM

The JMIM algorithm chooses applicable attributes based on mutual information (Bennasar, Hicks, & Setchi, 2015). It utilizes joint information over a greedy methodology by accepting ascribes as a feed-forward structure (Hacilar, Nalbantoglu, Aran, & Bakir-Gungor, 2020). From this information, it very well may be reasoned that firmness quality evaluator is uttermost noticeably utilized in filter. This evaluator retains just those ascribes in the image dataset, which is the justification for accomplishing always predictable arrangement. Mutual information and monotonicity are additionally broadly utilized in filter feature selection algorithms.

7.3.2 WRAPPER FEATURE SELECTION METHODS

This algorithm needs a rigid data mining technique and utilizations its exhibition as a computation basis satisfies some conditions with enhancing properties, and is profoundly computationally effective.

7.3.2.1 SFS ALGORITHM

This method utilizes heuristic inquiry in a forward way. It is known as Successive Forward Selection method. It begins with a vacant set and computes one component in the main emphasis (Marcano-Cedeno et al., 2010). In the following emphasis, some existing and new features are used to make a complete set and the perfect one is chosen. It gives the best outcomes with little information (Bouatmane, Roula, Bouridane, & Al-Maadeed, 2011).

7.3.2.2 SBS ALGORITHM

This algorithm is known as sequential backward selection algorithm and entire features are at first combined to the list of capabilities (Liu, Jiang, & Zheng, 2016). An assessment measure is utilized, which gradually eliminates features individually based on some rule. The exhibition of this algorithm might be ideal (Tsymbal, Cunningham, Pechenizkiy, & Puuronen, 2003).

7.3.2.3 SFFS ALGORITHM

It is known as the Sequential floating forward search method. It starts compute with one property to the rundown of abilities as SFS method. The credits are taken out dependent on appraisal measures. SFFS ensures that no huge quality is cleared out from the feature set (Ververidis & Kotropoulos, 2008).

7.3.2.4 SFBS ALGORITHM

This algorithm is known as sequential floating backward search method that begins with a total rundown of features. In each cycle, forward and backward following are in process until the fulfillment of the most extreme target work. The worldwide list of capabilities is advantageously gotten because it investigates of highlights which are adding to accomplishing target work (Sun, Babbs, & Delp, 2006).

7.3.2.5 ASFFS ALGORITHM

This method is named as Adaptive Sequential Forward Floating Selection method (Somol, Pudil, Novovicová, & Paclık, 1999) is a granular perspective. It is high coordinated inquiry that entirety the forward way. The previous one advises the number of features that should be included in the incorporation stage, and the horizontal one advises the number of traits to be taken out in the prohibition stage. This method closes in a lesser number of repetitive features and additionally works precisely. It is more unpredictable, and it functions admirably on more modest datasets.

This wrapper method has some advantages, which are as follows – (i) it is smooth in nature and collaborates with the classifier; (ii) It is less computationally comprehensive than the randomized method; and (iii) It is limited prostrate to local optima.

Some disadvantages are – (i) It has high risk of over fitting; and (ii) it is more prostrate than randomized algorithms to getting fastened in a limited optimum.

7.3.3 EMBEDDED FEATURE SELECTION METHODS

This method attempts to exploit both filter method and wrapper method by the modification in assessment allowance at various stages. This strategy needs to lessen the overabundance computational season of covering method. A few calculations of the implanted technique are examined underneath.

7.3.3.1 BOOSTED DECISION STUMP FEATURE SELECTION ALGORITHM

It utilizes data acquire standards to choose features. The selected number of various features is allowed to an irregular parameter k. At each progression, it disregards every one of the features that were chosen beforehand. Thus, the main target is on inspected features with the most noteworthy data acquire (Levner, 2005). The characteristic of this method effectively forecast class worth with most noteworthy weight. It is an endless supply of a trait. It gets credits that have more data by playing out the avaricious pursuit.

7.3.3.2 SVM-RFE ALGORITHM

This method is based on the linear discriminate function of weights (Zhou & Tuck, 2007). The new target work is arranged by utilizing SVM to perform recursive element end. It is utilized for twofold class grouping. It utilizes standardization for the minimization of the SVM issue. Discovering ideal target work is troublesome in SVM-RFE.

7.3.3.3 LAZY FEATURE SELECTION ALGORITHM

This algorithm chooses and wipes out selected features from include space with some expectations. It utilizes the K-NN technique to anticipate the benchmark of a given trait, for the most part utilized in the text arrangement issue. Dissipating properties are taken advantage of and utilized as a component choice strategy. It is a classifier subordinate (Peng, Xuefeng, Jianyong, & Yumhong, 2009).

These methods have some advantages. It is used to interact with the basic classifier (b). It is less computationally complex in nature as compared to the wrapper method. There is a drawback is to select classifier dependent.

From these methods, it is clear that the embedded method has superior computational complexity as compared to wrapper. This method is not dependent on the classifier, but the wrapper method and also embedded both are dependent. These feature selection methods can be applied on the segmented images using ISIC and PH2 datasets (Garg & Jindal, 2020).

7.4 RESEARCH DESCRIPTION AND FUTURE CHALLENGES

The feature selection accepts a pivotal part in the clinical imaging space for characterization tasks. Considering the ongoing focuses, the feature selection strategies will be deliberated in future examinations:

- The method to election the robust features is used to improve the conjecture precision level and also for life span calculation of different skin lesion characterization.
- Nevertheless, in light of commotion and a couple of different elements, the eliminated features don't design the particular

sore domain. Consequently, the segment decision techniques are employed to pick simply significant features to identify exact injury or lesion identification.

- The usage of feature optimization method provides a great success depending on scarcely any new articles.
- The determination of deep learning feature to know capabilities likewise upholds higher acknowledgment exactness with inside the clinical area, especially for a big measure of imaging information.
- As entropy-primarily established totally techniques restrained the computational time of CAD framework and also improves acknowledgment exactness.
- The overall enforcement of each determination strategy relies upon at the fitness characteristics. The fitness characteristic may be more suitable using higher strategies such as Neural Network or many more.

7.5 DISCUSSION AND CONCLUSION

Health service Institutes are massing widespread quantities of information because of the speedy increase of information prices in scientific information repositories and fitness care provision. These information repositories maintain a wealth of information that may assist medical doctors make better diagnostic selections and enhance the quality of patient care. The information, on the opposite hand, is multidimensional, with feature numbers weighing in greater closely than samples. In the scientific field, rapid and specific machine learning structures are needed. Most machine learning algorithms want a massive range of samples for training due to the fact small samples can lessen generalization functionality and cause overfitting. Due to the motives related to the curse of dimensionality, the use of traditional strategies to cope with such information isn't a good idea. To attain higher accuracy, the choice of vital functions is required. The choice of vital feature is essential to reap higher accuracy. Before sample classification, feature selection is one of the maximum a success techniques for doing away with redundant and unimportant functions. Different strategies to scientific feature selection implementations are studied in this chapter. It has been proven that the feature selection technique now no longer reduces the range of features however additionally improves the accuracy rate, supporting within side the comprehension of diseases' root causes. Feature extraction, feature selection, dimensionality reduction, and characteristic choice have been all discussed. Filter, wrapper and embedded techniques are general strategies to feature selection techniques which might be described in detail, along with their algorithms. The measures for characteristic assessment are covered, in addition to a few beneficial facts approximately the pros and cons of these feature selection procedures. Both the feature selection method and reduction algorithm techniques are to be discovered for the benefit of improving the accuracy and execution time of scientific imaging. Furthermore, it is concluded that deciding on suitable features reduces a CAD system's universal complication.

KEYWORDS

- **attribute selection**
- **data mining**
- **feature selection**
- **imaging data**
- **medical challenges**
- **medical imaging**

REFERENCES

Agnihotri, D., Verma, K., & Tripathi, P., (2017). Variable global feature selection scheme for automatic classification of text documents. *Expert Systems with Applications, 81*, 268–281.

Alaoui, A., & Elberrichi, Z., (2020). Enhanced ant colony algorithm for best features selection for a decision tree classification of medical data. In: *Critical Approaches to Information Retrieval Research* (pp. 278–293). IGI Global.

Alirezanejad, M., Enayatifar, R., Motameni, H., & Nematzadeh, H., (2020). Heuristic filter feature selection methods for medical datasets. *Genomics, 112*, 1173–1181.

Batool, F. E., Attique, M., Sharif, M., Javed, K., Nazir, M., et al., (2020). Offline signature verification system: A novel technique of fusion of GLCM and geometric features using SVM. *Multimedia Tools and Applications*, 1–20.

Bellman, R., (1957). *Dynamic Programming* (p. 95). Princeton, NJ: Princeton University Press.

Bennasar, M., Hicks, Y., & Setchi, R., (2015). Feature selection using joint mutual information maximization. *Expert Systems with Applications, 42*, 8520–8532.

Bischl, B., Lang, M., Kotthoff, L., Schiffner, J., Richter, J., Studerus, E., Casalicchio, G., & Jones, Z. M., (2016). mlr: Machine learning in R. *J. Mach. Learn. Res., 17*(170), 1–5.

Bolón-Canedo, V., & Remeseiro, B., (2020). Feature selection in image analysis: A survey. *Artif. Intell. Rev., 53*, 2905–2931.

Bolón-Canedo, V., Sánchez-Maroño, N., & Alonso-Betanzos, A., (2010). On the effectiveness of discretization on gene selection of microarray data. In: *2010 International Joint Conference on Neural Networks* (pp. 1–8). Barcelona, Spain: IEEE.

Bolón-Canedo, V., Sánchez-Maroño, N., & Alonso-Betanzos, A., (2013). A review of feature selection methods on synthetic data. *Knowledge and Information Systems, 34*, 483–519.

Bommert, A., Sun, X., Bischl, B., Rahnenführer, J., & Lang, M., (2020). Benchmark for filter methods for feature selection in high-dimensional classification data. *Computational Statistics & Data Analysis, 143*, 106839.

Bouatmane, S., Roula, M. A., Bouridane, A., & Al-Maadeed, S., (2011). Round-Robin sequential forward selection algorithm for prostate cancer classification and diagnosis using multispectral imagery. *Machine Vision and Applications, 22*, 865–878.

Bugata, P., & Drotar, P., (2020). On some aspects of minimum redundancy, maximum relevance feature selection. *Science China Information Sciences, 63*, 1–15.

Chen, X., Xia, X., & Liu, B., (2014). An analysis of hybrid feature selection for improved medical data classification. *Journal of Computational Intelligence and Electronic Systems, 3*, 44–49.

Chen, Y. X., Tadikamalla, P. R., Shang, J., & Song, Y., (2020). Supply allocation: Bi-level programming and differential evolution algorithm for natural disaster relief. *Cluster Computing, 23*, 203–217.

Eisen, M. B., Spellman, P. T., Brown, P. O., & Botstein, D., (1998). Cluster analysis and display of genome-wide expression patterns. *Proceedings of the National Academy of Sciences of the United States of America, 95*, 14863–14868.

Fakoor, R., Ladhak, F., Nazi, A., & Huber, M., (2013). *Using Deep Learning to Enhance Cancer Diagnosis and Classification.* New York, USA: ACM.

Fontana, R. S., Sanderson, D. R., et al., (1986). Lung cancer screening: The Mayo program. *J. Occup. Med., 28*(8), 746–750.

Galeotti, T., Borrello, S., Palombini, G., Masotti, L., Ferrari, M. B., Cavatorta, P., Arcioni, A., et al., (1984). Lipid peroxidation and fluidity of plasma membranes from rat liver and Morris hepatoma 3924A. *FEBS Lett., 169*, 169–713.

Garg, S., & Jindal, B., (2020). Skin lesion segmentation using k-mean and optimized fire fly algorithm. *Multimedia Tools and Applications, 80*(1), 1–14.

Goldberg, D. E., (1989). *Genetic Algorithms in Search, Optimization, and Machine Learning.*

Gopi, K., & Ramashri, T., (2016). A GUI-based hybrid clustering technique for brain tumor segmentation and validation in MRI images. *Aust. J. Basic Appl. Sci., 10*(8), 84–96.

Habib, M., Aljarah, I., Faris, H., & Mirjalili, S., (2020). Multi-objective particle swarm optimization: Theory, literature review, and application in feature selection for medical diagnosis. *Evolutionary Machine Learning Techniques* (pp. 175–201). Singapore: Springer.

Hacilar, H., Nalbantoglu, O. U., Aran, O., & Bakir-Gungor, B., (2020). *Inflammatory Bowel Disease Biomarkers of Human Gut Microbiota Selected Via Ensemble Feature Selection Methods*. arXiv preprint arXiv: 200103019.

Handels, H., Ross, T., Kreusch, J., Wolff, H. H., & Pöppl, S. J., (1999). Feature selection for optimized skin tumor recognition using genetic algorithms. *Artif. Intell. Med., 16*(3), 283–297.

He, Y., Zhou, J., Lin, Y., & Zhu, T., (2019). A class imbalance-aware relief algorithm for the classification of tumors using microarray gene expression data. *Computational Biology and Chemistry, 80*, 121–127.

Hira, Z. M., & Gillies, D. F., (2015). A review of feature selection and feature extraction methods applied on microarray data. *Advances in Bioinformatics, 13*.

Hussain, N., Khan, M. A., Sharif, M., Khan, S. A., Albesher, A. A., et al., (2020). A deep neural network and classical features based scheme for object recognition: An application for machine inspection. *Multimedia Tools and Applications*. doi: 101007/s11042-020–08852-3.

Kancherla, K., & Mukkamala, S., (2012). Feature selection for lung cancer detection using SVM based recursive feature elimination method. *EvoBIO'12: Proceedings of the 10th European Conference on Evolutionary Computation, Machine Learning and Data Mining in Bioinformatics* (pp. 168–217).

Khaire, U. M., & Dhanalakshmi, R., (2019). Stability of feature selection algorithm: A review. *Journal of King Saud University-Computer and Information Sciences*.

Kharrat, A., & Mahmoud, N., (2019). Feature selection based on hybrid optimization for magnetic resonance imaging brain tumor classification and segmentation. *Applied Medical Informatics, 41*, 9–23.

Kharrat, A., & Neji, M., (2018). *A Hybrid Feature Selection for MRI Brain Tumor Classification*. 10.1007/978-3-319-76354-5_30.

Kushwaha, P., & Welekar, R., (2016). Feature selection for image retrieval based on genetic algorithm. *International Journal of Interactive Multimedia and Artificial Intelligence*, *4*, 16. 10.9781/ijimai.2016.423.

Levner, I., (2005). Feature selection and nearest centroid classification for protein mass spectrometry. *BMC Bioinformatics, 6*(1), 68.

Li, X. C., Malikić, S., Schäffer, A. A., Gertz, E. M., Day, C., Pérez-Guijarro, E., & Sahinalp, S. C., (2020). *PhISCS-BnB: A Fast Branch and Bound Algorithm for the Perfect Tumor Phylogeny Reconstruction Problem*. https://europepmc.org/article/ppr/ppr111948 (accessed on 13 October 2023).

Liu, H., Jiang, H., & Zheng, R., (2016). The hybrid feature selection algorithm based on maximum minimum backward selection search strategy for liver tissue pathological image classification. *Computational and Mathematical Methods in Medicine, 9*.

López-Pineda, A., Rodríguez-Moran, M. F., Álvarez-Aguilar, C., Valle, S. M. F., Acosta-Rosales, R., et al., (2018). Data mining of digitized health records in a resource-constrained setting reveals that timely immunophenotyping is associated with improved breast cancer outcomes. *BMC Cancer, 18*, 1–10.

Majid, A., Khan, M. A., Yasmin, M., Rehman, A., Yousafzai, A., et al., (2020). Classification of stomach infections: A paradigm of convolutional neural network along with classical features fusion and selection. *Microscopy Research and Technique, 83*, 562–576.

Manonmani, M., & Balakrishnan, S., (2020). Review of optimization-based feature selection algorithms on healthcare dataset. *Emerging Research in Data Engineering Systems and Computer Communications* (pp. 239–245). Singapore: Springer.

Marcano-Cedeno, A., Quintanilla-Domínguez, J., Cortina-Januchs, M., & Andina, D., (2010). *Feature Selection Using Sequential Forward Selection and Classification Applying Artificial Metaplasticity Neural Network* (pp. 2845–2850). IEEE.

Nanda, K., McCory, P., Myers, E., et al., (2000). Accuracy of the PAP test in screening for and follow up of cervical cytologic abnormalities a systematic review. *Annals of Internal Medicine, 132*(10), 810–819.

Nandi, G., (2011). An enhanced approach to Las Vegas filter (LVF) feature selection algorithm. In: *2nd National Conference on Emerging Trends and Applications in Computer Science (*pp. 1–3).

Othi, G., (2016). Hybrid tolerance rough set-firefly based supervised feature selection for MRI brain tumor image classification. *Applied Soft Computing, 46*, 639–651.

Parmar, C., Barry, J. D., Hosny, A., Quackenbush, J., & Aerts, H. J., (2018). Data analysis strategies in medical imaging. *Clinical Cancer Research, 24*, 3492–3499.

Pavithra, V., & Jayalakshmi, V., (2019). *A Review on Predicting Cardiovascular Diseases Using Data Mining Techniques* (pp. 374–380). Springer Cham.

Peng, Y., Xuefeng, Z., Jianyong, Z., & Yumhong, X., (2009). Lazy learner text categorization algorithm based on embedded feature selection. *Journal of Systems Engineering and Electronics, 20*, 651–659.

Rana, M. M., & Ahmed, K., (2020). *Feature Selection and Biomedical Signal Classification Using Minimum Redundancy Maximum Relevance and Artificial Neural Network* (pp. 207–214). Berlin: Springer.

Saeys, V., Inza, I., & Larrañaga, P., (2007). A review of feature selection techniques in bioinformatics. *Bioinformatics*, *23*(19), 2507–2517.

Saeys, Y., Inza, I., & Larrañaga, P., (2007). A review of feature selection techniques in bioinformatics. *Bioinformatics, 23*, 2507–2517.

Schroder, M., Bogdan, M., Hinterberger, T., & Birbaumer, N., (2003). Automated EEG feature selection for brain-computer interfaces. In: *First International IEEE EMBS Conference on Neural Engineering* (pp. 626–629).

Selvaraj, S., & Natarajan, J., (2011). Microarray data analysis and mining tools. *Bioinformation, 6*, 95.

Shah, S., & Chauhan, N. C., (2016). Techniques for detection and analysis of tumors from brain MRI images: A review. *J. Biomed. Eng. Med. Imaging, 3*(1), 9–20.

Shaheen, M., & Khan, M. Z., (2016). A method of data mining for selection of site for wind turbines. *Renewable and Sustainable Energy Reviews, 55*, 1225–1233.

Shaheen, M., Shahbaz, M., & Guergachi, A., (2013). Context based positive and negative spatio-temporal association rule mining. *Knowledge-Based Systems, 37*, 261–273.

Shao, W., Peng, Y., Zu, C., Wang, M., Zhang, D., et al., (2020). Hypergraph based multi-task feature selection for multimodal classification of Alzheimer's disease. *Computerized Medical Imaging and Graphics, 80*, 101663.

Solorio-Fernández, S., Carrasco-Ochoa, J. A., & Martínez-Trinidad, J. F., (2020). A review of unsupervised feature selection methods. *Artificial Intelligence Review, 53*, 907–994.

Somol, P., Pudil, P., Novovicová, J., & Paclık, P., (1999). Adaptive floating search methods in feature selection. *Pattern Recognition Letters, 20*, 1157–1163.

Sun, Y., Babbs, C., & Delp, E., (2006). *A Comparison of Feature Selection Methods for the Detection of Breast Cancers in Mammograms: Adaptive Sequential Floating Search vs. Genetic Algorithm* (pp. 6532–6535). IEEE.

Tadist, K., Najah, S., Nikolov, N. S., Mrabti, F., & Zahi, A., (2019). Feature selection methods and genomic big data: A systematic review. *Journal of Big Data, 6*, 79.

Taxdal, S. R., Ward, G. E., & Figge, F. H. J., (1955). Fluorescence of human lymphatic and cancer tissues following high doses of intravenous hematoporphyrin. *Surg. Forum., 5*, 619–624.

Too, J., Abdullah, A. R., Mohd, S. N., & Tee, W., (2019). EMG feature selection and classification using a pbest-guide binary particle swarm optimization. *Computation, 7*(1), 12.

Tsymbal, A., Cunningham, P., Pechenizkiy, M., & Puuronen, S., (2003). Search strategies for ensemble feature selection in medical diagnostics. In: *16th IEEE Symposium Computer-Based Medical System* (pp. 124–129).

Ververidis, D., & Kotropoulos, C., (2008). Fast and accurate sequential floating forward feature selection with the Bayes classifier applied to speech emotion recognition. *Signal Processing, 88*, 2956–2970.

Wang, H., Jing, X., & Niu, B., (2017). A discrete bacterial algorithm for feature selection in classification of microarray gene expression cancer data. *Knowledge-Based Systems, 126,* 8–19.

Wang, L., Wang, Y., & Chang, Q., (2016). Feature selection methods for big data bioinformatics: A survey from the search perspective. *Methods, 111*, 21–31.

Wang, S. H., Xie, S., Chen, X., Guttery, D. S., Tang, C., et al., (2019). Alcoholism identification based on an AlexNet transfer learning model. *Frontiers in Psychiatry, 10*, 205.

Wang, S., Tang, C., Sun, J., & Zhang, Y., (2019). Cerebral micro-bleeding detection based on densely connected neural network. *Frontiers in Neuroscience, 13*, 422.

WHO, (2004). *Deaths by Cause, Sex and Mortality Stratum*. World Health Organization.

Yu, X., Kang, C., Guttery, D. S., Kadry, S., Chen, Y., et al., (2020). ResNet-SCDA-50 for breast abnormality classification. *IEEE/ACM Transactions on Computational Biology and Bioinformatics.*

Zhang, Y. D., Govindaraj, V. V., Tang, C., Zhu, W., & Sun, J., (2019). High performance multiple sclerosis classification by data augmentation and AlexNet transfer learning model. *Journal of Medical Imaging and Health Informatics, 9*, 2012–2021.

Zhou, X., & Tuck, D. P., (2007). MSVM-RFE: Extensions of SVM-RFE for multiclass gene selection on DNA microarray data. *Bioinformatics, 23*, 1106–1114.

Zhou, X., Gao, X., Wang, J., Yu, H., Wang, Z., & Chi, Z., (2017). Eye tracking data guided feature selection for image classification. *Pattern Recogn., 63*, 56–70.

CHAPTER 8

Starch-Based Biodegradable Packaging Materials: Characterization and Multifarious Applications in the Pharmaceutical and Food Industries

PRATAP KALITA,[1,2] ABDUL BAQUEE AHMED,[3] SAIKAT SEN,[4] SATYABRAT SARMA,[4] and SATYENDRA DEKA[1]

[1]*Pratiksha Institute of Pharmaceutical Sciences, Guwahati, Assam, India*

[2]*PhD Research Scholar, Assam Science and Technology University, Guwahati, Assam, India*

[3]*Girijananda Chowdhury Institute of Pharmaceutical Science, Tezpur, Assam, India*

[4]*Faculty of Pharmaceutical Sciences, Assam Down Town University, Guwahati, Assam, India*

ABSTRACT

Synthetic and semisynthetic plastic materials are being used widely worldwide to create a diversity of useful products due to its multipurpose and advantageous nature but remarkable damage to the planet is increasing day by day due to their widespread use. Because of the non-biodegradable nature of plastic materials is creating huge loss to the planet. Starch is

Advances in Manufacturing and Processing of Materials: Characterization and Applications.
Narendra Kumar, PhD, Mohit Tyagi, PhD, Dilbagh Panchal, PhD & Ravi Pratap Singh, PhD (Eds.)

a native biopolymer or novel material, which is available abundantly in nature. From the literature it was observed that starch will be a good candidate and alternative to substitute for synthetic materials due to its more biodegradable and environmentally-friendly properties. Starch is abundantly available in nature with a higher level of availability, biologically compatible, enough mechanical properties, no toxicity, and the compatible nature of starch makes it suitable to replace synthetic plastics materials. For starch substances, for utilizing in the fabrication of starch-based packaging materials many challenges need to overcome. However, starch is used as a biopolymer to substitute synthetic polymers in pharmaceutical preparations, where starch is used as a binding agent, disintegrating agent, carrier of drug molecules, excipient in several formulations, etc. Starch-based materials or polymers have significant potential to be applied in different food packaging products or pharmaceutical excipients, which will reduce environmental pollution, too. Therefore, this chapter is designed to display in detail the characterization of starch-based biodegradable materials and their applications in pharmaceutical and food industry.

8.1 INTRODUCTION

Bakelite is the first synthetic polymer fabricated by the two most effective solvents includes phenol and formaldehyde. The scientist behind this invention was Leo Baekeland in the year of 1907. From that time, plastic is an unavoidable material in people's lives. In today's worlds, the protection and preservation of the substances like foods, cosmetics, medicines, and chemicals is mainly fulfilled by plastic materials. The application of plastic in various industries is increasing day by day, and the demand is remarkable, but the main disadvantage of plastics is poor biodegradability. The chemical nature and formed chemical bonds within the plastic materials are the main reasons to show the poor biodegradability of plastics. The way of disposal of plastics on the earth is making unsuitable situations like pollution because the natural land cannot degrade the plastic waste due to the presence of various chemical bonds like hydrogen bonds, hydrocarbon bonds, etc. At the present time, pollution of land, air, and water is mainly adulterated by the end product of plastic materials. The end product of the plastic substances occurred due to exposing in open environment or burning of the plastic substances in air, which also put the negative impact on human health (Crespy et al., 2008). A report generated by Global Plastic Production displays that around 348 million tons of plastic were produced in the year 2017. If we compare the current data with the previous report generated in 1950, it was a drastic change, and through the statistical analysis, the current data is 174 times higher than the previous data obtained in 1950. Petrochemical engineering is considered as the beginning science in order to generate the plastic materials and introduce in the earth. Fabricated plastic materials completely manufactured by using non-renewable petroleum resources. The utility of these synthetic plastic materials and their co-products are tremendously harming the nature in serious manner, because of which the current global environment is completely polluted. Currently, scientists are also reporting that within a couple of years, there will be a depletion of petroleum resources (Gibb, 2019).

Several research works demonstrated on plastics and reported that plastics significantly contaminated soil, and ultimately that put a bad impact on green crops and other livestock crops. Therefore, there is depletion of food security as human health primarily based on the green crops. Microplastic contaminants are dangerous factor that can spread through flood and other ways and contaminant the soil and depletion of soil properties used take place. Microplastics are found to have harmful effect on the soil microorganisms. Microplastics can easily penetrate the wall of the soil microorganisms (worms, insects) and destroy them (Scheurer & Bigalke, 2018). The smallest micro-plastic substances can easily invade and enter to the food chain, which further causes the serious destruction of microorganism (such as birds and rodents) available in soil after consumption of those type of foods (Chae & An, 2018). There is a strong bond between human health and environment, but due to severe utilization of plastic materials causing drastic effect on this bond as well spoiling the nature. Availability of the plastics in food materials also affect the human health and depleted the immunity system. Plastics and various additives present in fish, green vegetables, and crops can significantly affect the human immunity system, and finally, there is an imbalance in the human physiological system (Rochman et al., 2013).

A list of degenerative disorders like metabolic disorders, reproductive problems, obesity, oxidative stress, cardiac problems, respiratory problems, and nephron degenerative disorders are used to observe human health due to the consumption of plastics through food or eatables. So, it is very important to develop impactful strategies to produce functional and biodegradable materials which can easily replace the plastics. Application of the biodegradable product has positive and significant effect on environmental section, human health care system, soil pollution, etc. In earlier research works, it was observed that scientists and researchers are conducting several scientific works on synthetic and natural polymer to construct biodegradable products which are widely used in daily life. A bunch of natural polymers are being used in the preparation of several biodegradable products due to easy availability, low cost, renewability, biocompatibility, no toxicity, and can form thin films (Limonta et al., 2019). Modification of starch is also being a significant strategy to introduce several functional active groups in the starch skeleton, and finally produced effective and functional materials which are fit for production of several biodegradable products. Application of native starch studied in several fields like pharmaceutical industries, food industries, biomedical sector, packaging of materials, etc. (Kumar, 2018).

Chemically, starch is a polymer abundantly synthesized by green plants. Three major synthesized components by plant are cellulose, starch and pectin. Among natural polymers, starch is found to have potential and functional applications in the fields of biomedical, food as well as non-food discipline, building materials, consumer goods, and packing (Hemamalini & Giri, 2018). The predominant application of starch in current times is due to their low density, renewability, biocompatibility, biodegradability, non-toxicity, non-inflammatory, easy of availability and relatively low cost (Pawar et al., 2015). There are several methods to isolate the starch from the natural sources, and it was studied that isolation procedure also put impact on the granular structure as well as physicochemical attributes of the starch (Maniglia & Tapia-Blácido, 2016). A bunch of significant physicochemical properties like gelatinization, retrogradation, syneresis, morphological characteristics, solubility index, water holding capacity, micromeritics properties, crystalline structure, etc., make starch suitable and fit to use in production of biodegradable products (Verwimp et al., 2004). Native and modified starches have wide applications. In several aspects starch and modified starches are being used as gelling agent, a thickening agent, an emulsifying agent, an encapsulating agent (Morrison et al., 1984). The applicability of modified and non-modified starch substances has been demonstrated in diverse fields. Lots of new strategies applied on starch to develop starch-based paper by paper mill systems. Incorporation of the anionic trash and electrolytes in the starch system by the grafting process developed highly starch-based tensile substances, which led to higher demands on the rheological and water-holding properties. Fabricated starch-based materials are also absorbing as adhesive or other co-agents in several food formulation and packaging industries (Maurer et al., 1998).

The concentration of amylose and amylopectin is varying in starch granules. This level of concentration of amylose and amylopectin in starch granules are responsible to form biofilms. Due to the excellent pasting and flow properties and the mechanical strength of starch are increasing demand in the food packaging and food industries (Przetaczek-Rożnowska, 2017). Starch is abundantly available in green plants. The ratio of amylose/amylopectin is varying in starch granules. This ratio is responsible for forming biofilms, but depending on the source, this ratio used to vary. Alpha 1–4 and 1–6 linkage used to present in the amylose and amylopectin system, respectively. The branching characteristics of starch system is depending on the relative crystallinity or amorphous region of the starch moieties. The crystallinity behavior of the starch can be easily evaluated by the XRD analyzer. Double helical structure of the amylopectin region is also indicating the branch-chain length. Amorphous and crystalline regions of the starch granules are mainly dependent on the content of the amylopectin. Crystalline nature of the starch is important parameter to generate the significant mechanical attributes. The long consistency structural chain involved in the starch system is mainly made up of amylopectin, and different types of chains are generated, including chains A, B, & C (Jane et al., 1999; Bogracheva et al., 1998). The surface morphology of the starch granules studied by several researchers and reported starch granules are often produced spherical, oval, polygonal and sometimes irregular in shape. The surface morphology of starch system is typically depending on the starch origin or starch sources. Based on the starch sources, the granular size and shape used to vary. The starch system is insoluble in cold water but soluble in warm or hot water. At a specific temperature, starch granules swell and expand their volume. The percentage of amylopectin present in starch system significantly effect on the swelling index of the starch (Li et al., 2017; Das et al., 2015). Therefore, this chapter is designed to profile the different applications of starch-based materials in several aspects and their properties.

8.2 CHARACTERIZATION

8.2.1 MORPHOLOGICAL ATTRIBUTES OF STARCH

The morphological behavior of starch granules is an important attribute that provide the macroscopic and functional properties of it. There are so many sophisticated instruments to study the morphological behavior of the starch-based materials. Currently, sophisticated instruments like SEM, AFM, and TEM are used to study the morphological behavior of starch granules in order to see their structural characteristics. SEM stands for scanning electron for microscope, AFM stands for atomic force microscope, and TEM stands for transmission electron microscope. The nature and shape of the particles can be also easily performed through these instruments. The behavior of the particle including cracking or pores on the surface, any kind of lamination on the particle surface, homogeneity or heterogeneity behavior of the starch-based materials and roughness of the particles can be profiled through SEM images. The appearance of the starch-based materials and starch granules determined by the help of SEM. Starch granules are available in different shapes like polygonal, hexagonal, oval, elongated, spherical, etc., which can be evaluated by SEM study (Reddy et al., 2018).

The topological properties of the starch can be easily determined by AFM. The arrangement of the materials used to find out by the help of AFM. Topological behavior like deformation of the materials, arrangement order of the materials, and interrelation of the starch-based substances can be easily concluded by AFM. The outer layer characteristics like surface roughness and the effect of excipients on the starch's structure are usually studied by AFM. Whenever fabrication of new starch-based product or materials by incorporation of the other additives than the final product studied under AFM to view their structural, arrangement behavior and constitutional behavior. Transmission electron microscopy (TEM) is an instrument that provides the internal structural behavior. From the TEM analysis, some smooth indent appearances were used to observe the starch-based materials. Surface behavior used to change for the starch-based materials due to the methods applied during manufacturing process. Smoother surface observed in the TEM micrographs of citrated starch, which may be suggested due to acid hydrolysis of glycosidic bonds of the starch granules. TEM micrographs also provide the information like polymer interaction within the materials, dispersion of the starch granules, morphological appearance, complete information of size and shape. The integrity of the starch-based materials can be also evaluated by TEM studies (Domene-López et al., 2019; Zhang et al., 2017).

X-ray diffraction (XRD) is used to study the internal structural behavior in the crystalline structure of the starch granules. The compact or packing behavior also can be easily analyzed by application of XRD. The crystallinity index and crystal structure of starch film as well as different starch-based products can be quantified by using XRD. XRD also provide the interlinked group details when native starch used to modify some process like citration, Carboxymethylation, hydrolysis, etc. The modification of starch was carried out by several research studies and characterized by XRD (Cheng et al., 2019; Suriyatem et al., 2018). During the formulation or fabrication time of starch-based materials, different types of additives or excipients are used to obtain the final products. Therefore, the compatibility study is an important parameter to investigate the compatible nature of the starch with other excipients. The compatibility study was performed with the help of FTIR, which provides information regarding the compatible nature of the starch-based materials. Present of different functional groups and different regions like saturation or unsaturation carbon chain can be also easily find out by FTIR (Dankar et al., 2018). FTIR also generates the different properties related to the amorphous region or crystalline region of the starch and starch-based products. The compatibilities of starch with other additives can be studied. The functional groups introduced in the starch skeleton due to modification can be easily identified. The presence of single bonds, double bonds, carbonyl groups, etc., visualized in the FTIR spectra and produced by several mechanisms like stretching, bending, twisting, oscillating, etc. (Pachuau et al., 2017, 2018).

8.2.2 GELATINIZATION AND RHEOLOGICAL ATTRIBUTES

Gelatinization and rheological properties of starch granules is considered as the main functional characteristic in terms of utilizing them in several packaging fields, especially in packaging of food and other materials. The practical application or use of the starch-based materials mainly based on the mechanical properties of it. The feasibility of starch to use as packaging material, it is very important to have the strong mechanical properties which can be determine through numerous methods. The dynamic mechanical properties also revealed the functional characteristics of the starch and starch-based products (do Evangelho et al., 2019; Santana et al., 2019). The starch long chain with high amylopectin branches is responsible to produce the significant rheological property. Parallelly rigid helical lipid-amylose complexes are also associated to display significant mechanical properties. Natural starch and modified starches displayed a strong viscosity profile, which directly put a strong impact on packaging materials. Peak viscosity, pasting temperatures and higher shear stability margin profile were very significant for the natural starches (Jan et al., 2017). Amylose and amylopectin content rapidly influenced on the pasting behavior of the starch and starch-based materials. Due to low content of amylose in the starch system revealed lower peak and higher breakdown viscosities, which is unable to produce amylose-lipid complexes and their findings also revealed that higher peak viscosities of starch isolated from corn, millet, sorghum and quinoa (Li & Zhu, 2018a).

The rheological behavior of the starch and starch-based biofilms or starch-based thin films facilitates the functional properties of the starch. The functional properties of the starch-based films and materials are also measured by their pasting characteristics. A higher level of amylose content was seen in pineapple starch, which produced beneficial pasting properties with the lowest peak viscosity. It was started that highest pasting temperature was observed in pineapple starch due to a higher level of amylose contents. Therefore, ratio of amylose and amylopectin in starch system is considered as remarkable parameter to introduce starch moieties in different starch-based products like thermostable or thermo-labile packaging materials (Nakthong et al., 2017). Any starch substances displayed the significant mechanical attributes due to their various potential and strong profiles in terms of starch integrity, amylose and amylopectin content, re-ordering of long and short chain, network like bridges within the starch molecules (Zhang et al., 2018). Chemical modification of the native starch always provides the more functional products due to refining of starch sample. It was reported that chemical modification can improve the pasting characterization of the starch granules due to shifting or changes in the starch chain linkage. It was observed in several research studies that higher values of peak viscosity, breakdown viscosity, and lower values of set-back viscosity of starch-based materials can form a very thin network-like structure in the starch system, which is beneficial for generating typical mechanical attributes. The chain length and the content of amylose and amylopectin of the starch system are also rapidly influenced by the sources of the raw materials from where starch used to isolate. The method of isolation process also influenced the integrity of the starch molecules and effect on the system chain length (Hoyos-Leyva et al., 2017; Verwimp et al., 2004).

The functionality and the application of the starch are mainly based on gelatinization and rheological properties. These properties are considered as the major physicochemical attributes among the others. The process of formation of gel of the starch system while heating known as the gelatinization process. Alpha 1–4 linkage in amylose system and alpha 1–6 linkage in amylopectin system is started to destruct once the starch granules absorb water in warm condition (Tabilo-Munizaga & Barbosa-Cánovas, 2005). The formation of gel from starch powder is also depends on the type of chains present in the starch system. Higher content of amylopectin in starch granules starts to form gel at slower rate. Starch contain more amylose content undergoes rapid gelatinization process. Therefore, gelatinization of starch is considered as the crucial factor to profile their rheological or pasting properties of the starch or starch-based products. Different modification methods on starch granules were found to be very effective, which make the starch system useful in diverse field. The incorporation of several additives in the starch system also provides the effective gelatinization and rheological properties applicable to food packaging field. Physical, chemical and enzymatic modification of starches possesses unique properties (Liu et al., n.d.).

8.2.3 BIODEGRADABLE ATTRIBUTES

Starch is a natural polymer which possesses many unique properties, among them biodegradability of starch also one of that. Starch-based bio-films or starch-based materials are biocompatible to use in biomedical aspects or other relevant fields. Biodegradable nature of packaging materials is the most important part in current time. The biodegradability of natural sources like starch, cellulose, and other biopolymers is considered a significant and important index to use in the packaging field. Due to the biodegradability nature of starch-based films, the utility is abundantly increasing in the current time and also starch is using as biopolymer in different pharmaceutical products development due to its biodegradable nature in biological fluids (Colussi et al., 2017). Ease of digestion on starch-based films or starch-based packaging products by microorganisms present in soil is considered as determining factor for biodegradability. The degradation of the starch-based materials used rapidly degraded by the microorganism like fungi, bacteria, algae. Due to the biodegradable nature of the starch-based materials, currently utility of the starch from natural sources are getting effective demand in diverse field. Several research works reported that biodegradable materials used decomposed by the microorganisms and produce CO_2 and H_2O as final products under presence of O_2 and hydrogen and methane as final products under absence or air or O_2. In the current time, modification of starch is a novel technique to improve the starch quality and also its biodegradability for using in diverse field like food packaging, pharmaceutical dosage form development, and biomedical applications (Kaur et al., 2019).

8.2.4 THERMAL ATTRIBUTES

Thermal behavior of the starch system used to characterize by DSC and TGA. DSC stands for differential scanning colorimeter, and TGA stands for thermo-gravimetric analysis. DSC analysis has been considered to be the preferred tool and the method of choice for analyzing endothermic gelatinization of starch. The thermal behavior of the starch granules is considered as a crucial parameter because by knowing the thermal characteristics of any starch granules, the preparation of starch-based materials or starch-based films can be designed and fabricated. The gelatinization behavior of starch, also influenced by heat applied to form a gel, characterized by DSC and TGA thermal curves. Properties of starch system or starch-based film materials like glass transition temperature (Tg) and melting point (Tm) are easily determined. The barrier properties of the starch also depend on the Tg of starch system. Sign of well compatibilities, miscibility's and dispersibilities of the starch components assured by Tg. It was reported that starch-based materials can produce one Tg or more than one Tg. Detected with one Tg for the starch-based materials, it was concluded that there was a good dispersion of the substances and that they were compatible with each other. Detected with more than one Tg for the starch-based materials concluded that there is no well dispersion of the substances and not compatible with each other. Barrier properties of the starch-based materials also depend on the value of Tg, and it was concluded that higher Tg value-containing materials showed better barrier properties. The melting point (Tm) is defined as the temperature that required for a starch components or starch crystals to change from solid to liquid (Jan et al., 2017; Li & Zhu, 2018a; Oleyaei, Zahedi et al., 2016; Oleyaei, Almasi et al., 2016).

DSC thermograms of natural and modified starch granules are considered as the typical parameter to profile their gelatinization temperature. The utility of the native starch and modified starch in current time is getting high demand due their typical DSC thermograms. The typical DSC thermogram of the starch granules make them suitable to use in several fields including food packaging (Oleyaei, Almasi et al., 2016). The highest gelatinization characteristics displayed by the native starch system followed by modified starch. High amylose content, rigid crystalline structure, chain length of the starch granules are the main responsible factor to show highest gelatinization properties of the native starch system (Oleyaei, Zahedi et al., 2016). Earlier research work stated that modification of the starch with citric acid cause the destruction of the starch skeleton and introduced the carbonyl group, resulted in the lower the gelatinization temperature compared to the native starch one. A typical thermogram generated by a DSC analyzer provides information like onset temperature, peak temperature, and conclusion temperature (Pachuau et al., 2018). TGA is considered as the main instruments to study the thermal degradation of the starch components. Degradation of the starch-based materials used to study by TGA analyzer. The starch-based materials or starch-based films kept under specific nitrogen air condition and samples used to heat for a specific period of time, and degradation of the materials used calculate. The decomposed substances of starch-based materials or films under TGA observed when the graphical representation used to figure out by taking the resulting mass versus temperature profiles (Malmir et al., 2018).

8.2.5 BARRIER ATTRIBUTES

Barrier properties are very important for starch-based materials to show the resistance activity against transport during packaging. The quality of the food packaging is depending on the barrier properties of starch.

The evaluation of the water transfer from inside to outside environment used to determine in terms of WVTR and WVP. WVTR stands for water vapor transmission rate, and WVP stands for water permeability rate. In term of food packaging the barrier properties are considered as the crucial parameter to study and demonstrate the barrier attributes of the starch and starch-based film or materials (Nandi & Guha, 2018). In the recent studies it was stated that the water barrier properties of the starch-based substances or materials studies by evaluating their water vapor transmission rate and water vapor permeability. These two parameters rapidly involved in the packing of the materials. The withstand capacity of a starch-based materials may be also concluded by the expressing VVTR and WVP. Lower value of VVTR and WVP containing starch-based materials may be good candidate for utilizing in diverse fields. In several studies the value of VVTR and WVP as barrier property of the starch-based materials and also revealed their potential application in the food packaging. The prepared starch biofilms, biocompatible thin films, and others starch based materials displayed significant barrier property which fit them to use in food packaging and other areas (do Evangelho et al., 2019; Santana et al., 2019).

Inhibition of the moisture uptake, oxidation of the starch-based films, and microbial growth or microbial degradation to the starch-based film materials are main countable properties in the field of packaging. Several research works demonstrated and suggested that implementation of the starch and cellulose nanoparticles to the natural starch (cassava) significantly reduced their WVP. The lower value of WVP also associated with the shelf life of the products that packed by the starch-based biomaterials. The significant value of WVP of any starch-based materials has the tendency to inhibit several factors to enter the inside of the package and save the life of packed products. The listed factors like deterioration, absorption behavior of moisture, erosion, microbial contamination, and decomposition are completely inhibited and not allowed to contaminate with the products inside the package, thus increasing the shelf life of the products (Santana et al., 2019). Increasing the oxygen barrier properties of starch-based films is a positive strategy to produce it effective in the field of food packaging. The oxygen barrier property of starch-based materials is considered as an important index/parameter for determining the safety of packaged foods. Earlier studies have been reported that OTR (oxygen transmission rate) and OP (oxygen permeability) are two parameters which rapidly influenced on the oxygen barrier property of starch-based materials (Kaur et al., 2019). The reduction in value of the OTR and OP is considered as functional and potential attribute to improve the oxygen barrier property of the starch-based films and make suitable to use them in packaging field. Xanthan gum-like hydrocolloids can be incorporated into the starch-based films to reduce the void spaces present in the starch molecules and improve the oxygen barrier property (Nagar et al., 2020).

8.3 APPLICATION IN FOOD INDUSTRY

Basically, in the food industries, the starch-based materials are being used in two main categories which are including food packaging and bio-film. Minimum safety measurements like prevention of oxidation of food materials, deterioration of active substances, destruction of the quality of products, and contamination from microorganisms should be provided by the starch-based packaging materials (Zhao et al., 2008). It is well known to society or communities that the degradation of plastic materials is not easy by the digestible enzymes available in the soil, therefore, disposal of plastic packaging products harms the environment; especially soil pollution is the main example. The smoke produced during the burning time of the plastic materials also can produce the drastic air pollution and harm entire environment. The components used during the fabrication of the conventional packaging materials are not completely inert, example synthetic and semisynthetic plastic materials (Avella et al., 2005; Ozdemir & Floros, 2004). Therefore, the utility of the starch-based materials in food industries and food packaging will be positive and advantageous in current times. Starch-based packaging materials are available with no biocompatibility. These materials are degraded easily by bio-based microorganisms, and end products are easily miscible with the soil. Starch-based materials are available with no color, no taste, no odor, and no toxicity (Phan et al., 2009). Preparation foaming materials by incorporating starch or modified starch substances can easily replace the plastic products like polystyrene. Fabrications of disposable dishes or trays are right example to demonstrate in comparison to plastic dishes. Plastic dishes rapidly produced water insoluble or toxic substances which are not absorbed by soil or environment. Starch based preparations or starch-based biofilms, or disposable dishes rapidly digested by the microorganisms abundantly available in soil (Siracusa et al., 2008).

Non-conventional starch from different natural sources is abundantly utilized in several aspects. The unique physicochemical attribute of the starch is made fit to use in diverse fields. Basically, the thermal, morphological, rheological, and mechanical parameters of the starch rapidly influenced on the packaging behavior. Food packaging and food industry are two broad areas where modified and non-modified starches can be used abundantly to prepare potential packaging materials and the functional food products (Li & Zhu, 2018a). Biodegradability is considered as the most effective attribute of the starch-based films and starch-based materials. The application of the starch-based materials in food packaging aspect is abundant. The starch-based materials are not having any negative effect on the food products; apart from this it provides the safety features and improvement of the quality of the food substances (Chakraborty et al., 2019). Considering starch-based films as the functional materials in food packaging is due to their potential mechanical, barrier, antibacterial, and antioxidant properties. Focusing on starch-based packaging materials should have the enough property to hold the food safety, food behavior, food quality and shelf life of the products. Fabrication of the effective and functional starch-based biofilms or starch-based packaging materials can be easily prepared by incorporation of different materials like antioxidants (prevents the oxidation of substances), antimicrobial agents (prevents the growth of microorganisms and inhibit contamination with other foreign particles), etc. (Luchese et al., 2018). Fabricated starch-based packaging materials provide the bio-efficacy to the packed materials. In contrast with the plastic packaging materials, the starch-based materials are displaying tremendous applicability in several fields which are already mentioned. Ease of degradation and eco-environment nature of the starch-based materials are main reason to get the more applicability in current time as packaging materials, and due to these unique characteristics, commercial applicability of the starch-based materials is expanding day by day (Li & Zhu, 2018b). The earlier research also explained the new function of the starch-based materials when prepared by addition of several agents to increase their functionality and to increase the shelf life of the food substances. The additives or agents applied during the preparation of starch-based materials are act as enhancer to increase the physicochemical attributes of the prepared materials. The improved physicochemical properties of the starch-based products also facilitated better environment for the packed products and helps to increase the shelf life of the food substances (Figure 8.1) (Zhang et al., 2020).

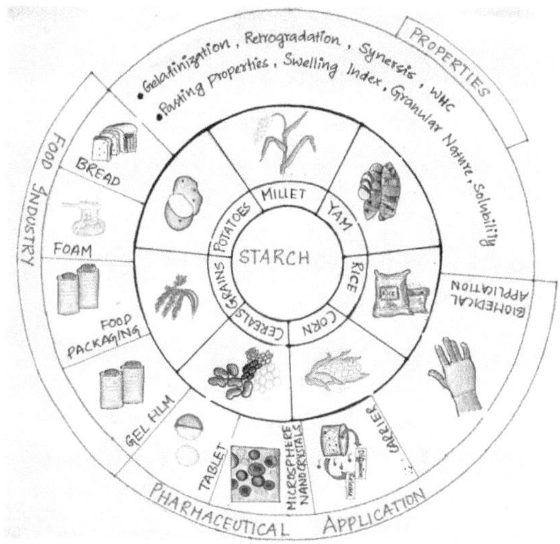

FIGURE 8.1 Graphical abstract.

The fabrications of starch based materials with incorporation of several co-materials rapidly provide to improve the mechanical behavior, so that utilization of the packaging materials can be increased abundantly in diverse fields. Involvement of the natural starch is also abundant in formulation of different food products. Utilization of native starch and modified starch in the preparation of different food products is considered as crucial ingredients. Starch is available in three different forms, including RDS, SDS, and RS, where RDS stands for rapidly digestible starch, SDS stands for slowly digestible starch, and RS stands for resistant starch. RDS and RS types of starch have lower levels of glycemic index, due to which the prepared food products also good candidates for consumption (Garcia et al., 2018). The preparation of thickening agents and thermolabile films by incorporating native starch is currently increasing, and the biocompatible nature of these prepared films with the biological and environmental microorganisms is remarkable (Li & Zhu, 2018b). From the previous report, chapatti was prepared by conjoining the finger millet with wheat millet. The prepared chapatti was having less rapidly digestible starch and higher amount of slowly digestible and resistant starch. Due to an increase of resistant starch content, this chapatti is a good candidate for the diabetic people. It is also reported that the composite chapatti showed the higher content of bioactive compounds like phenolic, flavonoid and antioxidant (Sharma et al., 2017).

Preparation of starch-based films like corn and cassava starch films are abundantly promoting in food application and also striking the customer demand in the market. The production of starch biofilms readily withdrawing the synthetic starch films from the market. Biopolymers like starch and its modified forms used to produce potential and functional attributes by enhancing their elasticity, elongation and tensile strength. Mechanical and barrier properties of the starch-based materials mainly depend on the starch content that used during the time of formulation. The concentration of the starch and their derivatives in the starch-based preparation provide the information regarding thickness of the materials (Luchese et al., 2017). During the manufacturing of thermoplastic films by using several modifiers along with native starch produced better product that produced more functional characteristics. In current research, several new approaches the packaging materials are prepared by crosslinking starch with different modifiers like glycerol, sorbitol, etc., which produces more thermo-labile packaging materials (Galdeano et al., 2009; Galdeano et al., n.d.). Application of the thermoplastic starch foam prepared by incorporating starch and other excipients able to replace the expanded polystyrenes in the food industry for food packaging purposes. Preparation of thermoplastic films based on the native starch is found to be very effective in order to provide the potential mechanical and barrier activity, which make the prepared films to fit for using in the food packaging or others packaging field. Compared to synthetic polymers or synthetic plastic packaging materials, starch-based products have beneficial and effective applications in food industries. Biocompatibility and biodegradability of natural and modified starches are considered as the main advantage over the synthetic plastics like polystyrene. Several studies were reported that starch-based packaging materials are usually having strong elongation property and enough tensile power to use in the field of food packaging industry (Bergel et al., 2017).

8.4 APPLICATION IN PHARMACEUTICAL FIELD

The applicability of natural starch in the pharmaceutical field is increasing tremendously due to its unique morphological and physicochemical behavior. Specifically, in the tablet formulation, starch is considered the choice of polymer that facilitates the release of active drugs from the tablet compact. In tablet formulation, the use of disintegrants always facilitates the faster breakdown of the tablet compact and the drug in solution form for the absorption process and display pharmacological response (Pachuau et al., 2017). The application of starch is also abundant in the field of drug discovery system. The excellent drug carrier properties of starch particles are used as effective biopolymers in the field of drug discovery and drug formulation and development. Significant applicability of starch in this field is due to its biodegradable nature and biocompatible in the biological system. Several studies displayed starches are being used in the pharmaceutical fields as binding agent, disintegrants agents, fillers or excipients (Chakraborty et al., 2018). Preparation of paracetamol tablet by using starch as disintegrants at different concentration level found to produce significant disintegration of the tablet. It was also reported that using starch and modified starch as disintegrant in tablet formulation produce significant improvement in dissolution of the tablets and increased the bio-efficacy of API (active pharmaceutical ingredients). Bio-compatibility and bio-degradability of the starch within the biological system or physiological system is unique and natural and modified starches are not generating any kinds of toxic effects in biological system. Dissolution efficiency and disintegration efficiency was found to be higher for natural starch isolated from the taro compared to corn standard starch (Pachuau et al., 2018).

A micromeritics property of starch powder is considered as the one of the finest property in order to achieve the unique formulation. Depending on the flow properties of the starch granules, compressibility and compatibility can be determined. To determine the flow properties of the starch granules, different parameters like Hausner's ratio, Carr's index, angle of repose, tapped density, bulk density, and porosity were used to measure. Morphological properties of the starch granules, like particle size, shape, elongation behavior, etc., are influenced by the flow properties of the starch granules. Granules are having enough good flow properties produce elegant tablet formulation, so that uniform disintegration and drug dissolution occurred to release the drugs in the medium (Manek et al., 2012; Pachuau et al., 2017). Formulation of micro- and nano-formulation by using starch is an important tool in current research field. The compatibility of starch with drug molecules is significant and also potential. Several natural starches like corn, maize starch is using to formulate nanocrystals, nanoparticles, nanowhishkers, nanotubes, etc. (Bel Haaj et al., 2016).

Targeted drug delivery system (DDS) current is considered an efficient DDS where natural starch granules are used to deliver the drug at the target site. In earlier research, natural starch was used as a drug carrier in the formulation of anticancer drug formulation, where starch was used as a polymer to release 5-fluorouracil at the target site of the body (Ahmed et al., 2012a). Ahmed et al. (2012b) reported that the compatibility index of native starch with other ingredient are very fruitful, therefore starch granules are compatible with the different types of pharmaceutical active ingredients and do not interfere with the quality of the drugs. Even the compressibility index is also effective for native starch while studied mixing with other ingredients (Ahmed et al., 2012b). The bio-adhesion property of native starch used to be very potential in terms of increase the bioadhesive characteristics of drugs molecules. Prepared bioadhesive microsphere with natural starch produced significant release of drug from the formulation. The entrapment efficiency of the prepared microsphere was significant and release study was also effective (Ahmed et al., 2012c).

8.5 CONCLUSION

The usefulness of starch based biodegradable materials in current time is attacking the researcher's interest and also customer compliance. Utilization of the starch-based materials and starch system is abundant due to its several functional aspects. Relation of excipients with starch system is very compatible, and provides the beneficial effect to the foods while packaging. In future research researchers can put their interest to build various functional materials and their proper use to replace the synthetic and semisynthetic plastic materials. The pollution and the negative impact of the plastic materials in worldwide is observed and considering as the dangerous factor. Starch is a native biopolymer or novel material, which is available abundantly in nature, and also provides the efficacy and safety measurement to the food substances. In current times, the starch may be subjected to a combination of different modifications to achieve desirable functional properties. Physical, chemical, and enzymatic modification starches will attract more attention in the future because the modification processes are chemical-free or environment-friendly, and the modified starches possess unique properties.

KEYWORDS

- **biodegradable materials**
- **biopolymer**
- **disintegrants**
- **gelatinization**
- **packaging materials**
- **pharmaceutical applications**
- **pollution**
- **starch**

REFERENCES

Ahmad, F., & Williams, P., (1999). Effect of salts on the gelatinization and rheological properties of sago starch. *Journal Of Agricultural and Food Chemistry*, *47*(8), 3359–3366. doi: 10.1021/jf981249r.

Ahmad, M. Z., Akhter, S., Ahmad, I., et al., (2012c). *In vitro* and *in vivo* evaluation of Assam bora rice starch-based bioadhesive microsphere as drug carrier for colon targeting. *Expert. Opin. Drug. Deliv., 9,* 141–149. https://doi.org/10.1517/17425247.2012.633507.

Ahmad, M. Z., Akhter, S., Anwar, M., et al., (2012a). Assam bora rice starch-based biocompatible mucoadhesive microsphere for targeted delivery of 5-fluorouracil in colorectal cancer. *Mol. Pharmaceutics, 9*, 2986–2994. https://doi.org/10.1021/mp300289y.

Ahmad, M. Z., Akhter, S., Anwar, M., et al., (2012b). Compatibility and compressibility studies of Assam Bora rice starch. *Powder Technol., 224*, 281–286. http://dx.doi.org/10.1016/j.powtec.2012.03.004.

Avella, M., De Vlieger, J., Errico, M., Fischer, S., Vacca, P., & Volpe, M., (2005). Biodegradable starch/clay nanocomposite films for food packaging applications. *Food Chemistry*, *93*(3), 467–474. doi: 10.1016/j.foodchem.2004.10.024.

Bajer, D., Janczak, K., & Bajer, K., (2020). Novel starch/chitosan/aloe vera composites as promising biopackaging materials. *Journal Of Polymers and The Environment*, *28*(3), 1021–1039. doi: 10.1007/s10924-020-01661-7.

Bel, H. S., Thielemans, W., Magnin, A., & Boufi, S., (2016). Starch nanocrystals and starch nanoparticles from waxy maize as nano reinforcement: A comparative study. *Carbohydrate Polymers*, *143*, 310–317. doi: 10.1016/j.carbpol.2016.01.061.

Bergel, B., Da Luz, L., & Santana, R., (2017). Comparative study of the influence of chitosan as coating of thermoplastic starch foam from potato, cassava, and corn starch. *Progress in Organic Coatings*, *106*, 27–32. doi: 10.1016/j.porgcoat.2017.02.010.

Bogracheva, T., Morris, V., Ring, S., & Hedley, C., (1998). The granular structure of C-type pea starch and its role in gelatinization. *Biopolymers, 45*(4), 323–332. doi: 10.1002/(sici)1097-0282(19980405)45:4<323::aid-bip6>3.0.co;2-n.

Chae, Y., & An, Y., (2018). Current research trends on plastic pollution and ecological impacts on the soil ecosystem: A review. *Environmental Pollution, 240*, 387–395. doi: 10.1016/j.envpol.2018.05.008.

Chakraborty, R., Kalita, P., & Sen, S., (2019). Natural starch in biomedical and food industry: Perception and overview. *Current Drug Discovery Technology, 16*(4), 355–367. doi: 10.2174/1570163815666181003143732.

Cheng, J., Wang, H., Kang, S., Xia, L., Jiang, S., Chen, M., & Jiang, S., (2019). An active packaging film based on yam starch with eugenol and its application for pork preservation. *Food Hydrocolloids, 96*, 546–554. doi: 10.1016/j.foodhyd.2019.06.007.

Colussi, R., Pinto, V., El Halal, S., Biduski, B., Prietto, L., & Castilhos, D., et al., (2017). Acetylated rice starches films with different levels of amylose: Mechanical, water vapor barrier, thermal, and biodegradability properties. *Food Chemistry, 221*, 1614–1620. doi: 10.1016/j.foodchem.2016.10.129.

Crespy, D., Bozonnet, M., & Meier, M., (2008). 100 years of Bakelite, the material of a 1000 uses. *Angewandte Chemie International Edition, 47*(18), 3322–3328. doi: 10.1002/anie.200704281.

Dankar, I., Haddarah, A., Omar, F., Pujolà, M., & Sepulcre, F., (2018). Characterization of food additive-potato starch complexes by FTIR and X-ray diffraction. *Food Chemistry, 260*, 7–12. doi: 10.1016/j.foodchem.2018.03.138.

Das, D., Jha, S., & Kumar, K., (2015). Isolation and release characteristics of starch from the rhizome of Indian Palo. *International Journal of Biological Macromolecules, 72*, 341–346. doi: 10.1016/j.ijbiomac.2014.08.009.

Do Evangelho, J., Da Silva, D. G., Biduski, B., El Halal, S., Kringel, D., & Gularte, M., et al., (2019). Antibacterial activity, optical, mechanical, and barrier properties of corn starch films containing orange essential oil. *Carbohydrate Polymers, 222*, 114981. doi: 10.1016/j.carbpol.2019.114981.

Domene-López, D., García-Quesada, J., Martin-Gullon, I., & Montalbán, M., (2019). Influence of starch composition and molecular weight on physicochemical properties of biodegradable films. *Polymers, 11*(7), 1084. doi: 10.3390/polym11071084.

Galdeano, M., Mali, S., Grossmann, M., Yamashita, F., & García, M., (2009). Effects of plasticizers on the properties of oat starch films. *Materials Science and Engineering: C, 29*(2), 532–538. doi: 10.1016/j.msec.2008.09.034.

Galdeano, M., Wilhelm, A., Mali, S., & Grossmann, M., (2013). Influence of thickness on properties of plasticized oat starch films. *Brazilian Archives of Biology and Technology, 56*(4), 637–644. doi: 10.1590/s1516-89132013000400014.

Garcia, V., Borges, J., Maciel, V., Mazalli, M., Lapa-Guimaraes, J., Vanin, F., & De Carvalho, R., (2018). Gelatin/starch orally disintegrating films as a promising system for vitamin C delivery. *Food Hydrocolloids, 79*, 127–135. doi: 10.1016/j.foodhyd.2017.12.027.

Gibb, B., (2019). Plastics are forever. *Nature Chemistry, 11*(5), 394–395. doi: 10.1038/s41557-019-0260-7.

Hemamalini, T., & Giri, D. V., (2018). Comprehensive review on electrospinning of starch polymer for biomedical applications. *International Journal of Biological Macromolecules, 106*, 712–718. doi: 10.1016/j.ijbiomac.2017.08.079.

Hoyos-Leyva, J., Alonso-Gomez, L., Rueda-Enciso, J., Yee-Madeira, H., Bello-Perez, L., & Alvarez-Ramirez, J., (2017). Morphological, physicochemical and functional characteristics of starch from *Marantha ruiziana* Koern. *LWT – Food Science and Technology*, *83*, 150–156. doi: 10.1016/j.lwt.2017.05.019.

Jan, K., Panesar, P., Rana, J., & Singh, S., (2017). Structural, thermal, and rheological properties of starches isolated from Indian quinoa varieties. *International Journal of Biological Macromolecules*, *102*, 315–322. doi: 10.1016/j.ijbiomac.2017.04.027.

Jane, J., Chen, Y., Lee, L., McPherson, A., Wong, K., Radosavljevic, M., & Kasemsuwan, T., (1999). Effects of amylopectin branch chain length and amylose content on the gelatinization and pasting properties of starch. *Cereal Chemistry Journal*, *76*(5), 629–637. doi: 10.1094/cchem.1999.76.5.629.

Kaur, K., Jindal, R., Maiti, M., & Mahajan, S., (2019). Studies on the properties and biodegradability of PVA/Trapa natans starch (N-st) composite films and PVA/N-st-g-poly (EMA) composite films. *International Journal of Biological Macromolecules*, *123*, 826–836. doi: 10.1016/j.ijbiomac.2018.11.134.

Kraak, A., (1992). Industrial applications of potato starch products. *Industrial Crops and Products*, *1*(2–4), 107–112. doi: 10.1016/0926-6690(92)90007-i.

Kulshreshtha, Y., Schlangen, E., Jonkers, H., Vardon, P., & Van, P. L., (2017). CoRncrete: A corn starch based building material. *Construction And Building Materials*, *154*, 411–423. doi: 10.1016/j.conbuildmat.2017.07.184.

Kumar, P., (2018). Role of plastics on human health. *The Indian Journal of Pediatrics*, *85*(5), 384–389. doi: 10.1007/s12098-017-2595-7.

Leon-Bejarano, M., Durmus, Y., Ovando-Martínez, M., & Simsek, S., (2020). Physical, barrier, mechanical, and biodegradability properties of modified starch films with nut by-products extracts. *Foods*, *9*(2), 226. doi: 10.3390/foods9020226.

Li, C., Oh, S., Lee, D., Baik, H., & Chung, H., (2017). Effect of germination on the structures and physicochemical properties of starches from brown rice, oat, sorghum, and millet. *International Journal of Biological Macromolecules*, *105*, 931–939. doi: 10.1016/j.ijbiomac.2017.07.123.

Li, G., & Zhu, F., (2018). Effect of high pressure on rheological and thermal properties of quinoa and maize starches. *Food Chemistry*, *241*, 380–386. doi: 10.1016/j.foodchem.2017.08.088.

Li, G., & Zhu, F., (2018). Quinoa starch: Structure, properties, and applications. *Carbohydrate Polymers*, *181*, 851–861. doi: 10.1016/j.carbpol.2017.11.067.

Limonta, G., Mancia, A., Benkhalqui, A., Bertolucci, C., Abelli, L., Fossi, M., & Panti, C., (2019). Microplastics induce transcriptional changes, immune response, and behavioral alterations in adult zebrafish. *Scientific Reports*, *9*(1). doi: 10.1038/s41598-019-52292-5.

Liu, H., Ramsden, L., & Corke, H., (1999). Physical properties and enzymatic digestibility of hydroxypropylated Ae, Wx, and normal maize starch. *Carbohydrate Polymers*, *40*(3), 175–182. doi: 10.1016/s0144-8617(99)00052-1.

Luchese, C., Abdalla, V., Spada, J., & Tessaro, I., (2018). Evaluation of blueberry residue incorporated cassava starch film as pH indicator in different simulants and foodstuffs. *Food Hydrocolloids*, *82*, 209–218. doi: 10.1016/j.foodhyd.2018.04.010.

Luchese, C., Spada, J., & Tessaro, I., (2017). Starch content affects physicochemical properties of corn and cassava starch-based films. *Industrial Crops and Products*, *109*, 619–626. doi: 10.1016/j.indcrop.2017.09.020.

Malmir, S., Montero, B., Rico, M., Barral, L., Bouza, R., & Farrag, Y., (2018). Effects of poly (3-hydroxybutyrate-co-3-hydroxyvalerate) microparticles on morphological, mechanical, thermal, and barrier properties in thermoplastic potato starch films. *Carbohydrate Polymers*, *194*, 357–364. doi: 10.1016/j.carbpol.2018.04.056.

Manek, R., Builders, P., Kolling, W., Emeje, M., & Kunle, O., (2012). Physicochemical and binder properties of starch obtained from cyperus esculentus. *AAPS Pharmscitech*, *13*(2), 379–388. doi: 10.1208/s12249-012-9761-z.

Maniglia, B., & Tapia-Blácido, D., (2016). Isolation and characterization of starch from babassu mesocarp. *Food Hydrocolloids*, *55*, 47–55. doi: 10.1016/j.foodhyd.2015.11.001.

Maurer, H., & Kearney, R., (1998). Opportunities and challenges for starch in the paper industry. *Starch – Stärke*, *50*(9), 396–402. doi: 10.1002/(sici)1521-379x(199809)50:9<396::aid-star396>3.0.co;2-8.

Morrison, W., Milligan, T., & Azudin, M., (1984). A relationship between the amylose and lipid contents of starches from diploid cereals. *Journal Of Cereal Science*, *2*(4), 257–271. doi: 10.1016/s0733-5210(84)80014-4.

Nagar, M., Sharanagat, V., Kumar, Y., & Singh, L., (2019). Development and characterization of elephant foot yam starch–hydrocolloids based edible packaging film: Physical, optical, thermal and barrier properties. *Journal of Food Science and Technology*, *57*(4), 1331–1341. doi: 10.1007/s13197-019-04167-w.

Nakthong, N., Wongsagonsup, R., & Amornsakchai, T., (2017). Characteristics and potential utilizations of starch from pineapple stem waste. *Industrial Crops and Products*, *105*, 74–82. doi: 10.1016/j.indcrop.2017.04.048.

Nandi, S., & Guha, P., (2018). Modeling the effect of guar gum on the physical, optical, barrier, and mechanical properties of potato starch-based composite film. *Carbohydrate Polymers*, *200*, 498–507. doi: 10.1016/j.carbpol.2018.08.028.

Nashed, G., Rutgers, R., & Sopade, P., (2003). The plasticization effect of glycerol and water on the gelatinization of wheat starch. *Starch – Stärke*, *55*(34), 131–137. doi: 10.1002/star.200390027.

Oleyaei, S., Almasi, H., Ghanbarzadeh, B., & Moayedi, A., (2016). Synergistic reinforcing effect of TiO_2 and montmorillonite on potato starch nanocomposite films: Thermal, mechanical and barrier properties. *Carbohydrate Polymers*, *152*, 253–262. doi: 10.1016/j.carbpol.2016.07.040.

Oleyaei, S., Zahedi, Y., Ghanbarzadeh, B., & Moayedi, A., (2016). Modification of physicochemical and thermal properties of starch films by incorporation of TiO_2 nanoparticles. *International Journal of Biological Macromolecules*, *89*, 256–264. doi: 10.1016/j.ijbiomac.2016.04.078.

Ozdemir, M., & Floros, J., (2004). Active food packaging technologies. *Critical Reviews in Food Science and Nutrition*, *44*(3), 185–193. doi: 10.1080/10408690490441578.

Pachuau, L., Dutta, R. S., Devi, T. B., Deka, D., & Lalhlenmawia, H., (2018). Taro starch (*Colocasia esculenta*) and citric acid modified taro starch as tablet disintegrating agents. *International Journal of Biological Macromolecules, 118*(Part-A), 397–405. https://doi.org/10.1016/j.ijbiomac.2018.06.086.

Pachuau, L., Dutta, R. S., Roy, P., Kalita, P., & Lalhlenmawia, H., (2017). Physicochemical and disintegrant properties of glutinous rice starch of Mizoram, India. *International Journal of Biological Macromolecules*, *95*, 1298–1304. doi: 10.1016/j.ijbiomac.2016.11.029.

Pachuau, L., Dutta, R., Devi, T., Deka, D., & Hauzel, L., (2018). Taro starch (*Colocasia esculenta*) and citric acid modified taro starch as tablet disintegrating agents. *International Journal of Biological Macromolecules*, *118*, 397–405. doi: 10.1016/j.ijbiomac.2018.06.086.

Pawar, H., (2014). An overview of natural polysaccharides as biological macromolecules: Their chemical modifications and pharmaceutical applications. *Biology And Medicine*, *7*(1). doi: 10.4172/0974-8369.1000224.

Phan, T. D., Debeaufort, F., Voilley, A., & Luu, D., (2009). Biopolymer interactions affect the functional properties of edible films based on agar, cassava starch and arabinoxylan blends. *Journal of Food Engineering*, *90*(4), 548–558. doi: 10.1016/j.jfoodeng.2008.07.023.

Przetaczek-Rożnowska, I., (2017). Physicochemical properties of starches isolated from pumpkin compared with potato and corn starches. *International Journal of Biological Macromolecules*, *101*, 536–542. doi: 10.1016/j.ijbiomac.2017.03.092.

Reddy, C., Choi, S., Lee, D., & Lim, S., (2018). Complex formation between starch and stearic acid: Effect of enzymatic debranching for starch. *Food Chemistry*, *244*, 136–142. doi: 10.1016/j.foodchem.2017.10.040.

Rochman, C., Hoh, E., Kurobe, T., & Teh, S., (2013). Ingested plastic transfers hazardous chemicals to fish and induces hepatic stress. *Scientific Reports*, *3*(1). doi: 10.1038/srep03263.

Santana, J., De Carvalho, C. É., Rodrigues, P., Correia, P., Cruz, R., & Druzian, J., (2018). Morphological, barrier, and mechanical properties of cassava starch films reinforced with cellulose and starch nanoparticles. *Journal Of Applied Polymer Science*, *136*(4), 47001. doi: 10.1002/app.47001.

Sarko, A., & Wu, H., (1978). The crystal structures of A-, B- and C-polymorphs of amylose and starch. *Starch – Stärke*, *30*(3), 73–78. doi: 10.1002/star.19780300302.

Scheurer, M., & Bigalke, M., (2018). Microplastics in Swiss floodplain soils. *Environmental Science & Technology*, *52*(6), 3591–3598. doi: 10.1021/acs.est.7b06003.

Sharma, B., Gujral, H., & Solah, V., (2017). Effect of incorporating finger millet in wheat flour on mixolab behavior, chapatti quality and starch digestibility. *Food Chemistry*, *231*, 156–164. doi: 10.1016/j.foodchem.2017.03.118.

Siracusa, V., Rocculi, P., Romani, S., & Rosa, M., (2008). Biodegradable polymers for food packaging: A review. *Trends in Food Science & Technology*, *19*(12), 634–643. doi: 10.1016/j.tifs.2008.07.003.

Suriyatem, R., Auras, R., & Rachtanapun, P., (2018). Improvement of mechanical properties and thermal stability of biodegradable rice starch–based films blended with carboxymethyl chitosan. *Industrial Crops and Products*, *122*, 37–48. doi: 10.1016/j.indcrop.2018.05.047.

Tabilo-Munizaga, G., & Barbosa-Cánovas, G., (2005). Rheology for the food industry. *Journal Of Food Engineering*, *67*(1-2), 147–156. doi: 10.1016/j.jfoodeng.2004.05.062.

Verwimp, T., Vandeputte, G., Marrant, K., & Delcour, J., (2004). Isolation and characterization of rye starch. *Journal of Cereal Science*, *39*(1), 85–90. doi: 10.1016/s0733-5210(03)00068-7.

Zhang, H., Zhang, J., Gao, Y., Wang, W., Dong, H., Hou, H., & Liu, X., (2017). Effect of modification extent of montmorillonite on the performance of starch nanocomposite films. *Starch – Stärke*, *69*(11, 12), 1700088. doi: 10.1002/star.201700088.

Zhang, K., Huang, T., Yan, H., Hu, X., & Ren, T., (2020). Novel pH-sensitive films based on starch/polyvinyl alcohol and food anthocyanins as a visual indicator of shrimp deterioration. *International Journal of Biological Macromolecules*, *145*, 768–776. doi: 10.1016/j.ijbiomac.2019.12.159.

Zhang, Y., Hu, M., Zhu, K., Wu, G., & Tan, L., (2018). Functional properties and utilization of *Artocarpus heterophyllus* Lam seed starch from new species in China. *International Journal of Biological Macromolecules*, *107*, 1395–1405. doi: 10.1016/j.ijbiomac.2017.10.001.

Zhao, R., Torley, P., & Halley, P., (2008). Emerging biodegradable materials: Starch- and protein-based bio-nanocomposites. *Journal of Materials Science*, *43*(9), 3058–3071. doi: 10.1007/s10853-007-2434-8.

CHAPTER 9

Semiconductor/Semiconductor Core/Shell Nanomaterial: A Versatile Novel Material

PALLABI BORO, S. RIJUWANA HAQUE, and SUPARNA BHATTACHARJEE

Department of Applied Sciences, Gauhati University, Guwahati, Assam, India

ABSTRACT

The present work includes a brief discussion regarding Semiconductor/Semiconductor Core/Shell Nanomaterials, which is an important class of Novel Materials in recent times. As an example of such type of nanomaterials, a discussion on the synthesis, characterization, and application of CdS/ZnS core/shell nanostructure is included in the chapter. Novel material refers to materials such as superconducting materials, smart materials, new-type composite materials, magnetic materials, etc., which can exhibit unique physical, chemical, optical, and other properties. Nowadays, researchers in the field of Science and Technology are concentrating on developing newer and newer novel materials so as to apply these materials for the progress of human civilization as well as for industrial upgrading. Semiconductor/Semiconductor Core/Shell Nanomaterials are one class of such novel materials which find widespread applications in varied fields. Whenever a core (inner material) is surrounded by a shell (outer layer material), then the structure is known as the core/shell structure. If both the

Advances in Manufacturing and Processing of Materials: Characterization and Applications.
Narendra Kumar, PhD, Mohit Tyagi, PhD, Dilbagh Panchal, PhD & Ravi Pratap Singh, PhD (Eds.)

inner and the outer materials are semiconducting nanomaterials, then the resultant structure is known as Semiconductor/Semiconductor Core/Shell Nanostructure. Core/shell nanomaterials may be of different shapes such as spherical, prism, octahedron, wire, rod, cube, tube, etc. Due to the shell boundary over the simple core material, the core/shell material exhibits improved optical properties over the core material which increases the stability and photoluminescence efficiency of the material. Moreover, the core/shell nanomaterials exhibit higher quantum yields and greater stability compared to the core material. CdS/ZnS is one example of semiconductor/semiconductor core/shell material belonging to group II-VI. CdS/ZnS materials can be synthesized by many synthesizing techniques such as chemical precipitation, sol-gel technique, microemulsion, polymerization, inverse micelle formation, etc. The characterization techniques which are used to study the optical properties of the CdS/ZnS core/shell materials are Optical Absorption Spectroscopy (OAS), Energy Dispersive X-ray Spectroscopy (EDS), Photoluminescence (PL) spectroscopy, Raman Spectroscopy, Fourier Transform Infrared Spectroscopy (FTIR), etc. Moreover, to study the structural properties, some techniques are used which are X-ray Diffraction (XRD) study, Scanning Electron Microscopy (SEM), Transmission Electron Microscopy (TEM), Small angle X-Ray scattering (SAXS), Atomic Force Microscopy (AFM), etc. CdS/ZnS core/shell nanomaterials can also be used in designing solar cells of low cost, long life, and good stability. The designed solar cells based on CdS/ZnS core/shell nanomaterials have been widely concerned in recent years. Moreover, in many other fields, CdS/ZnS nanomaterials are widely used, such as light-emitting diodes, cell imaging, biosensors, and many others. A wide range of colors can be studied combining the modularity of core/shell semiconductor nanocrystals and stability of organic polymer. For example, CdS/ZnS core/shell semiconductors are used to generate violet to blue colors.

9.1 INTRODUCTION

Novel material is a term that refers to a certain class of materials which have already been developed or are in the developing state. Such materials exhibit unique physical, optical, chemical, and other properties (https://nanomaterials.alfa-chemistry.com/products/novel-materials.html).

Nanomaterials are one of the best examples of novel materials, which are extensively studied nowadays. Among the nanomaterials, quantum dots (QDs) have many applications in varied fields. The materials whose size lies in a range of 1–100 nm are called nanomaterials. The term QD refers to semiconductor nanocrystals with a volume smaller than the volume defined by the exciton Bohr radius of a particular semiconductor. Hence, QDs possess distinctive properties that are dissimilar to their respective bulk materials. Core/shell QDs are developed structures of QDs for further engineering (Mirnajafizadeh & Stride, 2020). Because of the shell boundary over the simple QD core, the core/shell QD exhibits improved optical properties over simple QDs, which increase the stability and photoluminescence (PL) efficiency of the material. A QD has unsaturated bonds or dangling bonds on the surface, and as a result, they possess a large surface area to volume ratio. If an appropriate material is deposited over a simple QD, then the structure of core/shell QDs is formed, and these final materials exhibit higher quantum yield and greater stability than the core QD. Furthermore, the shell protects the core material from oxidation and chemical degradation (Mirnajafizadeh & Stride, 2020). This is why core/shell structures are being increasingly studied day by day.

Semiconducting novel nanomaterials belonging to group II-VI and exhibits wide bandgap. This gap becomes even wider in the nano-scale dimension. These materials are useful for application in devices which operate in the optical range from UV to visible region. Therefore, now a day, the synthesis of core-shell nanostructures is finding much attention amongst there searchers. Core-shell semiconducting materials exhibit properties which are different from the properties of small molecules and also massive crystalline semiconductors. These core/shell nanocrystals are the structures where one QD core is coated with a distinct semiconductor shell material. The resulting core/shell structures are typically composed of Type II-VI, IV-VI, and III-V semiconductors that have configurations like CdS/ZnS, CdSe/ZnS, CdSe/CdS, etc. For example, CdS/ZnS core/shell nanomaterials of type II-VI have wide applications in varied fields. The bandgap value for Core CdS is 2.42 eV and shell ZnS nanoparticles is 3.68 eV. When ZnS is deposited as a shell material over the CdS core, a core/shell structure possessing a definite band alignment is resulted. In this chapter, as an example of semiconductor/semiconductor core/shell nanomaterial, we have presented a discussion on the synthesis, characterization, properties, and applications of CdS/ZnS core/shell nanoparticles.

As such, the properties of zinc sulfide (ZnS) and cadmium sulfide (CdS) are briefly discussed in subsections.

9.1.1 CDS NANOPARTICLES

CdS is a semiconductor material which belongs to group II-VI of periodic table, and it has a direct bandgap of 2.42 eV. It is yellow in color. The main application of CdS nanoparticles is as a pigment. CdS nanoparticles exist in two changeable structures, namely cubic hawleyite and hexagonal greenockite. It finds applications in the production of solar cell and is used as the buffer layer in the manufacture of CIGS (Copper–Indium–Gallium–Selenide) (https://bradscholars.brad.ac.uk/handle/10454/16720). CdS nanoparticles can be used in electronic and opto-electronic devices (Salata et al., 1996).

9.1.2 ZNS NANOPARTICLES

ZnS is also a group II-VI semiconductor material possessing a direct bandgap of 3.68 eV. Because of its chemical stableness compared to the chalcogenide materials, these nanoparticles are greatly used as phosphor in PL, in optical sensors and in electroluminescence (EL). Also, it exhibits more optical transparency from the visible region to the infrared region and due to this reason ZnS semiconductor nanoparticles are known as the most promising material among II-VI groups (Vishwakarma, 2013).

9.1.3 CORE/SHELL NANOPARTICLES

The characteristics of semiconductor nanocrystals vary with the relative alignment of conduction and valence band edges for both the core and the shell materials. Based on the properties of semiconductor nanomaterial core/shell nanoparticles are classified into as Type-I, Type-II, and reverse Type-I semiconductors. In type-I semiconducting core/shell nanomaterials, the core comprises of electrons and holes. In Type-II semiconducting core/shell nanomaterials, one type of carriers is confined to the shell while the other is confined to the core. In the formation of Type-I core/shell nanoparticles, the core material's bandgap is smaller as compared to the shell material's bandgap. The offsets between the conduction band and the valence band are such that the energy of the shell's conduction band is lower than that of the core. This effectively traps the electrons and holes in the core material (Dabhane et al., 2021). Examples of materials in this class are CdSe/CdS, CdSe/ZnS, InAs/CdSe, etc. CdS and ZnS nanomaterials have a direct band gap of 2.42 eV and 3.68 eV, respectively. CdS/ZnS nanomaterial corresponds to Type-I core/shell nanomaterials due to the reason that the bandgap of ZnS shell nanomaterial is larger than the band gap of the core CdS nanomaterial. In the reverse Type-I core/shell structure, the energy band gap value of the core is wider than the energy band gap value of the shell material, and for the shell material, the conduction and the valence band edges of the shell lie within those of the core. The emission wavelength can be adjusted by changing the shell thickness of the material. For example, CdS/HgS, CdS/CdSe, and ZnSe/CdSe belong to this class of materials. In Type-II core/shell structure both the energy of both conduction band and the valence band of the core are lower in energy (or higher) than those of the shell. Example of this class of material is ZnTe/CdSe, CdTe/CdSe, CdS/ZnSe, etc. (https://en.wikipedia.org/wiki/Core%E2%80%93shell_semiconductor_nanocrystal).

9.1.4 BAND GAP ALIGNMENT OF TYPE-I, TYPE-II CORE/SHELL NANOMATERIALS

Figure 9.1(a) shows the alignment of Type-I semiconductor nanomaterials' bandgap. In type-I semiconductor core/shell nanomaterial, the band gap of shell material is larger than the core material. Moreover, the edges of the conduction and valence bands of the core material are within the energy gap of the shell material, resulting in the captivity of electrons and holes in the inner region. This arrangement is important for confining electrons and holes in the core material. As a result, the band gap energy (Eg1) of the core material determines the emission energy (ħωPL) of the core material. Moreover, the valence band of the shell material is at lower energy level compared to the core material. As such, when a layer of the shell material is deposited on the top of the core material, the overall optical properties of the core material increase. However, as the thickness of the shell material increases, the surface activity of the core reduces. This also reduces quantum yield. Some Type-I semiconductor/semiconductor nanomaterials are such as CdTe/CdS, CdSe/CdS, CdSe/ZnS, etc. In Type-II semiconductor/semiconductor nanomaterials, both the valence and conduction band of the core material can be lower or higher than that of the shell material. Due to this reason, one carrier is primarily confined to the shell, and the other one is primarily confined to the core material. This Type-II core/shell nanomaterials is expected to have many more novel properties like – the decay time of exciton a bit longer as compared to Type-I core/shell nanomaterial due to electron and hole separation. The band gap energy Eg12 of the material can be determined between the valence band edge of one semiconductor and the conduction band edge of another. If Uc and Uv are the band energy offsets at the juncture then, Eg12 can be related by the equation as follows:

$$Eg12 = Eg1 - Uv = Eg2 - Uc \tag{1}$$

where; Eg1 and Eg2 represents energy bandgap value of each semiconductor.

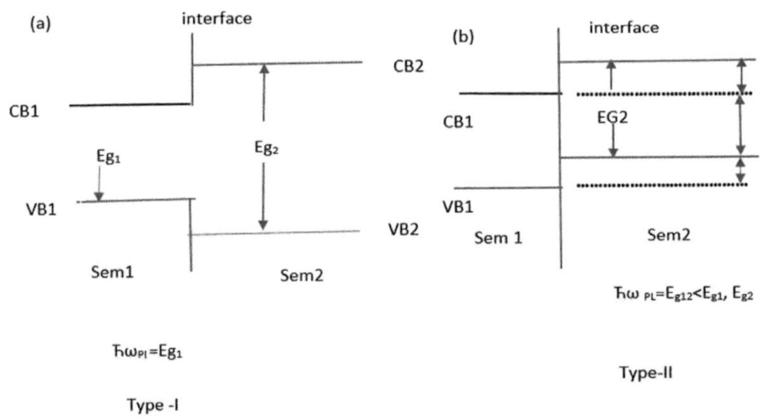

FIGURE 9.1 Bandgap alignment of (a) Type-I; and (b) Type-II core/shell semi-conductor nanomaterials.

9.2 CLASSIFICATION OF CORE/SHELL NANOPARTICLES

Based on material properties core/shell nanoparticles are classified into four main groups, which are – (i) inorganic/inorganic; (ii) inorganic/organic; (iii) organic/inorganic; and (iv) organic/organic.

9.2.1 INORGANIC/INORGANIC CORE/SHELL NANOPARTICLES

Inorganic/inorganic nanoparticles are the most important class among all other types of core/shell nanoparticles. Some inorganic/inorganic nanoparticles are CdS/PbS, CdS/CdSe, CdS/HgS, CdS/ZnS, CdS/HgS/CdS, CdSe/CdS, CdSe/ZnS, etc. These nanoparticles find wide applications in information storage, semiconductor efficiency, optical imaging, biological imaging, etc.

9.2.2 INORGANIC/ORGANIC CORE/SHELL NANOPARTICLES

Whenever a shell is a polymer or any other massive material, then to form this type of core/shell structure this material can be deposited over a core material which may be metallic compound, metal, or metal oxide. The advantage of organic coating over the inorganic coating is that the surface atoms of the inner material can be oxidized to the metal oxide in an NTP. From catalysts to additives, these materials have many applications, such as in pigments treatment, paints, cosmetics, inks, etc. Examples of inorganic/organic core/shell nanoparticles are SnO_2/graphene, CdS/thiol, Mn-doped CdSe/hexa decylamine, CdS/PMMA, etc.

9.2.2.1 ORGANIC/INORGANIC CORE/SHELL NANOPARTICLES

In these type of nanomaterials cores is polystyrene, poly(ethyleneoxide), polyurethane, poly(vinyl benzyl chloride), poly(vinyl pyrrolidone), dextrose, surfactant, and others various copolymers such as acrylonitrile-butadiene-styrene, poly(styrene-acrylic acid) and styrene-methyl methacrylate. These core/shell nanoparticles can show the dual properties of both organic and inorganic materials. The metal oxide which coats inorganic materials over an organic material can increase the strength of entire material and also increase the thermal and colloidal stability. Examples of organic/Inorganic core/shell materials are PS/Ag (PS-polystyrene), PS/SiO_2, PS/TiO_2, PS/Fe_3O_4, PS/ZnS, PVP/SiO_2, etc.

9.2.2.2 ORGANIC/ORGANIC CORE/SHELL NANOPARTICLES

Whenever both the core and the shell materials are made of organic materials then these structures are known as the organic/organic core/shell materials, and they find applications in many fields. Examples: biomaterials, chemical sensing, chemical separation, etc. Some Organic/Organic core/shell nanoparticles are such as PS/NIPA, PS/PTBA, TdG_2/PS-PEO, PNIPAM/CMCS, etc.

9.3 SYNTHESIS PROCESS OF CDS/ZNS CORE/SHELL (TYPE-I) NANOMATERIALS

As an example, we discuss below the synthesis processes, characterization techniques and application areas of CdS/ZnS nanoparticles briefly: The core/shell CdS/ZnS nanomaterial can be synthesized by various wet chemical methods such as chemical precipitation, sol-gel technique, microemulsion, polymerization, inverse micelle formation, etc. On the basis of the properties of the core and shell material, the synthesis process, properties, and morphology of the material can be modified. For example, in the year 2005, Oladeji et al. CdS/ZnS core/shell nanomaterials were synthesized using chemically and thermally activated cold diffusion method (Oladeji & Chow, 2005). From secondary ion mass spectrometry (SIMS), they found that Cd is more likely to diffuse into the ZnS film than Zn is to diffuse into the CdS film. CdS/ZnS nanoparticles were synthesized by Ethayaraja et al. in the year 2007, in a series of water in oil microemulsion solution with the increase of microemulsion drop sizes (Ethayaraja et al., 2007). They confirmed that the shell material was formed on the core CdS nanoparticles based on the redshift of the UV-visible absorption spectrum. The diameter of the synthesized CdS/ZnS nanoparticles was in arrange of 3–8 nm. They confirmed that the size of core/shell nanoparticles increases as the water to surfactant molar ratio (R) increases. Soltani et al. in the year 2012 synthesized ZnS/CdS and CdS/ZnS nanoparticles via two steps using microwave irradiation (Maleki et al., 2007). Core nanoparticles were prepared first. In the second step, shell nanoparticles were added over the core nanoparticles.

They controlled the thickness of shell material by adjusting the concentration of core material. From the UV-vis spectrum, they show a shifting towards lower wavelength in the absorption edge of CdS/ZnS nanoparticles compared to CdS (core) nanoparticles while for absorption edge of ZnS/CdS nanoparticles shows red shift compared to ZnS(core) nanoparticles. From PL analysis they observed that the emission peaks for CdS/ZnS and ZnS/CdS nanoparticles were in the 400–650 nm range. In the year 2013, S

Agarwal et al. synthesized CdS/PS and ZnS/PS nanoparticles using solution casting method varying wt.% of (CdS) and ZnS (Agarwal, Patidar, & Saxena, 2013). The Effective thermal conductivity of polymer nanocomposites was measured from room temperature to 150°C using transient plane source method. They found that thermal conductivity increased up to 4 wt.% for CdS/ZnS nanoparticles but found to decrease for 6 and 8 wt.% for CdS/ZnS nanoparticles. H Kumar et al. in the year 2014 (Kumar, Barman, & Raj, 2014) synthesized CdS and ZnS individually and finally ZnS was added to CdS Nanomaterials to get the final CdS/ZnS core/shell semiconductor materials. CV Reddy et al. also synthesized some core-shell nanoparticles in the year 2017 (Venkata, Shim, & Cho, 2017). They studied the optical, structural, and also photocatalytic applications of these nanoparticles. Many other characterization techniques were used to characterize the CdS/ZnS nanoparticles such as PXRD, TEM, HRTEM, FT-IR, Raman spectroscopy, BET analysis, etc. From BET analysis, they confirmed that CdS/ZnS nanoparticles have a larger surface area and pore diameter than individual CdS and ZnS nanoparticles. They studied degradation of methyl Orange (MO) using CdS/ZnS nanoparticles and found that CdS/ZnS nanoparticles exhibited more catalytic activity compared to pure CdS and ZnS nanoparticles. They confirmed the vibrational modes of CdS and ZnS NPs from RAMAN and FT-IR analysis.

9.4 DIFFERENT CHARACTERIZATION TECHNIQUES OF CDS/ZNS CORE/SHELL NOVEL NANOMATERIALS

There are many characterization techniques to study the optical, physical, electrical, etc., properties of core/shell novel materials. To characterize the synthesized II-VI semiconductor/semiconductor core/shell nanomaterials as for example CdS/ZnS, the following characterization techniques may be considered. Some of the techniques may be discussed briefly in subsections.

9.4.1 CHEMICAL CHARACTERIZATION TECHNIQUES

1. **Optical Absorption Spectroscopy (OAS):** One of the main characteristics of materials is that they absorb some portion of light

incident upon it. The amount of absorbed light and wavelength depends upon the chemical and the electronic structure of the material. The amount of absorption of the sample with respect to the wavelength is recorded by the absorption spectrometer. From this, we can calculate the value of bandgap, absorption coefficient and absorbance of the material. The following relation given by Tauc is used to estimate the bandgap value of a nanomaterial (https://onlinelibrary.wiley.com/doi/full/10.1002/pssb.201552007).

$$(\alpha hv)1/n = K(hv - Eg) \tag{2}$$

where α represents the absorption coefficient; h is Planck's constant; and v indicates the frequency of incident light; K is a constant; and Eg is the band gap of the material. For direct bandgap semiconductor nanomaterial n = 1/2. In our case, to estimate the band gape value of CdS/ZnS nanoparticles Eqn. (2) is used.

2. **Energy Dispersive X-Ray Spectroscopy (EDS):** One of the electron spectroscopic techniques is EDS which depends on the emission and ejection of photons (may be X-ray) or the electrons from the atoms. The chemical compositions of a material can be determined by estimating the energy of X-rays emitted by the corresponding material. An EDS spectrum is obtained from the EDS characterization, and it is a plot showing how frequently each energy level emits the X-ray. Corresponding to the energy level, this EDS range overall shows a few peaks for which beams of X-rays had been produced, and hence the single element corresponding to the peak is identified.
3. **Photoluminescence (PL) Spectroscopy:** It is a process where the material is excited with photon followed by photon emission. In this measurement, visible light is made to be incident on the material, and this light stimulates the material into an electronic state. In this photo-excitation process, excited state electrons have more energy than the electrons in the ground state. Electrons from the ground state of the valence band jump to the upper state of the conduction band during the photo-excitation process. These electrons have excess energy, and they lose it during the non-radiative relaxation process. These excited electrons have the probability of recombining with the holes in the valence band. In a pure sample,

one pair of electron and hole is known as the exciton. In impure or doped samples, there may be some recombination via acceptor or donor levels. Three possible transitions may occur in the case of doped samples, which are from donor level to valence band, conduction band to acceptor level, and from the donor to acceptor.

Few other techniques of chemical characterization of these materials are listed below:

- Raman spectroscopy;
- Fourier transform infrared spectroscopy (FTIR);
- Electron spectroscopy;
- Ionic spectrometry;
- Rutherford backscattering spectrometry (RBS);
- Secondary ion mass spectrometry (SIMS).

9.4.2 STRUCTURAL CHARACTERIZATION TECHNIQUES

1. **X-Ray Diffraction (XRD) Study:** XRD is an important characterization technique which is used to study the structural information of a crystallite material. Average crystallite of a material can be estimated using this technique. As we know X-rays are the electromagnetic radiation having the photon energy ranging from 100 eV to 100 keV and their wavelength lies in a range (0.1–1A°) which is about the size of an atom and hence it is ideally suitable for identification of the crystal structure at the atomic level.
   The average crystallite size D can be calculated using the Debye-Scherrer's formula which is as follows (Kumar et al., 2014):

$$D = \frac{K\lambda}{\beta Cos\theta} \tag{3}$$

   where; K is known as Scherrer's constant and it has a value equal to 0.9; λ represents the X-ray wavelength, full width at half maximum (FWHM) is indicated by β (in radians); and θ represents the Bragg's diffraction angle (in degrees).
2. **Scanning Electron Microscopy (SEM):** It is an electron microscopy technique. This technique creates an image of the sample,

scanning the sample with a focused electron beam and also detecting the backscattered or the secondary electrons. The resolution of the scanning electron microscope is a few nanometers, and it can operate at magnification ranges from ~10 to 1,00,000. It can give the topographical information of the sample, and moreover, it can also provide chemical composition information near the surface of the sample.

3. **Transmission Electron Microscopy (TEM):** It is an electron microscope which is very crucial to visualize a specimen and to get highly magnified image of the sample. Although TEM and light microscope operate on the same basic principle, the main difference between the two is that in order to magnify the images of the specimen, TEM use electrons rather than using light. It can magnify the specimen up to 2,000 times than normal light microscope. As the frequency of electron beam is larger, the images of the specimen with higher resolution can be obtained from TEM analysis. It can magnify something up to 2 million times.

Few other techniques of structural characterization of the materials are listed below:

- Small angle X-ray scattering (SAXS);
- Scanning probe microscopy (SPM);
- Scanning tunneling microscopy (STM);
- Atomic force microscopy (AFM).

9.5 DISCUSSION OF CHARACTERIZATION RESULTS TAKING CDS/ZNS NANOMATERIAL AS AN EXAMPLE

The results of UV-vis analysis, PL analysis, and XRD analysis of CdS/ZnS nanoparticles are discussed in subsections.

9.5.1 UV ANALYSIS

CdS is one of the semiconductor materials of group of II-VI semiconductor possessing a bulk band gap value of 2.42 eV. When reduced to nanoparticle form, the band gap of the material increases, which in turn confirms the formation of nanoparticles of CdS. This bandgap enhancement can be confirmed from UV visible spectroscopy results. In the UV visible spectra, a sudden abrupt change in the value of absorbance of the material occurs at a wavelength value corresponding to the energy bandgap of the material. This wavelength at which the abrupt change occurs is known as the absorption edge. The absorption edge for the nanomaterials is always blue-shifted corresponding to that of the bulk, which confirms the formation of nanomaterials. However, when a shell material is deposited over the core material, the particle size increases, and due to this, the absorption edges are red-shifted resulting in a corresponding decrease in band gap value. Such results were obtained by Reddy et al., Baláž et al., T Geremew et al., etc. We, the authors P Boro, S Rijuwana Haque, and S Bhattacharjee, synthesized the CdS/ZnS by chemical reaction method, adopting a one-pot synthesis technique. To synthesize CdS/ZnS nanoparticles first we synthesized CdS nanoparticles and then the precursor of ZnS nanoparticles was added to the CdS nanoparticles, under constant magnetic stirring so as to form CdS/ZnS nanoparticles. We used cadmium chloride ($CdCl_2$), sodium sulfide (Na_2S) and $ZnCl_2$ as Cd, S and Zn source, respectively. As the capping agent, PVA was used throughout the reaction. We studied the UV-visible absorption spectra of the synthesized sample CdS, ZnS, and CdS/ZnS using a UV-vis spectrophotometer. The values of absorption edges for all the three samples were obtained and we calculated the band gap values for each sample using Tauc's relation which was then compared with the corresponding values of the bulk materials (Table 9.1).

The spectra obtained from UV-analysis are shown in Figure 9.2.

TABLE 9.1 Different Parameters of CdS, ZnS, and CdS/ZnS Nanoparticles from UV-Vis Analysis and the Corresponding Bulk Value for the Same Materials

SL. No.	Sample Name	Absorption Edge for Bulk Material (nm)	Calculated Absorption Edge for Nanoparticles (nm)	Bandgap for Bulk Materials (nm)	Calculated Bandgap for Nanomaterials (eV)
1.	CdS	500 (Saraswathi et al., 2008)	316	2.42	4.13
2.	ZnS	335 (Saraswathi et al., 2008)	302	3.68	4.23
3.	CdS/ZnS	–	595	–	2.37

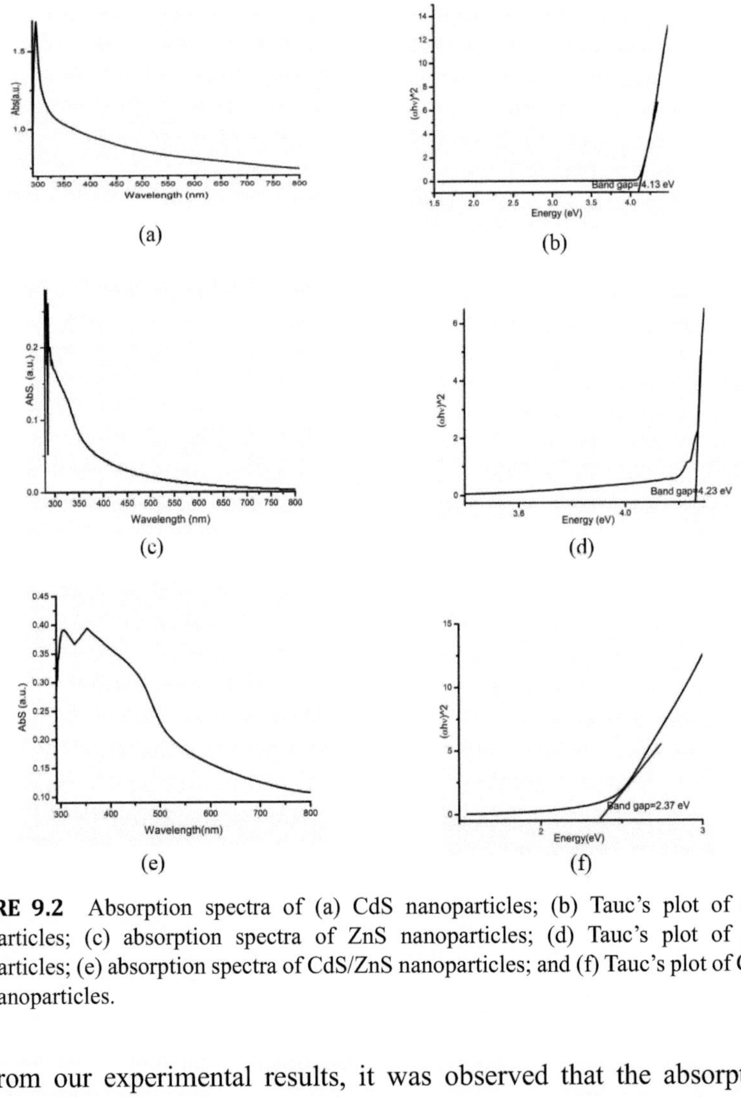

FIGURE 9.2 Absorption spectra of (a) CdS nanoparticles; (b) Tauc's plot of CdS nanoparticles; (c) absorption spectra of ZnS nanoparticles; (d) Tauc's plot of ZnS nanoparticles; (e) absorption spectra of CdS/ZnS nanoparticles; and (f) Tauc's plot of CdS/ZnS nanoparticles.

From our experimental results, it was observed that the absorption edge for CdS/ZnS nanomaterials was red-shifted compared to the core CdS nanoparticles. This can be ascribed to the potential well effect and to the overall particle size. When the CdS core is coated with a ZnS shell the size of the core/shell material becomes larger in comparison to that of the core, and as a result, the absorption edge of CdS/ZnS nanomaterials is redshifted (lower energy), and hence, the overall energy bandgap value for CdS/ZnS nanoparticles becomes small.

9.5.2 *PL RESULTS*

Maleki et al. studied the PL results of CdS nanoparticles in the year 2007 (Maleki et al., 2007). They dispersed the CdS nanoparticles in an ethyl alcohol solution, which was then subjected to PL analysis at an excitation wavelength of 220 nm. They observed that the PL spectra contained one strong emission at nearby 310 nm and another narrow emission at 340 nm. They concluded that the luminescence at 340 nm may be due to the higher-level energy transition in the CdS crystal. This may occur due to the slowly trapped excitons or due to the recombination of the excitons present in the sample. Similarly N Soltani et al. studied optical properties of CdS/ZnS nanoparticles in the year 2012. They observed two green emission peaks centered at wavelength ~542 and ~597 nm. They recorded the PL spectra of CdS/ZnS nanoparticles at room temperature under the excitation wavelength 410 nm. They reported that these peaks were due to the recombination excitons in sulfur and cadmium vacancies of core CdS nanoparticles. They observed that the intensity of emission peak decreased as the shell thickness of CdS/ZnS nanoparticles increased. Similarly, we P Boro, S Rijuwana Haque and S Bhattacharjee studied the PL spectra of CdS, ZnS and CdS/ZnS nanoparticles (Table 9.2). The spectra obtained by us are given in Figure 9.3.

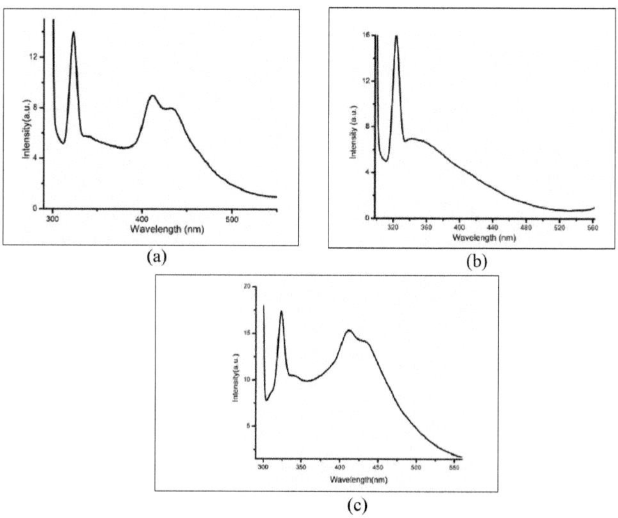

FIGURE 9.3 PL spectra of (a) CdS nanoparticles; (b) ZnS nanoparticles; and (c) CdS/ZnS nanoparticles.

TABLE 9.2 Position of Emission Peak of CdS, ZnS, and CdS/ZnS Materials

SL. No.	Name of Sample	Experimental Position of Emission Peak for Nanoparticles (nm)
1.	CdS	321.76
2.	ZnS	322.68
3.	CdS/ZnS	323.46

We observed from Figure 9.2 that when ZnS nanomaterial is coated over the CdS nanomaterial so as to form CdS/ZnS core/shell structure then the emission peak for CdS/ZnS nanomaterial is red shifted as compared to the core material. Such studies of PL spectra for CdS/ZnS nanomaterials are reported by Nayereh et al., Geremew et al., Kumar et al., Reddy et al., Saikia et al., etc., by different synthesis methods.

9.5.3 XRD ANALYSIS

XRD analysis of a sample plays a crucial role in identifying the reflection planes as well as in determining the average crystallite size of a crystallite sample. When a shell material is deposited over a core material, the reflection planes corresponding to the mixed structured material change from the reflection plane of a single core material. After the deposition of the shell material, the size of the material gets increased, which confirms the deposition of shell material successfully and hence confirms the formation of core/shell nanomaterial. XRD pattern of CdS/ZnS samples were studied by many researchers such as Soltani et al., Geremew et al., Usha, Bai et al., Kumar et al., etc.

The standard reflection planes for CdS nanoparticles are (111), (200), (311), (331) corresponding to 2θ values 2θ = 26.5°, 30.7°, 43.9°, 52.0°, etc. The standard JCPDS card no for CdS is (JCPDS File no: 42-1411). The reflection plane for ZnS nanoparticles is (111), (220), (311), etc., corresponding to 2θ values 2θ = 28.9°, 48.1°, and 57.1°, etc. The standard JCPDS card no for ZnS is JCPDS File no: 80-0020.

In the year 2012, N Soltani et al. synthesized CdS/ZnS nanomaterials and characterized the samples by XRD for structural information. They reported that the 2 theta values for core/shell CdS/ZnS nanoparticles were obtained at 26.6°, 28.30 and 52.0° which match with the hexagonal crystalline planes for ZnS and CdS nanomaterials [ICDD PDF 77-2306 and 02-1310] (Soltani et al., 2012). Similarly in 2012, H Kumar et al. synthesized CdS and ZnS nanoparticles using wet chemical method. When they performed XRD analysis of CdS and ZnS nanoparticles, the reflection planes for CdS obtained by them are (002, (200), (201), etc. For ZnS nanoparticles, the reflection planes obtained are (111), (220), (222), etc. For CdS/ZnS nanoparticles the reflection planes obtained are (111), (220), (222), etc. We Pallabi Boro, S Rijuwana Haque and Suparna Bhattacharjee obtained the following graphs of CdS/ZnS nanomaterial from XRD analysis.

From Figure 9.4, we observed that the XRD pattern exhibits a mixture phase of CdS and ZnS nanoparticles. We obtained the intense peak corresponding to angles $2\theta = 28.2°, 31.80°, 45.68°, 52.0°$, etc., which match with both crystalline planes of hexagonal structure of ZnS and CdS (ICDDPDF 77-2306 and 02-1310). The average crystallite size is estimated using the Debye Scherrer's formula and is gotten to be 11.87 nm. Similarly, various other characterizations can be carried out for CdS, ZnS, and CdS/ZnS nanoparticles, as we have listed in Sections 9.4.1 and 9.4.2, respectively.

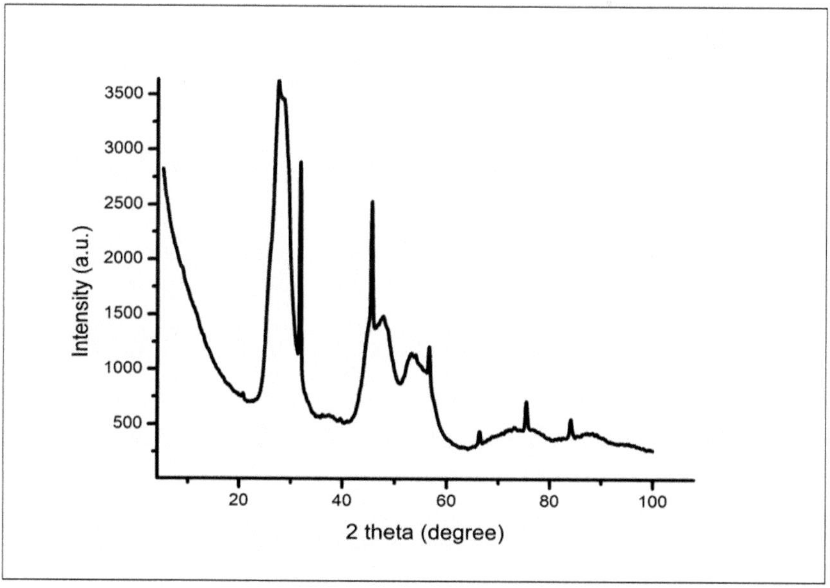

Figure 9.4 XRD pattern of CdS/ZnS nanoparticles.

9.6 APPLICATIONS OF CDS/ZNS CORE/SHELL NANOMATERIALS

Core/shell nanoparticles have many practical applications if we compare with single nanoparticles. They are specially used in biomedical and electronic fields. In the biological field, they are used for bioimaging, controlling drug delivery, cell imaging, biosensors, and many more. They can be used to design many electronic devices, for example, light-emitting diodes, solar cells, sensors, and so on.

CdS/ZnS nanomaterials are promising fluorescent probes compared to the traditional fluorescent dyes due to its characteristics which give them stable luminescence, good fluorescence, wide excitation spectrum, narrow emission spectrum, tunable fluorescence, etc. (Koneswaran & Narayanaswamy, 2012). For example, mercury is one of the most poisonous heavy metals. A fluorescent probe can be designed to detect mercury ions using the mechanism of energy transfer.

CdS/ZnS core/shell nanomaterials find applications in designing solar cells of low cost, long life, and good stability. The designed solar cells based on CdS/ZnS core/shell nanomaterials are of great concern in recent years. Moreover, in many other fields, CdS/ZnS nanomaterials are widely used, such as light-emitting diodes, cell imaging, sensors, and many others (Hernandez et al., 2017). A wide range of color can be studied by combining the modularity of core/shell semiconductor nanocrystals and the stability of the organic polymer.

9.7 CONCLUSION

Research on the improvement and application of novel materials plays a key role in the field of Science and Technology, in recent times. Nowadays, in order to develop human civilization as well as for industrial upgrading purposes, material innovation plays a very crucial role. Nanomaterials are an important class of novel materials. These materials are one of the most attractive branches of studying the present-day world. Amongst the nanomaterials, the Semiconductor/Semiconductor core/shell nanomaterials are one of the most widely investigated. There are various methods of synthesis and characterization of these materials. Such novel materials have crucial applications in varied fields, including electronics, optics, mechanics, and biological fields.

ACKNOWLEDGMENT

The authors sincerely acknowledge the Department of Applied Sciences for helping in availing the UV-visible and PL analysis, the Department of Instrumentation and USIC, Gauhati University for XRD facility. The first author (Pallabi Boro) also acknowledges CSIR, New Delhi for the financial support under JRF scheme [File no-09/059(0066)/2018-EMR-I, date: 22/02/2019].

KEYWORDS

- **core/shell nanomaterial**
- **light-emitting diodes**
- **nanomaterials**
- **novel materials**
- **semiconductor**
- **shell nanoparticles**
- **shell novel nanomaterials**
- **UV analysis**
- **XRD analysis**

REFERENCES

Agarwal, S., Patidar, D., & Saxena, N. S., (2013). *Heat. Mass Transf., 49*, 947–953.

Dabhane, H., Ghotekar, S., Tambade, P., Pansambal, S., Ananda, M. H. C., & Roza, V. M., (2021). *J. Environ. Chem. Ecotoxicol., 3*, 209–219.

Ethayaraja, M., Ravikumar, C., Muthukumaran, D., Dutta, K., & Bandyopadhyaya, R., (2007). *J. Phys. Chem. C., 111*, 3246–3252.

Ghosh, C. R., & Paria, S., (2012). *Chem. Rev., 112*, 2373–2433.

Hernandez, C. R., Acosta, M., Riech, I., Santana-Rodriguez, G., Mendez, G. J., Acosta, C., & Zambrano, M., (2017). *Optik, 148*, 95–100.

https://bradscholars.brad.ac.uk/handle/10454/16720 (accessed on 11 September 2023).

https://en.wikipedia.org/wiki/Core%E2%80%93shell_semiconductor_nanocrystal (accessed on 11 September 2023).

https://nanomaterials.alfa-chemistry.com/products/novel-materials.html (accessed on 11 September 2023).

https://onlinelibrary.wiley.com/doi/full/10.1002/pssb.201552007 (accessed on 11 September 2023).

Koneswaran, M., & Narayanaswamy, R., (2012). *Mikrochim Acta, 178*, 171–178.

Kumar, H., Barman, P. B., & Raj, S. R., (2014). *Int. J. Sci. Eng. Res., 5*, 40–53.

Kumar, H., Kumar, M., Barman, P. B., & Raj, S. R., (2014). *Appl. Phys. A., 117*, 1249–1258.

Maleki, M., Sasani, G. M., Mirdamadi, S., & Ghasemzadeh, R., (2007). *Semicond. Phys. Quantum Electron. Optoelectron., 10*, 30–32.

Mirnajafizadeh, F., & Stride, J A., (2020). *J. Nanotechnol. Mater. Sci., 2*, 121–126.

Oladeji, I. O., & Chow, L., (2005). *Thin Solid Films, 474*, 77–83.

Salata, O. V., Dobson, P. J., Sabesan, S., Hull, P. J., & Hutchison, J. L., (1996). *Thin Solid Films, 288*, 235–238.

Saraswathi, A. B., Manzoor, K., Ramakrishna, K., & Pattabi, M., (2008). *Mater. Chem. Phys., 112*, 789–792.

Soltani, N., Saion, E., Erfani, M., Bahrami, A., Navaseri, M., Rezaee, K., & Zobir, H. M., (2012). *Chalcogenide Lett., 9,* 379–387.

Venkata, R. C., Shim, J., & Cho, M., (2017). *J. Phys. Chem. Solids, 103*, 209–217.

Vishwakarma, K., (2013). *Asian J. Chem., 25*, S45–S47.

CHAPTER 10

Spiral Fluid Jet for Micro-Components Finishing

KELVII WEI GUO

Department of Mechanical and Biomedical Engineering, City University of Hong Kong, Kowloon Tong, Kowloon, Hong Kong

ABSTRACT

Microtechnology related to miniaturization and microfabrication is challenging with the development of various fields such as medicals, transportation, telecommunication, electronics, and optics. However, there always exists an issue during the microprocesses-post processing to make the relevant components meet with the correlated requirements. Consequently, the drawbacks (e.g., debris, burrs, scars, etc.) must be removed to avoid the mismatch from the practical applications.

Therefore, the relevant technologies potential for micro-components finishing are marked. Meanwhile, to meet with the requirements of the eco-friendly environment, a new approach for the post-processing (a spiral fluid jet) of micro-components was developed. Even though the achieved results are promising and relatively new in the field of microscale, it still needs further detailed research to improve the overall smoothness of the surfaces or to focus on particular locations to remove the marks, burrs and debris, and the correlated process mechanism should be investigated and characterized.

Advances in Manufacturing and Processing of Materials: Characterization and Applications.
Narendra Kumar, PhD, Mohit Tyagi, PhD, Dilbagh Panchal, PhD & Ravi Pratap Singh, PhD (Eds.)

10.1 INTRODUCTION

Micro-technology has emerged as an important discipline in recent years. It leads to miniaturization. Functionality of products is increased while the sizes remain small. The impact on new products and processes is significant. This enables new developments in telecommunication, biomedical systems, sensory devices, etc.

The demand for miniaturization leads to processes and equipment for manufacturing parts and products, which are measured in microns. Currently, photo-resist and/or micro-machining techniques are mainly taken for fabricating micro-devices. The finishing of the edges, side-walls, cavities, etc., may be poor due to factors like inhomogeneity of the material, imperfect cut, machine marks, tool wear and process stability.

Traditional mechanical finishing approaches to micro-feature finishing are impractical due to problems in efficiency and tool miniaturization. Chemical approaches to improve the finishing of micro-features are undesirable due to the use of environmentally unfriendly chemicals and difficulties in selective removal of material.

Surface finishing is performed to reduce marks and burrs resulting from primary manufacturing operations, and to improve the smoothness and sometimes the flatness of the surfaces. An important application is in the polishing of precision optics. There are also recent investigations in the polishing of precision molds and dies and artificial joints which contain complicated physical shapes. Fabrication of semi-conductor devices also imposes stringent demands in surface finishing like sub-micron surface flatness and nanometer level surface roughness (Biswas, Bayer, Biris, Wang, Dervishi, & Faupel, 2012; Kitsara, Kontziampasis, Agbulut, & Chen, 2019; Oliver, Henning, & Pieter, 2004; Mamalis, 2005).

Micro-holes and components from micro-fabrication processes like micropunching (Guo & Tam, 2012, 2013, 2014, 2015a, b, 2016, 2017a, b), drawing (Reichardt & Liewald, 2019; Raja & Ramesh, 2021; Takalkar, Babu, & Chinnapandi, 2019; Dwivedi & Agnihotri, 2017) and machining (Johnstone & Parameswaran, 2004; Mamalis, 2005; Janssen, Hoefnagels, De Keijser, Th. & Geers, 2008; Uhlmann et al., 2016) need secondary operations to improve the surface finishing, just like their macro-counter-parts.

To date, in order to realize polishing, the polishing tools are mostly contact with the needed part surface physically. They include the use of fixed abrasives in which the abrasive grit is embedded in binding materials like stone or plastic, loose abrasive or coated abrasive applied with soft polishing pads or sheets, filamentary brushes, etc. Because of the problems in tool miniaturization, wear, and positioning, these processes are NOT practical for the finishing of micro-components.

In the meantime, there were some investigations in non-contact polishing processes. Magnetorheological finishing (Arora & Singh, 2021) and electrorheological finishing (Arora & Singh, 2021) were targeted for the polishing of optical components. By activating a magnetic or electric field, the abrasive fluid is stiffened to promote the contact between the abrasive grit and the part surface. The main use is in the polishing of surfaces with gentle curvatures. Although these processes are non-contact methods, they are still NOT suitable for the micro-components contained the complicated forms and structures.

Using high-energy beam irradiation such as laser (Guo & Tam, 2012; Guo, Hua, Tse, & Mok, 2012; Guo, 2009) and electron beam (Thornton, 1980; Randolph, Hale, Guillorn, Rack, & Simpson, 2005; Bower, Gilchrist, Lueck, & Stoner, 2007; Choi et al., 2016) were also proposed for polishing. These are basically thermal processes as they involve melting and re-solidification of the surface material. The heat-affected zone can be large if the processes are not carefully controlled.

10.2 POLISHING METHODS

10.2.1 GRINDING-MECHANICAL POLISHING

Grinding is a mechanical method of surface polishing, which is using abrasives and a work wheel to scratch the surface directly. For different surface quality, there are different abrasive tools and wheels installed to the grinding. After the grinding, the rough surface will become smooth and shiny, and the roughness will be decreased. The mechanical polishing machines are used to grind materials such as metals, plastics, glasses, etc. This technique usually applies to the surface with high initial roughness and removes the high volume of imperfections on the surface.

Additionally, the grinding machine consists of a fixture to hold the work piece and a working wheel to rotate at a required speed. When the work piece is moved to contact the abrasive wheel, the abrasive wheel can scratch or cut the work piece. To move the workpiece, the operator can adjust the grinding area of the workpiece.

The advantage of grinding is that this technique is simple and low-cost. The disadvantage of grinding is that some inner surfaces and edges are difficult to grind, and the grinding machine is mainly used for flat surfaces, outer surfaces, and edges.

10.2.2 CHEMICAL MECHANICAL POLISHING (CMP)

Chemical mechanical polishing (CMP) is a planarization method with mechanical removal and chemical dissolution to create a flat and mirror surface. CMP can achieve the requirement of global planarization that the feature size smaller than 0.35 μm. In the semiconductor fabrication, this technique often applies in silicon wafer that removes excess material to obtain a smooth and flat surface. Normally, this technique often applies in semiconductor fabrication such as germanium, silicon, etc.

CMP consists of a carrier, a wafer, a polishing pad and the slurry. The wafer is the surface to be polished, which is fixed to the carrier. The carrier starts to rotate and press the wafer to the rotating polishing pad. The slurry is a mixture of abrasive particles and chemical additives, which is deposited on the pad so that the chemical reaction can increase the material removal rate.

The advantages of CMP are that CMP can provide excellent global planarization and apply to a wide range of wafer surfaces, and CMP can produce a planarized surface allows higher resolution of photolithography process. The disadvantages of CMP are that this technique is only suitable for semiconductor industries, and the complex shape surface is also not allowed for CMP.

10.2.3 ABRASIVE WATER JET (AWJ)

The abrasives used in a grinding wheel or blasting are synthetic and natural to realize the surface improvement because of their effective physical properties on size and shape stabilization, high toughness and excellent hardness.

Moreover, the artificial abrasives (Al_2O_3, $BaTiO_3$, B_4C, BN, SiC, TiC, WC, $ZrTiO_4$, etc.) are largely taken in the practical applications due to their outstanding qualities.

Abrasive water jet (AWJ) polishing is like the mechanical polishing. It is using the abrasive fluid to replace the abrasive wheel and scratch the surface directly to create a surface with good quality. The compressive water fluid jet is supplied to the AWJ machine and the abrasive is added to the water fluid when the water fluid is passing through the mixing chamber. After passing the mixing chamber, the water fluid is mixed with the abrasive to form an AWJ and pass through the nozzle. If the control parameters are changed, such as pressures, the abrasive size, and the nozzle diameter, the AWJ machine can be used in various applications. For example, an AWJ machine can be used for cutting materials if the jet pressure increases to form a high-power jet. In industries, this technique is often to polish optical materials and work on some detailed work pieces such as glasses, glass edges.

The advantages of AWJ are that AWJ is an inexpensive process due to no special machines and special consumables required. In the AWJ process, coolants and lubricants are not required to add on the work piece. Also, AWJ can create the high quality and finer surface depend on the abrasive size. To achieve different applications, various machine nozzles can be changed for varying spot size and removal rates. Specially, the nozzle of AWJ is flexible that the nozzle can be adjusted to different angles that the abrasive jet can focus to impact the detailed characteristics of the work piece such as inner surfaces, inner edges, and curved surfaces.

10.2.4 LASER POLISHING

Professional laser polishing is a fully automatic polishing process, which can avoid monotonous manual work. It polishes with laser radiation to make a thin surface layer melted and material flow from the peaks to the valleys under the effect of surface tension (Guo, 2012). Material is relocated while it is not removed. By comparison with the traditional abrasive polishing and grinding, the idea of laser polishing stimulates from the fundamentally different active principle (remelting). According to the initial roughness and the material, the roughness of the processed surface can be roughly up to 0.05 μm by laser polishing, which is attractive to a broad field of applications.

For the polishing process, it mainly includes two sub-variants: micro- and macro-polishing. It is well known that the continuous-wave (CW) laser radiation is often taken in macro-polishing. On the basis of the practical requirements, the processed surface with a roughness of several micrometers should be firstly removed by macropolishing after milling, turning or EDM-process.

The relevant processing parameters are taken in according to the initial roughness and the materials needed to be polished. Meanwhile, during the polishing, the remelting depths are strictly controlled from 20 to 200 μm. Nowadays, as an efficient solid-state laser-Nd:YAG laser, it can make the surface roughness from 5.0 μm down to 0.05 down with the optimal processing parameters. The results show that the beneficial characteristics superior to other techniques are: (i) The automated process can achieve complex 3D surface polishing, which manual polishing cannot provide; (ii) compared with the related manual polishing, less processing times along with high level automation are easily achievable; and (iii) flexibly selected the processing area and keep other areas untouched without any pollution, wastes and contaminations.

Up to now, many fabricated micro-components need secondary finishing processing as shown in Figure 10.1 (Figure 10.1 shows the relevant features of micro-components finishings (micro-channels and micro-turbine). It shows that some debris or burrs distributed in the fabricated structures which hinder the efficiency of the correlated applications afterwards.), just like their macro-counterparts. Therefore, finishing processes involving physical contact between the tool and the component are not feasible due to problems with tool miniaturization.

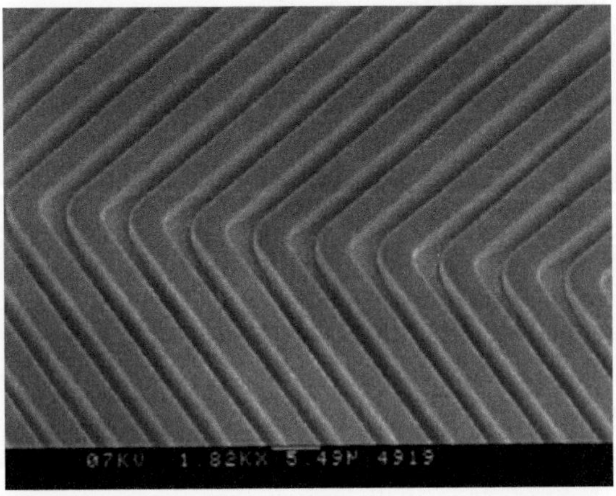

(a) Glass micro-channels formed by DRIE (deep reaction ion etching).

(Continued)

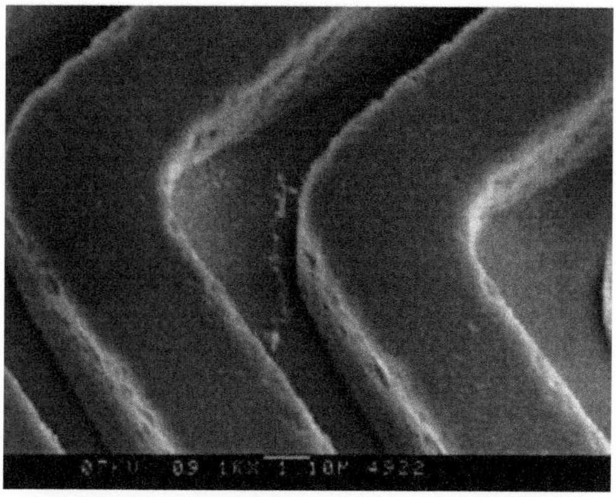

(b) Detailed micrograph of glass-micro-channels.

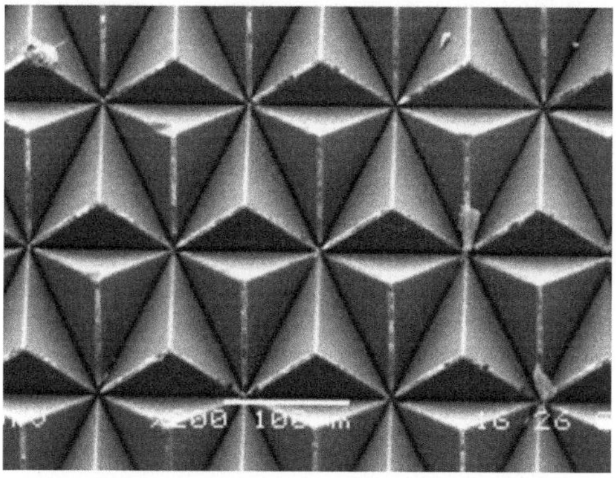

(c) Micro-prism array for total internal reflection of light by diamond cutting.

FIGURE 10.1 Features of micro-finishings.

10.2.5 ISSUES IN MICRO-FORMING

It is well known that because of low cost, high precision and efficiency, no pollution and mass production, the techniques related to microforming is fascinating to micro-parts fabrications. At the current situation of the urgent need of microcomponents and the trend to miniaturization, the techniques related to microfabrication are more and more important.

In according with the requirements of the microparts, the correlated techniques are taken in the processing. Up to now for the metallic parts, they are usually processed by the conventional methods, such as milling, grinding and turning. However, with the size down to the microscale, the research on the details of microprocessing is more crucial, which is significantly different from the traditional technologies. Due to the size effect, the results show that except for the flow behavior, the microproduction process and machine systems, the micro-tools have the significant influence on the accuracy and repeatability.

It should be noted that the property of the processed materials is relied on its microstructure and the correlated size effect. Meanwhile, the interactions during the processing are also additional effects such as spring-back, tribology, wear, forming forces, unstable results, etc.

Also, manufacturing of tools concerning microforming is a key to the ideal processes. According to the requirements of the surface quality and geometrical accuracy, EDM as an effective conventional method is usually taken to realize the destination. For the microfabrication, microtechniques are adapted to realize the relevant needs. However, with the size effect, the results of the processing on the microscale are mostly restricted by the processing parameters, tools and the concerning materials. Actually, the requirement roughness can be achievable in the microforming, there are many restrictions for the applied tools even taking ultrasonic vibrations and nitriding processes to ameliorate the relevant issues.

Meanwhile, the roughness of *R*z 0.1–0.3 μm is achieved by the usage of cemented carbides with the diameter of tools down to 50 μm. As an effective processing tool, wire EDM is often taken in microtool fabrication available with 300 μm in the width. Nevertheless, for making microtools with radii and contour on the microscale, the high quality of thin wires should be considered (for instance, the tensile strength). Moreover, during the processing, the relevant machining accuracy should be controlled with the practical requirements.

To date, laser ablation is taken in order to overcome the issues abovementioned. Due to the unique characteristics of laser beam, it can process high brittle materials and is a noncontact processing technology while the processed area is evaporated along with the noncontact area remains unaffected. Although some results are successfully reported, the additional post processing is often required for laser-machined surfaces. To date, the surface roughness *R*a can be realized by mechanical polishing from 0.75 μm down to 0.07 μm. However, it is not possible inside small structures. As for the chemical methods, the surface roughness can be obviously improved by electrolytically polishing except for the pollution.

It is noted that due to scaling down from macro- to micro-scale, the complicated problems (such as tool wear, tool defects, machine stability, size effect, material strength, etc.) should be fully solved. Therefore, the micro-forming technology, at low cost and in large quantities, applied for fabrication of micro-parts should be explored in detail. Though the present researchers are genuinely enthusiastic about the long-term future of improving the micro-tools, research on the micro-post-treatment for the drawbacks arising from the above-mentioned problems has not received much attention in the literature. Consequently, surface-micro-finishing technique applied to finish surface of micromachining will be a promising alternative technique for micromachining.

10.3 SPIRAL FLUID JET (SFJ)

10.3.1 WORK FLOW OF SFJ

According to the above-mentioned, a new approach (spiral fluid jet) is proposed which is suitably applied to finish processing of components in the micron range, which is potential for mass processing as well as selective processing, the correlated schematic diagram is shown in Figure 10.2. The explored SFJ with low pressure is so narrow and focused that the flow can reach various regions of the micro-components. It is strong enough for effective finishing the surfaces and edges but not so strong as to destroy the components.

The self-designed workstage is adopted to permit operations necessary to characterize the length of the individual movement included accurately up-down moving, twisting, tilting, and reciprocating at designated locations and not mechanically interfering with the robotic parts. Meanwhile, the operations are manipulated with the aid of video imaging attached with slow-scan mode with high-resolution photographic recording.

The workstage was designed with translation range from 0 mm ~ 100 mm, reasonable precision with 0.01~0.05 μm precision, rapid assembly and low-cost. To avoid interference with the robot, the *x*, *y*, and *z* motions are realized, which are anchored symmetrically on the robotic platform. Its precision is single-axis 360° rotational motion and coarse 3-axis linear motions up to 10 mm. For motion along the micromachining axis, a separate Z-axis stage and an integral X-Y stage guides motion parallel to the plane of the robotic stage are used, and rotational motion is mounted atop the Z-stage.

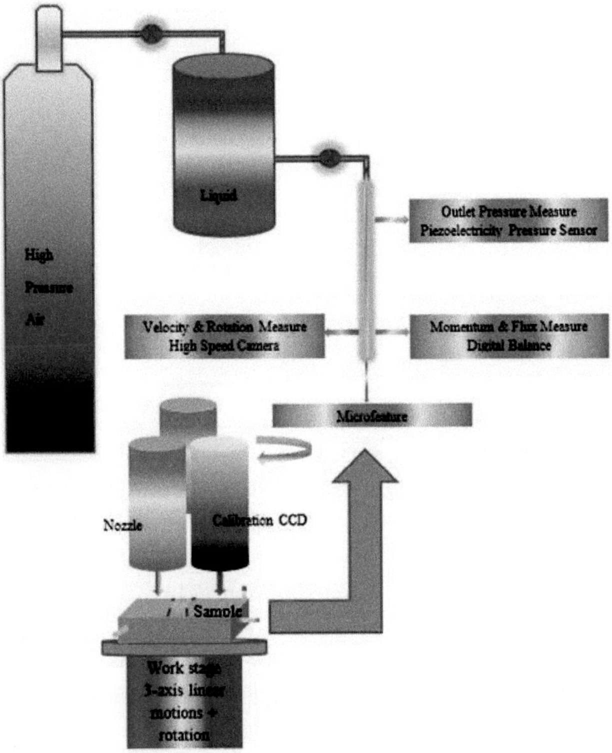

FIGURE 10.2 Schematic diagram of SFJ.

10.3.2 *MODELING OF FLOW*

Rotation of stream = Outlet velocity × (12/γ) × 60, γ: twist rate.

For continuous flow region:

$$V = \left(\frac{2P}{\rho} \right)^{1/2}$$

where V is the outlet velocity, P is the working pressure, and ρ is the liquid density.

$$Q = c \times d^2 \times \sqrt{P}$$

where; Q is the stream fluent; d is the nozzle diameter; and c is the coefficient (1.94~1.46).

10.3.3 MICRO-FEATURES OF MICROHOLE PROCESSED BY SFJ

Figures 10.3 and 10.4 show the variation of the micro-features of the punched microholes (included front side and back side). The debris attached to the inner of the microhole and distributed in the surrounding of the microhole can be removed clearly, as shown in Figure 10.3 (in red). The deburring on the surface of microhole can be polished partially (in green).

By comparison, as shown in Figure 10.4, the effect of SFJ processing is obvious. The debris distributed on the backside is removed entirely. The morphology of the processed microhole is clear and meets the finished requirements.

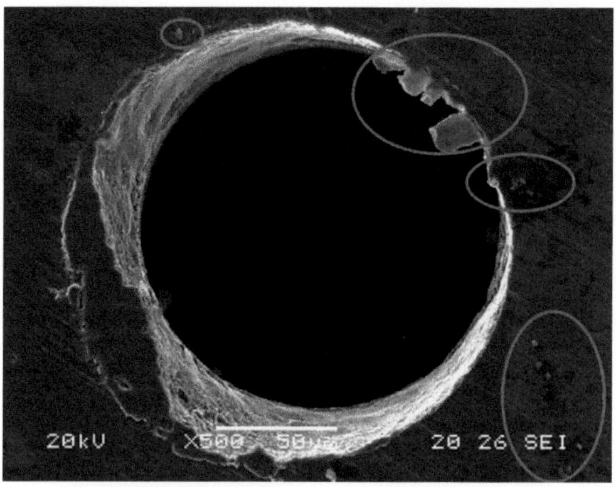

(a) Initial microhole A (front side)

(Continued)

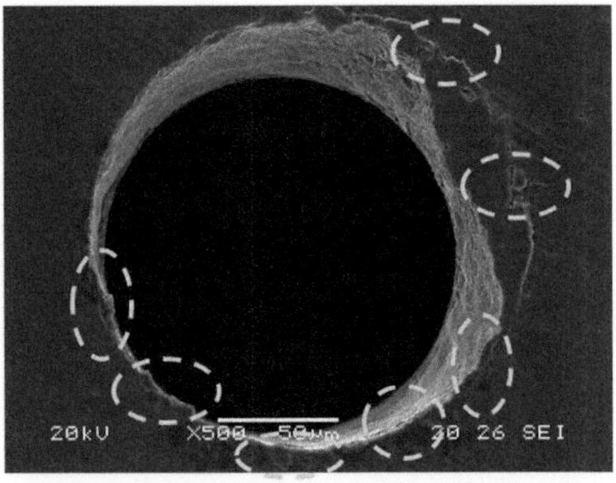

(b) Microhole A processed by spiral fluid jet (front side)

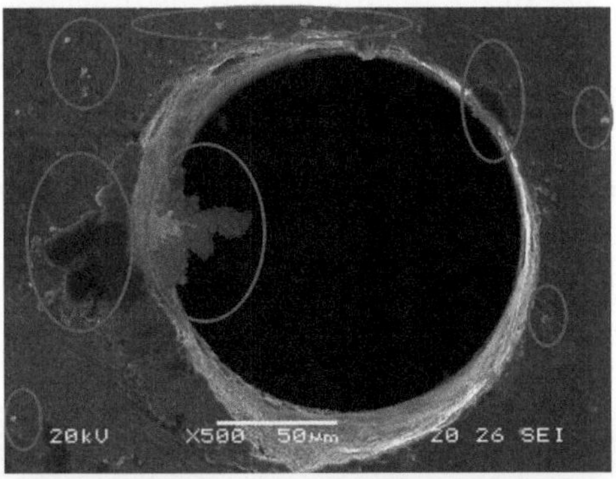

(c) Initial microhole B (front side)

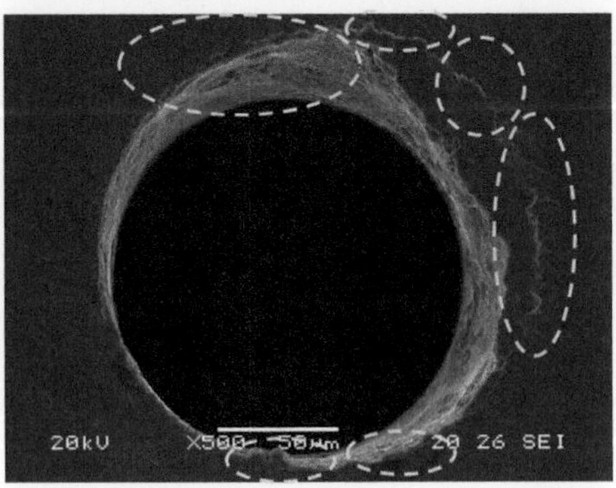

(d) Microhole B processed by spiral fluid jet (front side)

FIGURE 10.3 Morphology of microhole (front side).

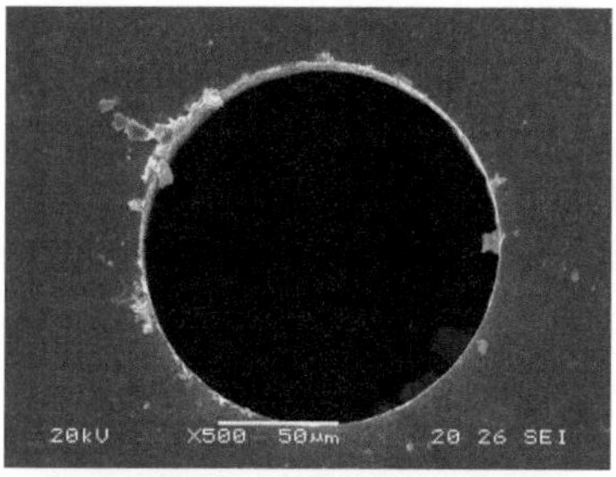

(a) Initial microhole A (backside)

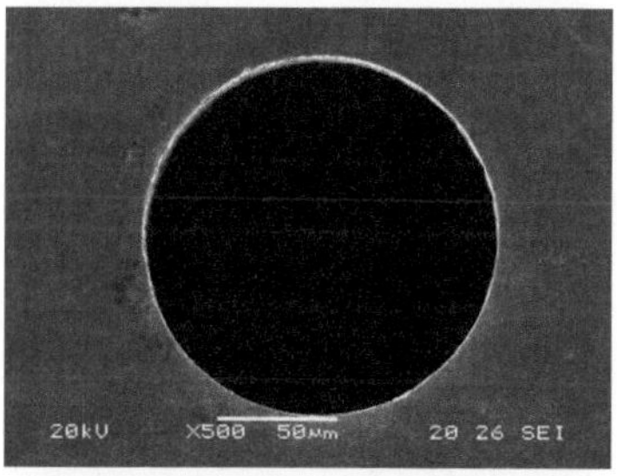

(b) Microhole A processed by spiral fluid jet (backside)

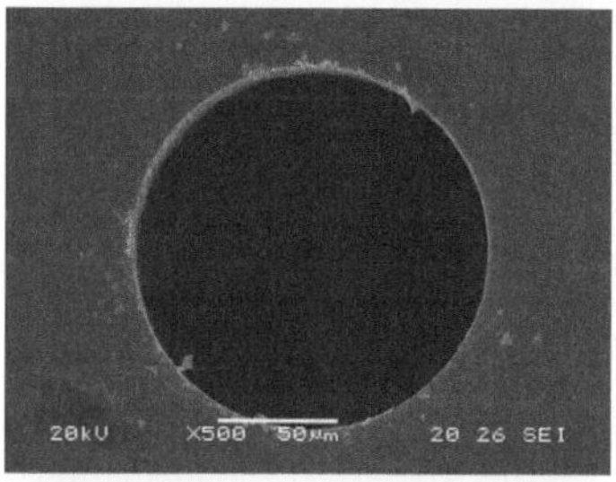

(c) Initial microhole B (backside)

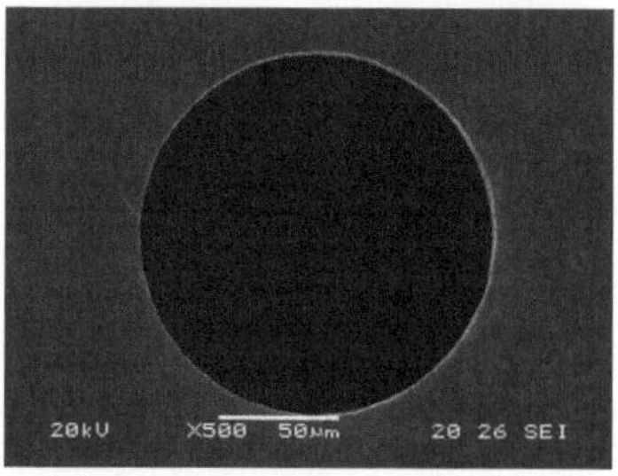

(d) Microhole B processed by spiral fluid jet (backside)

FIGURE 10.4 Morphology of microhole (back side).

10.4 CONCLUSION

Experiments demonstrate the SFJ can apply for the microparts finishing successfully. The high quality of the final microparts finishing relies on the optimal SFJ micro-finishing parameters achieved by studying the influence of manufacturing parameters on the morphology variation in detail along with the correlated mechanism extensively explored or revealed. Ultimately, it will be further propitious to increase the productivity of the process.

KEYWORDS

- **abrasive**
- **finishing**
- **grinding**
- **laser**
- **micro-components**
- **microhole**
- **polishing**
- **spiral fluid jet**

REFERENCES

Arora, K., & Singh, A. K., (2021). Theoretical and experimental investigation on surface roughness of straight bevel gears using a novel magnetorheological finishing process. *Wear, 476*, 203693.

Biswas, A., Bayer, I. S., Biris, A. S., Wang, T., Dervishi, E., & Faupel, F., (2012). Advances in top-down and bottom-up surface nanofabrication: Techniques, applications & future prospects. *Advances in Colloid and Interface Science, 170*(1, 2), 2–27.

Bower, C. A., Gilchrist, K. H., Lueck, M. R., & Stoner, B. R., (2007). Microfabrication of fine-pitch high aspect ratio Faraday cup arrays in silicon. *Sensors and Actuators A: Physical, 137*(24), 296–301.

Choi, S. S., Park, M. J., Han, C. H., Oh, S. J., Yamaguchi, T., Park, N. K., Yoo, J. H., et al., (2016). Formation of an Au membrane incorporated with carbon atoms under electron beam irradiations. *Surface and Coatings Technology, 306*(Part A), 113–118.

Dwivedi, R., & Agnihotri, G., (2017). Study of deep drawing process parameters. *Materials Today: Proceedings, 4*(2), Part A 820–826.

Guo, K. W., & Tam, H. Y., (2012). Effects of extended punching on wear of the WC/Co micropunch and the punched microholes. *The International Journal of Advanced Manufacturing Technology, 59*(9–12), 955–960.

Guo, K. W., & Tam, H. Y., (2012). Study on polishing DF2 (AISI O1) Steel by Nd: YAG laser. *Journal of Materials Science and Technology, 1*(1), 54–77. doi: 10.5539/jmsr.vln1p54.

Guo, K. W., & Tam, H. Y., (2013). Influence of the processing time on the finishing of punched micro holes by planetary stirring with natural sand grains. *Proceedings of the Institution of Mechanical Engineers Part B: Journal of Engineering Manufacture, 227*(6), 339–347.

Guo, K. W., & Tam, H. Y., (2014). Effects of carbon nanotubes on wear of WC/Co micropunches. *The International Journal of Advanced Manufacturing Technology, 72*, 269–275.

Guo, K. W., & Tam, H. Y., (2015a). TEM morphology of carbon nanotubes (CNTs) and its effect on the life of micropunch. In: Khan, M., (ed.), *The Transmission Electron Microscope Theory and Applications*. Springer, ISBN 978-953-51-2150-3.

Guo, K. W., & Tam, H. Y., (2015b). Carbon nanotubes (CNTs) for prolonging the life of micropunch. *Journal of Coating Science and Technology, 2*, 65–71.

Guo, K. W., & Tam, H. Y., (2016). Filtering brackish water by photovoltaic-powered membrane. *Journal of Membrane and Separation Technology, 5*(2), 69–76.

Guo, K. W., & Tam, H. Y., (2017a). Ti microholes potential for biomedical application punched by 75WC25Co micropunch. *Journal of Applied Biotechnology & Bioengineering, 4*(3), 572–576.

Guo, K. W., & Tam, H. Y., (2017b). Effect of cobalt on the morphology of microhole formed by micropunching. *Cobalt*. InTech.

Guo, K. W., (2009). Effect of polishing parameters on morphology of DF2 (AISI-O1) steel surface polished by Nd: YAG laser. *Surface Engineering, 25*(3), 187–195.

Guo, K. W., (2012). Pulsed Nd: YAG laser micromachined DF2 (AISI-O1) surface aided with confocal laser signal processing. *Journal of Optics Research, 14*(3, 4), 333–359.

Guo, W., Hua, M., Tse, P. W. T., & Mok, A. C. K., (2012). Process parameters selection for laser polishing DF2 (AISI O1) by Nd: YAG pulsed laser using orthogonal design. The *International Journal of Advanced Manufacturing Technology, 59*(9–12), 1009–1023. doi: 10.1007/s00170-0113558-1.

Janssen, P. J. M., Hoefnagels, J. P. M., De Keijser, Th. H., & Geers, M. G. D., (2008). Processing induced size effects in plastic yielding upon miniaturization. *Journal of the Mechanics and Physics of Solids, 56*, 2687–2706.

Johnstone, R. W., & Parameswaran, M., (2004). *An Introduction to Surface-Micromachining*. Kluwer Academic Publishers Netrherlands.

Kitsara, M., Kontziampasis, D., Agbulut, O., & Chen, Y., (2019). Heart on a chip: Micro-nanofabrication and microfluidics steering the future of cardiac tissue engineering. *Microelectronic Engineering, 203, 204*, 44–62.

Mamalis, A. G., (2005). Advanced manufacturing engineering. *Journal of Materials Processing Technology, 161*, 1–9.

Oliver, G., Henning, K., & Pieter, T., (2004). Microsystem engineering of lab-on-a-chip devices. *Journal of Angewandte Chemie, 43*(34), 4382–4390.

Raja, C. P., & Ramesh, T., (2021). Influence of size effects and its key issues during microforming and its associated processes: A review. *Engineering Science and Technology, an International Journal, 24*(2), 556–570.

Randolph, S. J., Hale, M. D., Guillorn, M. A., Rack, P. D., & Simpson, M. L., (2005). A microfabrication process for a vacuum-encapsulated microchamber. *Microelectronic Engineering, 77*(3, 4), 412–419.

Reichardt, G., & Liewald, M., (2019). Investigation on friction behavior of deep drawing radii using volatile media as lubricant substitutes. *Procedia Manufacturing, 29*, 193–200.

Takalkar, A. S., Babu, L., & Chinnapandi, M., (2019). Deep drawing process at the elevated temperature: A critical review and future research directions. *CIRP Journal of Manufacturing Science and Technology, 27*, 56–67.

Thornton, P. R., (1980). Electron physics in device microfabrication. II electron resists, x-ray lithography, and electron beam lithography update. *Advances in Electronics and Electron Physics, 54*, 69–139.

Uhlmann, E., Mullany, B., Biermann, D., Rajurkar, K. P., Hausotte, T., & Brinksmeier, E., (2016). Process chains for high-precision components with micro-scale features. *CIRP Annals – Manufacturing Technology, 65*, 549–572.

Zhu, W. L., & Beaucamp, A., (2020). Compliant grinding and polishing: A review. *International Journal of Machine Tools and Manufacture, 158*, 103634.

CHAPTER 11

Analysis of SG Iron Machining in Turning Operations

NIRVEDH UMESH SHIRKE,[1] APURVA RANE,[1] AFNAAN SAYYED,[1] ATUL KULKARNI,[2] and SATISH CHINCHANIKER[3]

[1]*Department of Mechanical Engineering, Vishwakarma Institute of Information Technology, Pune, Maharashtra, India*

[2]*Associate Professor, Department of Mechanical Engineering, Vishwakarma Institute of Information Technology, Pune, Maharashtra, India*

[3]*Professor, Department of Mechanical Engineering, Vishwakarma Institute of Information Technology, Pune, Maharashtra, India*

ABSTRACT

Due to its ductility nature, SG iron is widely used in heavy machinery and the construction industry. Mechanically, SG iron has groups of graphite called nodules, which makes it strong but also very difficult to work with. A common issue faced in SG iron is tool failure and production time loss. The primary objective of this research work is to assist in optimizing turning operations in SG iron.

11.1 INTRODUCTION

SG Iron, commonly known as nodular ductile iron, ductile iron, or compacted graphite iron, is largely used in heavy machinery manufacturing

Advances in Manufacturing and Processing of Materials: Characterization and Applications.
Narendra Kumar, PhD, Mohit Tyagi, PhD, Dilbagh Panchal, PhD & Ravi Pratap Singh, PhD (Eds.)

and construction industry. Due to its ductility, it is widely used in mechanical applications that require high fatigue resistance. Mechanically, SG Iron has groups of graphite in it called nodules, which makes it strong but also very difficult and unpredictable to work with. If a wrong tool is applied, it is highly likely that it will lead to tool failure and drop in surface roughing (Abele et al., 2022; Dearnley, et al., 1985; Bil et al., 2004; Ghani et al., 2015).

But the ductility that makes it so widely used also make SG Iron relatively challenging to machine accurately. The most common problem faced in SG.

Iron machining is tool failure and production time loss. The chemical and mechanical composition of SG Iron makes it challenging for parameters to be set and the question arises, "Which kind of tools and setups would be the best for machining SG Iron most efficiently?"

With this research work, we wish to provide enough evidence that will assist in optimizing turning operations on SG Iron. We will be reviewing simulation analysis of SG Iron, its mechanical properties, effects under various conditions and parameters, various tools, and coating setups; and attempt to conclude what might be the best way to machine SG Iron.

11.2 MICROSTRUCTURE ANALYSIS

SG IRON has nodular characteristics, which make it challenging to machine. The graphite matrix and ferrite matrix which result in uneven hardness distribution. The volatile hardness stresses the tools, thus reducing tool life and may result in failure. SG Iron exhibits morphological characteristics, which means that there is a strong adhesion present between graphite and steel matrix. This strong adhesion, along with round edges and irregular bumpy surfaces, results in a substantial reduction in crack formation and growth, which makes SG Iron's mechanical properties exponentially better than other forms of cast irons. These traits exhibited by SG Iron result in formation of fracture zones near nodules causing separation of chip (Wang et al., 2017; Nayyar et al., 2012; Ravindra et al., 1993; Kumar et al., 2021).

The chip formed is segmented and short. It is observed that when the tool approaches the cutting surface, it leads to plastic deformation. Major zones of impact are shattered zone, fracture zone and decohesion zone.

The presence of nodularity directly affects the tool life and machinability of SG Iron.

Finite elemental analysis and regression models were used to study and define the molecular composition and characteristics of SG Iron. The chemical compositions may vary for different experiments we have reviewed, but the mechanical properties observed were the same, thus giving a constantly similar output (Figure 11.1).

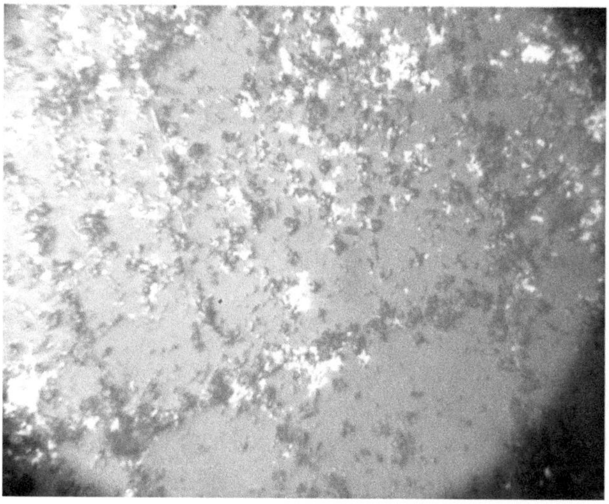

FIGURE 11.1 Microscopic image of AlCrN coated carbide cutting insert (TNMA) with HV = 2407.

11.3 TOOLS AND COATING

Tools are the main component in machining processes. Since it is widely known that SG Iron has irregular mechanical properties due to the presence of nodularity, nodules in the metal surface often forms hardened portions which are quite difficult to machine, and it is also observed that the increase in nodularity greatly affects tool life. For optimal productivity and better-quality surface roughness (surface finish) the tool must be able to sustain extremely high temperatures and speeds, while also preventing the degradation of cut as it is getting worn off (Heck et al., 2008; Martinez et al., 2017; Kalidasan et al., 2016; Kumar et al., 2017; Voigt et al., 1999; Li et al., 2021; Yigit et al., 2008).

Carbide tools with multilayer-tungsten coating was used in many experiments for testing the maximum tool life, tool wear, surface roughness, temperature differentials, and quality of chips.

Initially, we will discuss the tools. A tool is mounted on a holder in the machine. It interacts with the material at a single point. There are three main parameters that control the movement of tool. These parameters are – speed of cutting, feed rate and depth of cut. Variation in these parameters greatly affects the quality of work done as well as tool life. It is observed that Speed of cutting has the most significant effect on tool life, whereas the depth of cut and feed rate greatly effects the surface roughness and quality of finish. As the speed of cutting and feed rate is increased, the surface roughness decreased, which indicates their inverse relation.

After reviewing many papers, it is evident that the best tool for machining SG Iron is the Carbide tool (WC), and many coatings were suggested but the most commonly used one and also seemingly the most effective one is the Tungsten carbide coating ($TiCN+Al_2O_3+TiN$). The WC insert is of low Cutting speed, Feed rate and Depth of cut. CBN inserts gave the best surface roughness compared to all the experimented inserts.

There was also an experiment in which two-point cutting was used, which resulted in high productivity. Less force was observed by the rear tool, and friction on it was also reduced due to the cleansing effect from the first tool. This setup also resulted in a reduced ratio of feed to force, which means longer tool life. It was also observed that CBN coated tools performed better in dry turning as they work at a cooler temperature, gives good surface finish, and does not adhere to the workpiece when compared with TiAlN coated tool.

Geometry of tool has a substantial effect on surface finish and chip formation. The variation in parameters can be observed in the chips obtained after machining. It is observed in microstructure analysis that Some amount of damage occurs at the tip of the tool in the shear zone.

In case of high-speed turning operation, it is observed that the residual stress is observed on the surface of the part produced. It is mainly due to the thermal shock experienced on the surface of the part produced. Coating can reduce the stress experienced by the tool, while also improving the tool life and surface finishing. Tool coating acts as a thermal barrier and protects the substrate from overheating. Toughness of the substrate protects the tool from premature failure and provides good form stability (Boothroyd & Bailey, 1966; Fang & Jahawir, 1996; Fatahalla et al., 1996; Grzesik et al., 2009; Tu et al., 2020; Colding, 1991).

Coatings add certain properties to the tool, such as hardness, friction resistance, low thermal conductivity, chemical interference, etc.

Inserts can be coated in two main types: chemical vapor deposition (CVD) and physical vapor deposition (PVD). Each coating type provides different features. CVD coatings are thick, usually 9–20 microns, and highly wear-resistant, making them useful for steel and cast-iron machining and widely used in turning operations. However, such a thick coating can compromise edge toughness. PVD coatings are typically thin, usually 2–3 microns, yet tougher and smoother than CVD coatings. They are useful for machining materials, such as titanium alloys and hard-to-machine stainless steels. They typically notch or chip cutting edges.

11.3.1 EXPERIMENTAL SETUP

With the help of practical experiments involving three different types of pvd coating on carbide tools, there were inferences derived.

Three different coatings – (i) AlCrN, (ii) TiCN, and (iii) AlTi. The thickness of coating was kept between 4 and 5 microns. The Microhardness test was carried out results showed that Alti had HV: 2317, AlCrN had HV: 2407, TiCN had HV: 2407. The machining job chosen for the experiment was a SG Iron differential case. The operation chosen was Outer Diameter roughing, and the insert chosen was TNMA. Different sets of parameters were chosen for the experiment (Tables 11.1–11.3; Figure 11.2).

TABLE 11.1 AlCrN Trial

Tooling Grade	Operation	Number of Jobs Done by Tool Insert [Before Coating]	Number of Jobs Done by Tool Insert [After Coating]	Speed (mm/mm)	Feed (mm/min)
AlCrN	40	37	50	300	0.35
AlCrN	40	37	46	350	0.35
AlCrN	40	37	30	400	0.35
AlCrN	40	37	30	300	0.25
AlCrN	40	37	35	350	0.25
AlCrN	40	37	33	400	0.25
AlCrN	40	37	54	300	0.40
AlCrN	40	37	49	350	0.40
AlCrN	40	37	45	400	0.40

TABLE 11.2 AlTi Trial

Tooling Grade	Operation	Number of Jobs Done by Tool Insert [Before Coating]	Number of Jobs Done by Tool Insert [After Coating]	Speed (mm/mm)	Feed (mm/min)
AlTi	40	37	46	300	0.35
AlTi	40	37	53	350	0.35
AlTi	40	37	54	400	0.35
AlTi	40	37	48	300	0.25
AlTi	40	37	50	350	0.25
AlTi	40	37	53	400	0.25
AlTi	40	37	42	300	0.40
AlTi	40	37	55	350	0.40
AlTi	40	37	51	400	0.40

TABLE 11.3 TiCN Trial

Tooling Grade	Operation	Number of Jobs Done by Tool Insert [Before Coating]	Number of Jobs Done by Tool Insert [After Coating]	Speed (mm/mm)	Feed (mm/min)
TiCN	40	37	36	300	0.35
TiCN	40	37	42	350	0.35
TiCN	40	37	38	400	0.35
TiCN	40	37	45	300	0.25
TiCN	40	37	39	350	0.25
TiCN	40	37	34	400	0.25
TiCN	40	37	46	300	0.40
TiCN	40	37	42	350	0.40
TiCN	40	37	37	400	0.40

FIGURE 11.2 Experimental setup.

11.3.2 RESULTS OF EXPERIMENT

Thus, concluding from above trials:

1. AlCrN performed best at the speed of 300 rpm, feed rate of 0.40 and dept of cut at 0.55 mm.
2. AlTi performed best at the speed of 350 rpm, feed rate of 0.40 and dept of cut at 0.55 mm.
3. TiCN performed best at the speed of 300 rpm, feed rate of 0.40 and dept of cut at 0.55 mm.
4. From the above trails it was observed that AlCrN and AlTi gave the results, but AlCrN gave an increase of 2 seconds in production time. AlTi gave better surface finishing here as well.
5. With the new parameters, company was getting 34–36 jobs per insert, so we recommend AlTi coating along with speed 400 and feed 0.35. This setup will have the best tool life while also reducing production time and give a better surface finish.
6. Both AlCrN and TiCN tool inserts performed the worst at higher speed of 400 rpm, 0.35 feed rate and 0.55 dept of cut whereas AlTi gave one of its best performances at the same parameters.
7. Cutting speed and feed are the most influencing parameter for both cutting tool material while machining SG Iron within the range of levels of experimentation carried out (Figures 11.3–11.5).

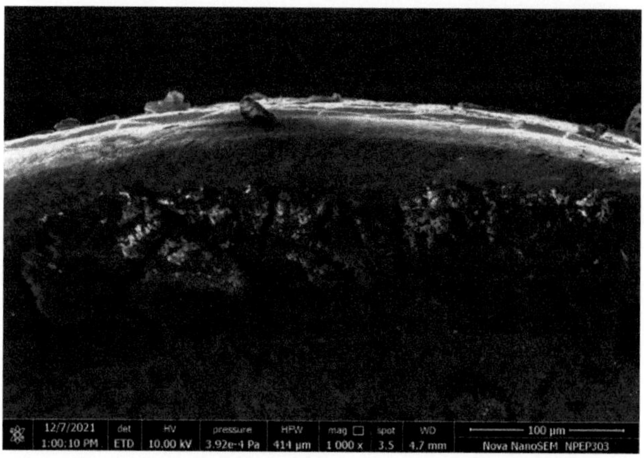

FIGURE 11.3 Microscopic image of AlCrN coated tool edge after machining.

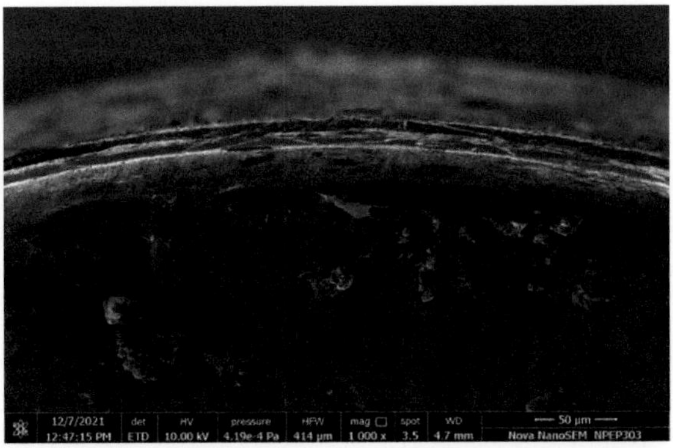

FIGURE 11.4 Microscopic image of AlTi coated tool edge after machining.

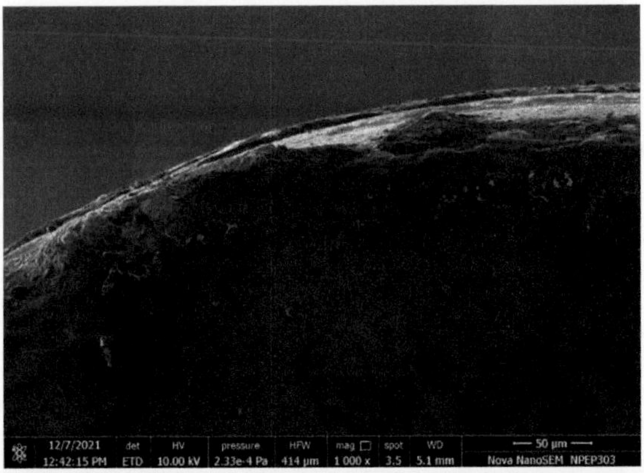

FIGURE 11.5 Microscopic image of TiCN coated tool edge after machining.

11.4 COOLANT/LUBRICATION

Coolant is a responsible component to keep the temperatures of tool and workpiece in check. The rate of flow of coolant also aids in reducing friction between the tool and the workpiece. The added benefit of reducing friction is increase in tool life. Many conditions and types of coolants are tested: The type of spray used, positions of nozzles for maximum reach of the coolant, various flow rates.

In dry turning operations, if the graphite content is high within the material composition, the graphite while machining, acts as a lubricating layer, reducing friction and improving tool life.

11.5 CHIP FORMATION – MECHANICS AND ANALYSIS

Chips are the pieces of metal shaved off by the tool from the workpiece in the machining process. Chips vary in multiple forms with various settings in the parameters of machining, as well as the tool used. A chip can help us learn a lot about the machining process and tool wear. Generally, when the speed of cutting is increased, the chips that are formed will be segmented as the workpiece is subjected to heavy forces. The forces result in premature fractures in chip, breaking it into smaller pieces. When the parameters are set correctly, we obtain continuous spiral chips, but for that to occur, the feed is generally kept slower. This causes productivity to get slowed down and is not used in industry. The nodularity of SG Iron also plays a crucial role in the formation of chips, as they tend to fracture where graphite flakes are present.

*Economic turning of CGI in continuous cut is only feasible (under the following conditions. At low cutting speeds and high feed speeds conventional coated carbide inserts can be used).

In all chip breaking models based on continuous chip formation, it is presumed that there is no contact between the chip and work surface. The other variables present in work that also effect chip formation is cutting forces, bending at chip root, shear angle, tool friction angle, tool contact length, chip thickness, chip velocity. Also, it is to be noted that the simulated chip formations may not be possible in real life because of various parameters and forces in action, as well as the composition of the material in workpiece itself.

11.6 CONCLUSION

SG iron is a tricky material to work with, but if we follow certain combination of parameters and limitations, then even with the nodularity present, we can achieve a minimum 55% increase in overall efficiency:

1. The best tool type for machining SG Iron is carbide based (WC) tool.
2. The suggested tool-coating combination for wet turning is Tungsten carbide coating (TiCN+Al_2O_3+TiN) on CNB tool.
3. For dry machining CBN tool is suggested, without any coating.

Parameters recommended by tool manufacturers are the best place to start working with. Simulations and data can be run to find optimal parameters for that particular operational and work-piece requirement. While working with SG Iron, it is suggested that the Depth of Cut should not exceed 0.15.

Coolant also plays an important role in tool life as it reduces friction and gives a cooling effect. Sufficient flow of coolant is needed in certain operations, if failed to provide may lead to work-piece as well as tool damage.

KEYWORDS

- **carbide tools**
- **cast iron**
- **chip formation**
- **lubrication**
- **machining**
- **microstructure analysis**
- **SG iron**
- **turning**

REFERENCES

Abele, E., Sahm, A., & Schulz, H., (2002). Wear mechanism when machining compacted graphite iron. *CIRP Annals, 51*(1), 53–56.

Bil, H., Kılıç, S. E., & Tekkaya, A. E., (2004). A comparison of orthogonal cutting data from experiments with three different finite element models. *International Journal of Machine Tools and Manufacture, 44*(9), 933–944.

Boothroyd, G., & Bailey, J. A., (1966). Effects of strain rate and temperature in orthogonal metal cutting. *Journal of Mechanical Engineering Science, 8*(3), 264–275.

Colding, B. N., (1991). A tool-temperature/tool-life relationship covering a wide range of cutting data. *CIRP Annals, 40*(1), 35–40.

Dearnley, P. A., (1985). A metallurgical evaluation of tool wear and chip formation when machining pearlitic grey cast irons with dissimilar graphite morphologies. *Wear, 101*(1), 33–68.

Fang, X. D., & Jawahir, I. S., (1996). An analytical model for cyclic chip formation in 2D machining with chip breaking. *CIRP Annals, 45*(1), 53–58.

Fatahalla, N., Bahi, S., & Hussein, O., (1996). Metallurgical parameters, mechanical properties, and machinability of ductile cast iron. *Journal of Materials Science, 31*, 5765–5772.

Ghani, J. A., Kian, Y. S., & Harun, C. H. C., (2015). Performance of commercial and palm oil lubricants in turning FCD700 ductile cast iron using carbide tools. *Jurnal Tribologi, 7*, 1–9.

Grzesik, W., Rech, J., Żak, K., & Claudin, C., (2009). Machining performance of pearlitic–ferritic nodular cast iron with coated carbide and silicon nitride ceramic tools. *International Journal of Machine Tools and Manufacture, 49*(2), 125–133.

Heck, M., Ortner, H. M., Flege, S., Reuter, U., & Ensinger, W., (2008). Analytical investigations concerning the wear behavior of cutting tools used for the machining of compacted graphite iron and grey cast iron. *International Journal of Refractory Metals and Hard Materials, 26*(3), 197–206.

Kalidasan, R., Yatin, M., Sarma, D. K., Senthilvelan, S., & Dixit, U. S., (2016). An experimental study of cutting forces and temperature in multi-tool turning of grey cast iron. *International Journal of Machining and Machinability of Materials, 18*(5–6), 540–551.

Kumar, K. M., & Hariharan, P., (2017). Prediction and investigation of surface roughness while turning sg iron with cubic boron nitride (cbn) and tungsten carbide inserts. *Transactions of the Canadian Society for Mechanical Engineering, 41*(1), 129–141.

Kumar, K. M., Arun, K., Sathishkumar, N., Narayanan, M. P., & Raviraj, E., (2021). Experimental investigation on the machinability of nodular ductile iron with cubic boron nitride and tungsten carbide inserts. *Materials Today: Proceedings, 39*, 1386–1389.

Li, S., Lin, H., Zhang, T., Sui, J., & Wang, C., (2021). High-speed machining of malleable cast iron by various cutting tools coated by physical vapor deposition. *Chinese Journal of Mechanical Engineering, 34*(1), 46.

Martinez, I., Tanaka, R., Yamane, Y., Sekiya, K., Yamada, K., Ishihara, T., & Furuya, S., (2017). Wear mechanism of coated tools in the turning of ductile cast iron having wide range of tensile strength. *Precision Engineering, 47*, 46–53.

Nayyar, V., Kaminski, J., Kinnander, A., & Nyborg, L., (2012). An experimental investigation of machinability of graphitic cast iron grades; flake, compacted and spheroidal graphite iron in continuous machining operations. *Procedia Cirp., 1*, 488–493.

Ravindra, H. V., Srinivasa, Y. G., & Krishnamurthy, R., (1993). Modelling of tool wear based on cutting forces in turning. *Wear, 169*(1), 25–32.

Tu, L., Tian, S., Xu, F., Wang, X., Xu, C., He, B., & Zhang, W., (2020). Cutting performance of cubic boron nitride-coated tools in dry turning of hardened ductile iron. *Journal of Manufacturing Processes, 56*, 158–168.

Voigt, R. C., Marwanga, R. O., & Cohen, P. H., (1999). Machinability of gray iron—mechanics of chip formation. *International Journal of Cast Metals Research, 11*(6), 567–572.

Wang, C., Lin, H., Wang, X., Zheng, L., & Xiong, W., (2017). Effect of different oil-on-water cooling conditions on tool wear in turning of compacted graphite cast iron. *Journal of Cleaner Production, 148*, 477–489.

Yigit, R., Celik, E., Findik, F., & Koksal, S., (2008). Effect of cutting speed on the performance of coated and uncoated cutting tools in turning nodular cast iron. *Journal of Materials Processing Technology, 204*(1–3), 80–88.

CHAPTER 12

Additive Manufacturing: A Look at the Current Technology, Materials, and Applications

ADITYA VERMA, PULKIT DUREJA, MANDEEP DHANDA, and R. S. WALIA

Department of Production and Industrial Engineering, Punjab Engineering College, Chandigarh, India

ABSTRACT

Additive manufacturing (AM) or 3D printing is a swiftly advancing technology that is a changing approach to industrial manufacturing. Today, 3D printing has become a flexible and cost-effective manufacturing method in various industries. It has gained widespread acceptance because of its capability to make intricate geometries with customized material properties. Over the past decades, several new technologies and processes for 3D printing have come out, leading to a decrease in manufacturing guidelines and standardization. The focus of this research paper is to organize the huge information available on AM into a concise paper. This chapter also sheds light on present state of various material used for AM and its challenges and uses. The wide uses of AM in bio-medics, aerospace, and electrical industry are also reported. Overall, this chapter gives a brief overview of AM, including its benefits and drawbacks as a criterion for future research and development.

Advances in Manufacturing and Processing of Materials: Characterization and Applications.
Narendra Kumar, PhD, Mohit Tyagi, PhD, Dilbagh Panchal, PhD & Ravi Pratap Singh, PhD (Eds.)

12.1 INTRODUCTION

3D printing comes under the branch of additive manufacturing (AM). It is the process that makes 3D objects shape when electronic data source is provided to it. Consecutive layers of materials are added one on top of one another by computer (Gokhare, Raut, & Shinde, 2017). 3D printing was discovered by Charles Hull, who designed the first 3D printer in the year 1984. According to Ahmed et al. (2021) 3D printing includes the following steps: Making of 3D CAD model, converting to STL file, STL file transfer to machine, machine buildup, printing of prototype, post processing, inspection of part and final delivery of the finished item. Different processes and techniques are used in 3D printing. Processes which melt or soften the material to produce the layers and the methods which cure liquid materials through divergent techniques. Polymer materials and different bio-materials are used for different medical application based on respective properties like ease of printing, biocompatible, morphologically mimicking living tissue, etc. 3D printing brings forth increased productivity, flexibility and customization according to user's need. The absence of compatible materials, highly skilled labor, and less research prevents 3D printing from becoming mainstream.

In recent times, 3D printing has fast become mainstream in the health field. It is finding its applications in various sectors of health, like the manufacturing of surgical equipment, anatomical models, prosthetics, and DDS. These have been evolving specially in Dentistry because of the customer specific needs in this field especially oral surgery and musculoskeletal system (Dement, Thompson, & Bergmann, 2017). Different AM techniques like Stereo-lithography, SLS and FDM are employed to make various dental services. Usage of CAD and CAM software is helping in development of new materials and procedures that are more economical and beneficial for the patients.

Mass manufacturing coupled with customization of drill guides and surgery instruments on an enormous scale are possible. Digital Dentistry being a new concept, of esthetic value and therefore providing superior innovation and therefore adding comfort for the patient (Yakut, Betawi, & Veldhuis, 2018). It is imperative that dentistry moves alongside so as to benefit both the patient as well as the dentist in the future. According to the current trends of AM techniques, the prospect for collaborative design is bright, and these designs, with progressing duplication, can develop affinity with the customer.

Author De Jong, & De Bruijn (2013) states that the emergence of open-source communities in 3D printing has helped in innovation which in turn has led to growth of the industry. These innovations are generally adopted by proactive companies, which can help them in price reduction and improving the overall efficiency of the manufacturing process. According to Kats, Speicher, Salonen, & Gers (2019), low-cost medical equipment is necessary, especially in the rural developing world. Even though there is shortage of trained CAD designers in the rural areas, downloading a STL file from the internet can help even untrained individuals to easily work with 3D printers. It is necessary to decrease the gap between the current 3D abilities and medical requirements. Presently, there are different techniques of 3D printing, but each one has its own drawbacks, which prevent them from becoming mainstream. Lack of suitable material, cytotoxicity, and post-processing difficulties make their adoption harder. Studying the mechanical properties of polymer material has somewhat helped our understanding of the material required for dental prosthetics, but attaining homogeneity is still an issue. According to Tartaglia et al. (2021) and Tunchel, Blay, Kolerman, Mijiritsky, & Shibli (2016) other dental utilizations like aligners, Implants and dental bridges are also currently under research. Long period observation is still required to confirm their non-toxicity, and dimensional accuracy says (Dikova, Dzhendov, Ivanov, & Bliznakova, 2018). By adding time, 3D printing is also giving way to 4D printing, especially for oral soft tissue regeneration, where the structure printed with cells changes itself after some time (Nesic et al., 2020). Again, the lack of biocompatible material prevents its mainstream usage and adoption. One thing is for certain that 3D printed dental utilities are more comfortable and efficient.

12.2 PRINCIPLE OF ADDITIVE MANUFACTURING (AM)

The process of making a part by AM starts in Computer. Various designing software like AutoCAD, SolidWorks, Autodesk, etc., are used to make design of the component. Then, this file is converted into STL format. Here, the STL file is dissected into several layers of thickness depending upon the requirement (for a high surface finish, more layers lead to thin thickness and vice versa). Every layer is generated through selective deposition of material layer by layer to form a primitive as shown in Figure 12.1.

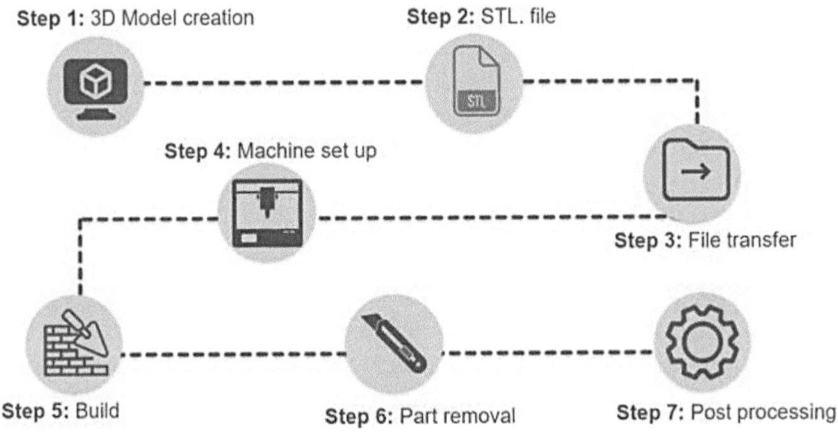

FIGURE 12.1 Steps involved in 3D printing a part

12.3 ADDITIVE MANUFACTURING (AM) TECHNIQUES AND PROCESSES

Different types of 3D Printers use different techniques to serve specific applications and purposes. Different techniques of AM have different resolution and ability to make parts (Ngo, Kashani, Imbalzano, Nguyen, & Hui, 2018). Different techniques have their own merit and demerits. Various techniques have different goal to build the part like somewhere strength of part can be the primary objective whereas somewhere accuracy may be more desirable. For example, Extrusion 3D Printing tends to have the highest layer thickness because of the large diameter of the extrusion nozzle. On the other hand, powder bed fusion and vat photo-polymerization techniques have much smaller layer thicknesses and a smoother surface finish. ASTM classifies AM technologies into following seven categories (Gao et al., 2015):

- Material extrusion;
- Powder bed fusion;
- Vat photopolymerization;
- Material jetting;
- Binder jetting;
- Sheet lamination;
- Directed energy depositions.

Details about each technique are listed in Table 12.1 and Figure 12.2.

TABLE 12.1 Classification of Various 3D Printing Techniques

Methods	Techniques Used	Materials Used	Source of Energy	Advantages	Disadvantages
Material extrusion	Fused deposition modeling (FDM)	• Acrylonitrile butadiene styrene (ABS) • Polylactide (PLA) thermoplastics	Thermal energy	• Produces complex items with high accuracy and low cost. • No special tooling required. • Simple printing of copy from normal inkjet printer.	• Restricted to small size product. • Raw material limitations. • FDM is a developing process.
Power bed fusion	Selective laser sintering (SLS)	Polyamides, thermoplastic elastomers (TPE), polystyrenes, polyaryletherketones (PAEK)	High power laser beam	• Uses materials like titanium and nylon. • Does not require any support materials.	• Post processing necessary. • Materials used for processing is limited.
	Direct metal laser sintering (DMLS)	Aluminum, stainless steel and titanium, as well as niche alloys like MONEL® K500 and Nickel Alloy 718.	High power laser beam	• Create geometrically complex parts. • Create strong metal parts. • Topology optimization, reduced weight and increased strength.	• Small build size. • Expensive process.
	Electron beam melting.	Cobalt chrome, titanium alloys, steel powder, and nickel alloy 719.	High energy electron beam.	• Rapid production. • Less need for supports. • Minimal waste.	• Restricted part size. • Low accuracy. • High material cost.

TABLE 12.1 *(Continued)*

Methods	Techniques Used	Materials Used	Source of Energy	Advantages	Disadvantages
Vat photopoly-merization	Stereolithography (SLA)	Resins (standard, engineering, dental and medical, or castable resins).	Ultraviolet radiation	• Fast • Accurate • Works with different materials. • Cost effective. • Sustainable	• Fragile • Expensive machinery unsuitable for mass production.
	Digital light processing (DLP)	Liqcreate resins	Ultraviolet radiation or visible light.	• Intricate and accurate designs. • Faster and more economical than SLA.	• Parts degrade under sun. • Parts susceptible to breaks or cracks. • Expensive costs of resins.
Material jetting	MultiJet fusion	Polyamide and thermoplastic polyurethane	Thermal energy photo curing.	• Fast build time. • Cost efficient. • Parts are strong and virtually isotropic. • Manufacturing of bigger parts. • Automated post-production process.	• Rough surface quality. • Lack of ceramic material.
	PolyJet	ABS materials, bio-compatible PolyJet photopolymer, simulated polypropylene.		• Objects with different levels of flexibility can be produced. • Versatile. • Multi-colored objects possible.	• Overexposure to UV lights can weaken the part. • Time consuming. • Costly. • Plumbing required for removing support material.

TABLE 12.1 *(Continued)*

Methods	Techniques Used	Materials Used	Source of Energy	Advantages	Disadvantages
Binder jet	Indirect inkjet printing (Binder 3DP)	Metals, sand, and ceramics that come in a powdered fashion.	Thermal energy	• Very large size capabilities. • Very high production speed. • Ability to work on a wide variety of items.	• High cost of systems. • No technical ceramics support. • Increased competition from HP, GE, desktop metal.
Sheet lamination	Laminated object manufacturing (LOM)	Polymer composites, ceramics, paper and metal rolls.	Laser beam	• Reduced tooling and production time. • Versatile • Lower cost. • Great for big structures.	• Poor surface quality and spatial accuracy. • Not suitable for complex shapes.
Direct energy deposition	Laser engineered net shaping (LENS)	Metals and alloys in the form of powder or wire, ceramics and polymers.	Laser beam	• Good granular structure. • Few material limitations. • Doesn't require secondary firing operation.	• Post processing is involved. • Part must be cut from build substrate. • Rough surface finish. • Low dimensional accuracy.
	Electron beam welding			• Clean process. • High quality and integrity. • Repeatable accuracy and quality control.	• Setup costs. • Size limitations. • Complex technical equipment.

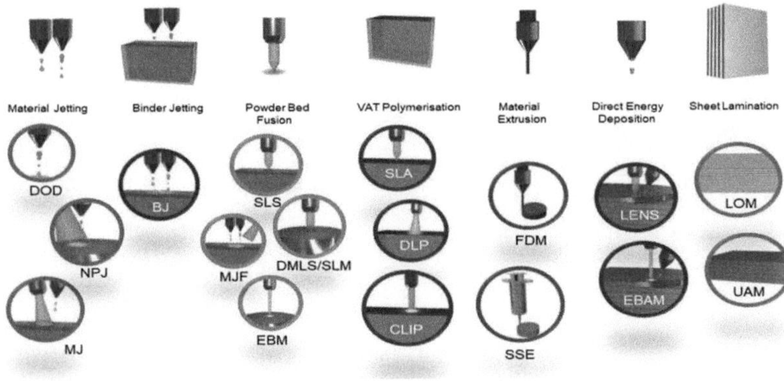

FIGURE 12.2 3D printing techniques.

12.4 MATERIALS FOR ADDITIVE MANUFACTURING (AM)

12.4.1 METALS AND ALLOYS

Metal AM mainly finds its modern applications in the field of aerospace industry. It also finds its uses in bio-medics, defense and automobile industries. The process involves melting metallic powder using laser or electronic beam. The melted material is modified layer by layer to form a solid part. Components manufactured have good mechanical properties and are highly accurate and precise. Hence, AM offers significant monetary superiority by producing more intricate structures at low costs with less waste.

Metal components produced using AM methods have finer microstructures and increased yield and ultimate strengths with anisotropic structure. Surface roughness and material defects greatly affect mechanical behavior as surface roughness can lead to concentration of stress and strain, and material defects decrease the fatigue resistance. Metal AM decreases cost of tooling, increases creativity of design and manufacturing of intricate and lightweight structures, and numerous part consolidation, which can get rid of part assemblies.

12.4.2 POLYMERS AND COMPOSITES

Polymers are extensively used materials in AM industry due to their manifoldness and acquisition. They are found in thermoplastic filaments, active monomers, powder or resins. However, they do not have strength and functionality, and so their use is often restricted to testing of prototypes.

The blend of 3D printing and a polymer amalgam offers great possibility for industrial manufacturing with great functionality and mechanical accomplishment. The use of polymers in mass production is growing steadily. Hence, appropriate materials and uses for 3D printing polymer composites should be explored and researched.

12.4.3 CONCRETE

This material makes use of terrace crafting, which is a process similar to inkjet printing (IJP). This method involves uses a bigger nozzle through which concrete paste is extruded at a very high pressure. 3D printing for concrete is still a new technology, and hence, its life cycle and performance are yet to be determined. Bonding between two layers is the main problem of 3D-printed concrete fabric. Moreover, a smaller setting time for the layers increases the chances of cold joints between layers. Challenges involved in 3D printing of wet concrete include structural properties problems such as toughness, improper layer bonding, distortion and building ability. Powder-bed fusion AM needs to be further researched and matured in order to create highly durable structures. The upcoming 3D printers have a vast prospective for these applications.

12.4.4 CERAMICS

Ceramics in 3D printing find large applications in nanomaterials and tissue engineering. Despite the accuracy of layer-wise printing, its appearance and few options of materials available make 3D printing of ceramics unfeasible. Extrusion of ceramic filament is called extrusion free-forming (EFF) of ceramics (EFF). Curing techniques can help in their finish. These include phase-changing, solvent evaporation and UV or heat curing. Direct ceramic stereolithography (CSL) involves a photocuring binder heated and cured to achieve the green body of the ceramic. CSL produces a smooth surface finish. But this process is highly expensive and is restricted to small number of materials available for this technique (Figure 12.3 and Table 12.2).

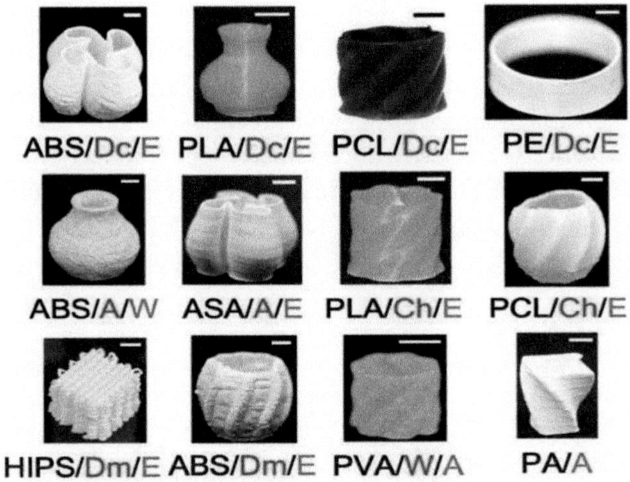

FIGURE 12.3 Several different porous materials 3D printed.

Source: Reprinted with permission from Karyappa et al., 2019

TABLE 12.2 Commonly Used Materials in Additive Manufacturing

Material	Techniques in which Used	Properties	Applications
ABS (acrylonitrile butadiene styrene)	Material extrusion-fused deposition modeling (FDM)	• High melting point (can withstand high heat). • Tensile strength about 4,850 psi.	Functional prototypes, conceptual modeling, and production of tools and a few end-use parts.
Poly-lactic acid (PLA)	Fused deposition modeling (FDM)	• Has an impressive tensile strength. • When exposed to high temperatures and pressure, the strength of PLA severely weakens.	Print plastic containers, medical implants, and packaging materials.
Carbon fiber reinforced polymer composite.	Powder bed fusion	• High specific stiffness strength. • Good corrosive resistance. • Superior fatigue performance.	Aerospace industry.
Glass fiber polymer composite	Powder bed fusion	• High Thermal conductivity. • Low coefficient of thermal expansion.	Cost-effective and high-performance make its uses diverse.

TABLE 12.2 *(Continued)*

Material	Techniques in which Used	Properties	Applications
TI6AL4V (titanium alloy)	Selective laser melting (SLM), direct laser metal sintering (DLMS), sheet lamination.	• High strength. • Low density. • Superior corrosion resistance.	Automotive, aerospace, medical/biomedical, chemical and military applications.
Resins	Stereolithography (SLA), digital light printing (DLP)	• High chemical resistance. • Material is rigid and delicate.	Highly useful in engineering and dental applications.
Ceramics	Stereolithography (SLA)	• Strong. • Durable. • Fire resistant.	Construction and building dental and aerospace applications.

12.5 ADVANTAGES AND CHALLENGES OF ADDITIVE MANUFACTURING (AM)

12.5.1 ADVANTAGES PROVIDED BY AM

3D printing is called AM because of its ability to create parts out of thin air by adding layer on layer. Since the material is added layer by layer, it enables the design and manufacturing of complex and intricate geometries. There are no problems with the need for fixtures or any other diverse tooling. Complexity of design is most affordable through AM. Printing can be stopped to put a foreign object inside the build, which later will be fully encapsulated by the part. For example, embedding sensors and circuits into the part. The tolerance and allowances of AM methods are generally way less than that of traditional manufacturing processes. AM can also manufacture parts that, otherwise, would require assembly of small parts to make it. It's on-demand on-location production can reduce the cost involved in inventory and transportation, thus saving both time and money.

12.5.2 BARRIERS AND CHALLENGES OF AM

3D printing, as of present, is not accustomed to mass production. Achieving profit through mass production by AM is still very difficult. Therefore, as of present, AM is generally being employed to cater to customer-specific needs in bio-medical or high-end automobiles and aircraft. the cost for achieving economies of scale through batch production of standardized part shapes using 3D printing is remarkably larger as compared to injection molding techniques due to the disparity in cycle time. There is always a conflict on what layer thickness must be there (Gao et al., 2015). As thick layer produces parts with poor surface finish while thin layer thin layer thickness increases the build time. Making heterogenous products with AM is presently difficult due to inter-layer deficiencies and uncertainty. Products resulting from AM systems frequently suffer from anisotropic mechanical properties due to interlayer bonding deficiencies. The enormous heterogeneity in machines, materials and processes also makes the growth of a constant standard for AM a challenging task. The 3D Printing industry is still an open-source community; hence, innovations by users can be adopted by big companies, leaving them helpless.

Furthermore, the environmental impact of 3D printing has not been properly studied, and more research needs to be done on it (Liu, Jiang, Zhang, Li, & Zhang, 2016) (Figure 12.4).

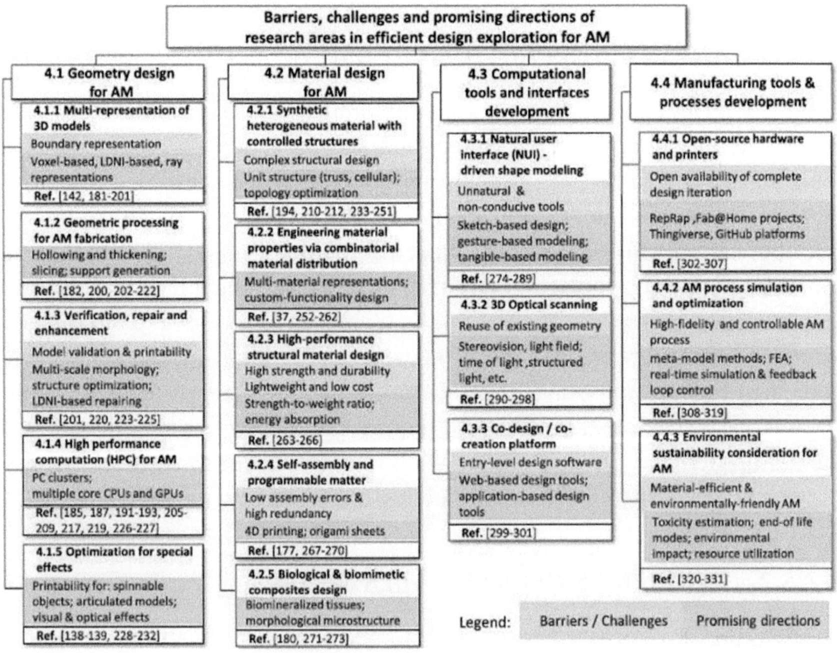

FIGURE 12.4 The hurdles, difficulties and promising directions of four research areas include: Design of geometry, material, development of computational tools and manufacturing processes in design exploration for AM.

Source: Adapted from Gao et al., 2015

12.6 DIMENSIONAL ACCURACY OF 3D PRINTING

Accuracy of finished part is essential in every manufacturing process and therefore a study was conducted to statically analyze dimensional accuracy in AM (Baturynska, 2018). This study involved predicting the dimensional features of part by two identical run of polymer laser sintering system with same design and process parameters. Very less study has been done on effect of controlling process parameters to achieve accuracy of 3D Printed part. STL model properties like number of mesh triangles and volume of CAD model are new parameters that need to be investigated further. Dimensional parameters were measured with the help of calipers and micrometers, and the final value of each measurement was taken as the mean of repeated measurement. Correlation between geometrical features and part placement and STL model properties were analyzed by Pearson correlation test which concluded that STL model properties influence some of the dimensional features depending on their orientation. It also must be noted that different techniques of AM have different accuracy of finished part. AM processes has the ability to produce precise shaped complex final parts in different sectors like aerospace, health, or automotive, however the processes must be carefully selected.

12.7 APPLICATION OF AM IN VARIOUS INDUSTRIES

3D printing has had an enormous impact in present day manufacturing and is opening new possibilities every day. Many important industries have found 3D printing useful and highly successful (Shahrubudin, Lee, & Ramlan, 2019). Some industries where this technique has been highly successful include Aerospace, Electrical, Automotive, and the biggest beneficiary of all, the healthcare and medical industry.

12.7.1 AEROSPACE INDUSTRY

3D has made possible for the potential to manufacture light-weight, improved and complex aircraft parts with minimized cost and high efficiency (Wang, Chen, & Yeh, 2019). Reduction in parts, in turn, can lead to fuel savings. 3D printing is extensively being employed in manufacturing spare parts of aircraft and also sometimes engines. Nickle-based alloys are preferred due to their higher tensile strength and corrosive resistance properties (Figure 12.5).

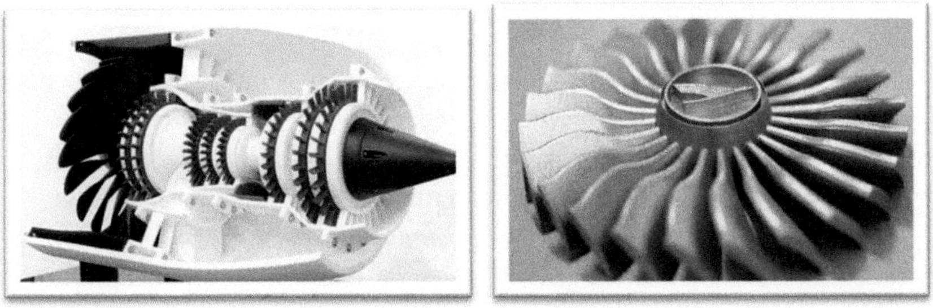

FIGURE 12.5 (a) 3D printed engine; and (b) propeller shaft.

Source: Reprinted from 3dprint.com. https://3dprint.com/17716/3d-printed-jet-engine/

12.7.2 AUTOMOTIVE INDUSTRY

In the automotive industry, 3D Printing technique allowed for lighter and more complex structures in the fast time. This has led to a surge in its usage.

Ford is a pioneer in the use of 3D printing technology. It also applies 3D printing technology to produce testing parts and engine parts while BMW has been using 3D printing technology to produce hand tools for automotive testing and assembly (Sreehitha, 2017). 3D printing technology reduces the wastage of the materials. Furthermore, 3D printing technology allows to test new designs in a very fast time period (Figure 12.6).

FIGURE 12.6 3D printing in car production.

Source: Reprinted from Formlabs. https://formlabs.com/asia/blog/3d-printed-car-how-3d-printing-is-changing-the-automotive-industry/

12.7.3 ELECTRICAL INDUSTRY

Various 3D printing technologies are used for structural electronic devices like active electrical materials, electrodes and devices with mass customization and adaptable design by implanting conductors into 3D printed devices (Espera, Dizon, Chen, & Advincula, 2019). FDM is being used to mass-produce electrodes. 3D printing is also being employed to produce circuits and microchips to high degree of precision (Figure 12.7).

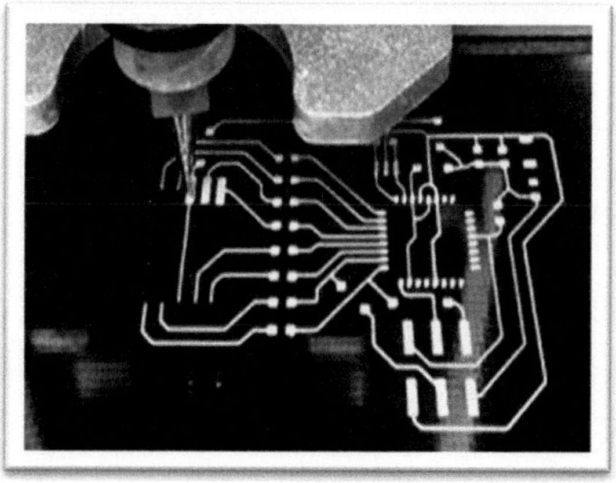

FIGURE 12.7 PCB fabricated using liquid metal printer.

Source: Sculpteo, 2015. https://www.sculpteo.com/blog/2015/12/08/electrifying-ideas-with-3d-printing/

12.7.4 MEDICAL INDUSTRY WITH FOCUS ON DENTISTRY

- Perhaps no industry has seen a revolutionized and drastic improvement in technology as had the medical industry. Such has been the growth of 3D printing in this sector that printing skin tissues, replacement body parts, organ are all possible now (Quan, Zhang, Xu, Luo, Nie, & Zhu, 2020; Dodziuk, 2016).
- 3D printing technology has achieved increased effectiveness, precise control of dose, high duplicability and to produce shot form with complex drug-release profiles (Prasad &Smyth, 2016).
- 3D printing has helped in tissue regeneration. The substitute tissues made by 3D printing technology has interconnected pore network, biocompatibility, suitable surface chemistry and good mechanical properties (A.-Do, Khorsand, Geary, & Salem, 2015).

- Prosthetics can be easily printed. 3D printing also makes it possible to customize them according to user's preferences (Manero et al., 2019).
- One area where 3D printing has become mainstream is the dentistry. Applications such as implants, aligners and dental bridges are produced by AM. Perhaps nowhere the benefit of customizability is felt as in here. Each individual has a different tooth structure and hence can have the dental equipment according to his own need. 3D Printed titanium implants are highly porous thus can greatly improve osseointegration (Tunchel, Blay, Kolerman, Mijiritsky, & Shibli, 2016). 3D Printed aligners have become the norm because of their ease and comfort (Maspero &Tartaglia, 2020). 3D printing is now capable of bioprinting, i.e., to print patient patient-tailored gingival tissue which are comfortable to patient. These can mimic complex tissue structure within hydrogel (Nesic et al., 2020) (Figure 12.8).

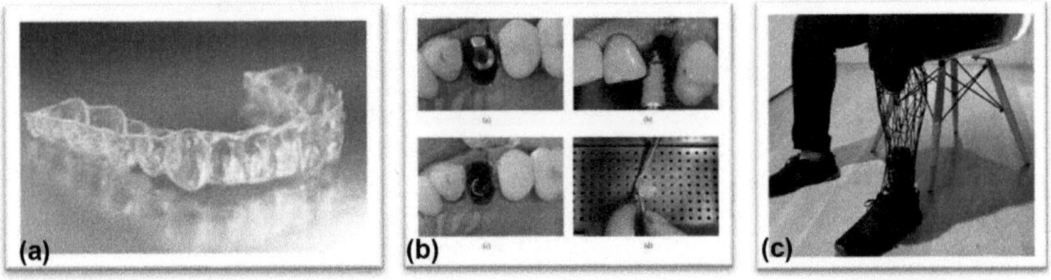

FIGURE 12.8 (a) Dental aligners; (b) titanium implants; and (c) 3D-printed prosthetic leg.

Source: a) Reprinted from AMFG (2019.)https://amfg.ai/2019/08/08/application-spotlight-3d-printing-for-digital-dentistry-clear-aligner-manufacturing/
b) Reprinted from Samy Tunchel et al., 2016. Copyright © 2016. https://creativecommons.org/licenses/by/4.0/
c) Reprinted from Parametrichouse. https://parametrichouse.com/3d-printed-prosthetics/

12.8 FUTURE POSSIBILITIES

The scope of future possibilities of 3D printing is vast. The possibility of metallic 3D printing will help in creation of rare replacement parts in a wide range of applications. Repair shops will be able to handle a larger amount of clientage, and online retailers can print unique and rare parts, thus providing more products through a JIT inventory way. More research needs to be carried out on suitable materials for different applications to make them profitable in mass production. AM technologies can greatly help in customization by increasing interaction with customers in the design stage. All these technologies are still at infant stage due to the fulsomeness linked with design and printing software. Many of the research still going on is fragmented; hence significant problems are still arising, preventing 3D printing from going mainstream.

12.9 CONCLUSION

The future of 3D printing is bright. 3D printing machines could be used in our homes, various businesses, disaster sites and also in the outer space. However, there are some challenges that must be solved, such as the lack of 3D Printing expertise and technical maturity. There are companies with narrow mindset that still proceed with traditional approach of selling copyrights instead of printing the actual parts. However, it is believed that the AM landscape will become more competitive in the coming future.

ACKNOWLEDGMENT

This study was performed under the guidance of the HOD of the Production Engineering department of PEC, Prof. R.S. Walia, and also Dr. Mandeep Dhanda. The beliefs, findings, results, and recommendations expressed in this material are the thinking and school of thought of the authors. The authors would also like to thank 3D Printing Department of PEC for allowing use of their resources and knowledge. The information was taken from various research papers, and references have been added below.

KEYWORDS

- **3D printing**
- **additive manufacturing**
- **aerospace**
- **automotive industry**
- **dimensional accuracy**
- **electrical industry**
- **osseointegration**

REFERENCES

Ahmed, A., Azam, A., Aslam, B. M. M., Khan, F. A., Aslam, R., & Tahir, Z., (2021). Discovering the technology evolution pathways for 3D printing (3DP) using bibliometric investigation and emerging applications of 3DP during COVID-19. *Clean. Environ. Syst.*, *3*, 100042. doi: 10.1016/j.cesys.2021.100042.

Anh-Vu, D., Khorsand, B., Geary, S. M., & Salem, A. K., (2015). 3D printing of scaffolds for tissue regeneration applications. *Adv. Healthc. Mater.*, *4*(12), 1742–1762. doi: Https://doi.org/10.1002/adhm.201500168.

Baturynska, I., (2018). *Statistical Analysis of Dimensional Accuracy in Additive Manufacturing Considering STL Model Properties* (pp. 2835–2849).

De Jong, J. P. J., & De Bruijn, E., (2013). Innovation lessons from 3-D printing. *MIT Sloan Manga. Rev.*, *52*(2), 43–52 [Online]. Available: http://sloanreview.mit.edu/files/2012/12/819a3bcb5e.pdf (accessed on 11 September 2023).

Dement, L. E., Thompson, M. S., & Bergmann, J. H. M., (2017). *Clinical Efficacy and Effectiveness of 3D Printing: A Systematic Review.* doi: 10.1136/bmjopen-2017-016891.

Dikova, T. D., Dzhendov, D. A., Ivanov, D., & Bliznakova, K., (2018). Dimensional accuracy and surface roughness of polymeric dental bridges produced by different 3D printing processes. *Arch. Mater. Sci. Eng.*, *94*(2), 65–75. doi: 10.5604/01.3001.0012.8660.

Dodziuk, H., (2016). Applications of 3D printing in healthcare. *Kardiochirurgia I Torakochirurgia Pol.*, *13*(3), 283–293. doi: 10.5114/kitp.2016.62625.

Espera, A. H., Dizon, J. R. C., Chen, Q., & Advincula, R. C., (2019). 3D-printing and advanced manufacturing for electronics. *Prog. Addit. Manuf.*, *4*(3), 245–267. doi: 10.1007/s40964-019-00077-7.

Gao, W., et al., (2015). The status, challenges, and future of additive manufacturing in engineering. *CAD Comput. Aided Des.*, *69*, 65–89. doi: 10.1016/j.cad.2015.04.001.

Gokhare, V. G., Raut, D. N., & Shinde, D. K., (2017). A review paper on 3D-Printing ASPECTS and various processes used in the 3D-printing. *Int. J. Eng. Res. Technol.*, *6*(6), 953–958. [Online]. Available: www.ijert.org (accessed on 11 September 2023).

Karyappa et al., 2019. Immersion Precipitation 3D Printing (ip3DP). 6(9). 1834-1844

Kats, D., Speicher, L., Salonen, B., & Gers, H. J., (2019). Paper 3D printing to supplement rural healthcare supplies-what do healthcare facilities want. *GHTC 2018 – IEEE Glob. Humanity. Technol. Conf. Proc.* doi: 10.1109/GHTC.2018.8601529.

Liu, Z., Jiang, Q., Zhang, Y., Li, T., & Zhang, H. C., (2016). *Sustainability of 3D Printing: A Critical Review and Recommendations.* doi: 10.1115/MSEC2016-8618.

Manero, A., et al., (2019). Implementation of 3D printing technology in the field of prosthetics: Past, present, and future. *International Journal of Environmental Research and Public Health, 16*(9). doi: 10.3390/ijerph16091641.

Maspero, C., & Tartaglia, G. M., (2020). 3D printing of clear orthodontic aligners: Where we are and where we are going. *Materials, 13*(22). doi: 10.3390/ma13225204.

Nesic, D., Durual, S., Marger, L., Mekki, M., Sailer, I., & Scherrer, S. S., (2020). Could 3D printing be the future for oral soft tissue regeneration? *Bioprinting*, *20*. doi: 10.1016/j.bprint.2020.e00100.

Ngo, T. D., Kashani, A., Imbalzano, G., Nguyen, K. T. Q., & Hui, D., (2018). Additive manufacturing (3D printing): A review of materials, methods, applications, and challenges. *Compos. Part B Eng.*, *143*, 172–196. doi: 10.1016/j.compositesb.2018.02.012.

Prasad, L. K., & Smyth, H., (2016). 3D printing technologies for drug delivery: A review. *Drug Dev. Ind. Pharm.*, *42*(7), 1019–1031. doi: 10.3109/03639045.2015.1120743.

Quan, H., Zhang, T., Xu, H., Luo, S., Nie, J., & Zhu, X., (2020). Photo-curing 3D printing technique and its challenges. *Bioact. Mater.*, *5*(1), 110–115. doi: 10.1016/j.bioactmat.2019.12.003.

Shahrubudin, N., Lee, T. C., & Ramlan, R., (2019). An overview on 3D printing technology: Technological, materials, and applications. *Procedia Manuf.*, *35*, 1286–1296. doi: 10.1016/j.promfg.2019.06.089.

Sreehitha, V., (2017). Impact of 3D printing in automotive industry. *International Journal of Mechanical and Production Engineering, 5*(2), 91–94.

Tartaglia, G. M., et al., (2021). Direct 3D printing of clear orthodontic aligners: Current state and future possibilities. *Materials (Basel)*, *14*(7), 1–11. doi: 10.3390/ma14071799.

Tunchel, S., Blay, A., Kolerman, R., Mijiritsky, E., & Shibli, J. A., (2016). 3D printing/additive manufacturing single titanium dental implants: A prospective multicenter study with 3 years of follow-up. *Int. J. Dent.*, *2016*. doi: 10.1155/2016/8590971.

Wang, Y. C., Chen, T., & Yeh, Y. L., (2019). Advanced 3D printing technologies for the aircraft industry: A fuzzy systematic approach for assessing the critical factors. *Int. J. Adv. Manuf. Technol.*, *105*(10), 4059–4069. doi: 10.1007/s00170-018-1927-8.

Yakut, M., Betawi, M. A., & Veldhuis, S. C., (2018). A review of metal additive manufacturing technologies. *Solid State Phenom: SSP*, *278*, 1–14. doi: 10.4028/www.scientific.net/SSP.278.1.

CHAPTER 13

Methods of Hardfacing of Aluminum Alloys

SHIVANK A. TYAGI,[1] NEHA BHADAURIA,[2] and
PRASHANT VASHISHTHA[2]

[1]Research Scholar, Department of Mechanical Engineering, NIT, Warangal, Telangana, India

[2]Department of Mechanical Engineering, KIET Group of Institutions, Ghaziabad, Uttar Pradesh, India

ABSTRACT

Due to poor wear resistance and hardness, Aluminum and its alloys are often avoided from being used in certain applications, specifically for those requiring good surface properties and involving dynamic loading conditions. However, being one of the lightest metals, Al gives a very high power-to-weight ratio in some significant applications such as aerospace and shipbuilding. Moreover, it has high corrosion resistance compared to steel, making it suitable for use in other applications bearing corrosion. This puts up a need to improve the surface properties of Al alloys, and the process of improving the hardness of the surface by modifying its surface by using some methodology is hard-facing. Several methods have been developed for performing hardfacing of Al alloys such as surfacing using welding processes, friction surfacing, laser cladding, etc. This chapter discusses the trials for hardfacing of Al alloys.

Advances in Manufacturing and Processing of Materials: Characterization and Applications.
Narendra Kumar, PhD, Mohit Tyagi, PhD, Dilbagh Panchal, PhD & Ravi Pratap Singh, PhD (Eds.)

13.1 INTRODUCTION

It has been observed that wear is among the prominent reasons for the failure of mechanical parts and reduced efficiency in machines and their parts. In general, wear is accompanied by corrosion in all environments, apart from vacuum and inert atmospheres. Wear can be defined as a continued deterioration through loss of material due to extended or overly frequent use (Gerard, 2013; Varenberg, n.d.). It degrades the condition of a part, leading to a loss of performance. The equipment used in several industries like construction, agriculture, rolling mills, etc., face severe wear problems. Though there are multiple modes of wear (Kato, 1992) that take place between two or more mating parts, the ones that cause major damage are abrasion, impact, corrosion and erosion (Varenberg, n.d.; Rigney, 1988). Aluminum (Al), being the most extensively used metal next to steel, has been a matter of research over the years in view of its low resistance to wear. The problem of wear in Al components occurs in many situations involving relative motion between different parts or when the component is subjected to corrosive and erosive environment. One such example is erosion on the wings of an aircraft due to the erosive action of air and suspended particulate in the air striking the wings (Bhadauria, Pandey, & Pandey, 2019). Several methods have been advanced for improving the wear properties of Al and its alloys. Some of these methods involve developing specific materials for enhancing wear resistance or by adding alloying elements with better wear resistant properties for enhancing the resistance to wear of the material at hand. Methods like stir casting for the preparation of specific metal matrix composites (MMCs) (Kok, 2005), friction surfacing (Vilaca, Gandra, & Vidal, 2012; Gandra et al., 2014) have been frequently used for improving wear resistance and other mechanical properties of Al alloys. Among many, one of the techniques often employed for improving the wear resistance of materials is surface modification (Shibe & Chawla, 2013). In the last few years, the results obtained by using hardfacing as a method for developing a wear resistant coating have been quite satisfactory. Hardfacing, also called "Hardsurfacing" (Digambar & Choudhary, 2014; Pradeep, Ramesh, & Prasad, 2010), is a method of developing certain specialized coatings of metals, ceramics or specific alloys by means of processes like welding, Thermal spraying, Laser cladding, Friction surfacing, etc., to resist deterioration of a part due to abrasive action of hard particles, corrosion, poor refractoriness or impact. Desired material can also be deposited on the surface, over the edges, or just at the points of the workpiece exposed to wear. Hardfacing can put the parts with damaged surfaces to use and reclaim equipment by improving their service life (Gualco, Svoboda, Surian, & Vedia, 2010). Several efficient methods are available for performing hardfacing such as the use of different welding techniques (TIG, MIG, SMAW, Gas welding, etc.), Thermal spraying, Laser cladding (Digambar & Choudhary, 2014). Also, use of composite coatings and friction stir processing (FSP) to add wear resistant particles on the surface of the substrate have been successfully used (Kuruveri, Panemangalore, Kuruveri, & John, 2022). Hardfacing has been used in diverse applications including surfacing of automobile parts like cylinder, cylinder head of IC Engine and piston, etc. (Digambar & Choudhary, 2014), surfacing of valve seat ring (Selvi, Sankaran, & Srivatsavan, 2008), surfacing of the heavy machine equipment like cranes crushers, agricultural equipment like chisel plowshare has been hardfaced for improving their service life (Bayhan, 2006), hardfacing of hot working tools such as die-casting, rolling, extrusion, forging, etc. (Ahn, 2013). The current chapter attempts to review some significant trials made for hardfacing of Al alloys with suitable processes and materials used.

13.2 HARDFACING USING WELDING TECHNIQUES

Yoshie et al. (1995) tried TIG welding for hardfacing of Al 5052 alloy. The idea was to allow for the formation and distribution of $CuAl_2$, which is an intermetallic compound in the microstructure which has a great impact on hardness. This was done by using a filler rod during TIG welding Al(20–51)%Cu. The welding parameters were found to affect the dilution and resulting hardness and microstructure. For constant current and welding speed, feed rate of 0.2 m/min of the filler rod gave higher hardness as compared to when it was 0.1 m/min. Further, it was observed that while using a filler rod of Al-51%Cu, there was a need to control dilution within 20% and also the amount of Cu in the deposited metal should be more than 40% for obtaining a hardened layer with a hardness of more than 300 HV. Similar experimentation was repeated by preheating the base metal to a temperature of 100°C. Abrasion characteristics of the samples were also studied and a comparison was made among AC8A, 5052, and 7075 Al alloy. A hard build-up layer was found to be efficient for improving the wear resistance, however as the hardness exceeded HV 300, there was a tendency for wear to accelerate due to the release of self-abrasive powder and development of microscopic defects, but a considerable influence on wear resistance was registered as the hardness exceeded HV 200. Wear resistance properties were expected to improve even further with increasing hardness but considering toughness, suitable composition was found to be for eutectic of around HV 250 (Yoshie & Yasuda, 1995).

Shivank et al. (2020) developed a surface composite layer on Al 6063 alloy plates using TIG welding. A layer of pre-prepared paste, mix of Sodium Silicate (Na_2SiO_3), lumina powder and Al metal powder in a ratio 1:1:0.2 was applied on the substrate. TIG arc heat led to melting of the surface, and a composite coating was obtained on the surface. SEM analysis showed a fairly uniform distribution of Al_2O_3 on the surface (Figure 13.1). A 30–40% increase in hardness was achieved (Tyagi, Bhadauria, Vashishtha, & Mishra, 2019). In a study, Dimitrova et al. (2017) developed a composite layer on the surface of pure Al using extruded filler wire and TIG welding process. For fabrication of wires, billets of Al reinforced with micro/nano-SiC particles were prepared by hot impression-die forging and then extruded into wires. Electroless nickel/copper coated micro/nano-SiC particles and a coating flux, with cryolite and three salts – sodium chloride, potassium chloride and magnesium chloride were prepared. Tungsten inert gas (TIG) welding with SiC metalized particles reinforced Al wire was employed for the deposition of coatings for enhancing the wear resistance of pure Al metal matrix. The filler wire was formed by extrusion of a pack (composed of several Al plates coated with films containing the reinforcing phase). A 15–31% rise in the resistance to wear of the coatings and a 30–40% increase in the hardness was registered, the reason for which was the inclusion of metalized micro/nano-SiC as reinforcing phase of particles in the layer (Dimitrova, Kandeva, & Kamburov, 2017).

Among all the welding processes TIG welding is found to be most popular for hardfacing Al alloys as it generates adequate temperature for melting and also efficient shielding is provided by an inert gas like Argon, Helium, etc. It also provides cathodic cleaning action and delivers high quality welds. Further, for hardfacing applications and for on-site repair works, the welding method is more suitable as compared to friction and laser-based hardfacing.

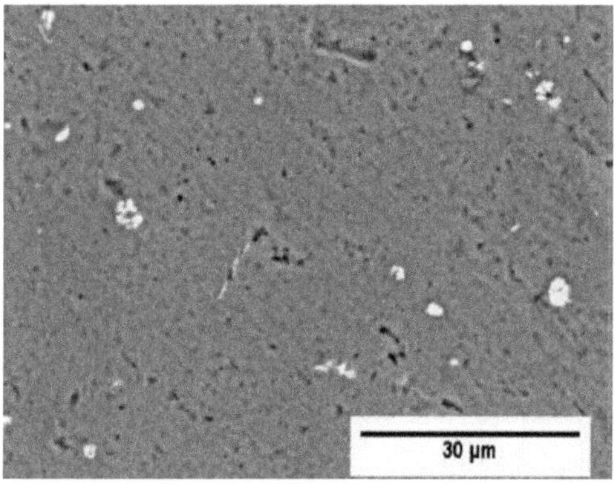

FIGURE 13.1 Micrograph showing the distribution of Al_2O_3 particles in Al-matrix.
Source: Reprinted with permission from Tyagi et al., 2019. Copyright © 2019 Elsevier Ltd

13.3 HARDFACING USING FRICTION-BASED TECHNIQUES

Reddy et al. (2009) used friction surfacing for deposition of coating of AA 2124 Al alloy composite reinforced with SiC particle over the surface of A356 Al alloy. The composite was taken in the form of wires of 15 mm diameter and was mounted on a drilling chuck rotated at a speed of 1,800 rpm. The base material 10 mm thick plate was fixed. Fairly even dispersion of particles was observed in Al–SiC particle composite layer. Figure 13.2 shows the microstructure of composite layer and adjoining base metal. For wear testing of the composite coating, a pin-on disc test was done, and considerable improvement in hardness leading to the rise in wear resistance of the composite coating was seen (Figure 13.2(b)). Microhardness of coating was also found to be marginally high (Figure 13.2(c)) (Reddy, Rao, & Mohandas, 2009).

Shafiei-Zarghani et al. (2011) developed an Al-Al_2O_3 nano-composite layer on an Al 6082 alloy workpiece using FSP. Al_2O_3 nanopowder was filled in a shallow groove on the substrate which was then subjected to 1 to 4 number of FSP passes. Increasing the number of passes led to the reduction of grain size, improved dispersion of Al_2O_3 particles (Figure 13.3), higher hardness and wear resistance. The average microhardness of the surface composite layer was found to be three-time as compared to that of the substrate (Shafiei-Zarghani, Kashani-Bozorg, & Hanzaki, 2011). Later, Britto et al. (2016) used the friction stir process (FSP) to add alumina powder for improving the surface hardness of Al. HSS tool with a feed rate of 30 mm/min, rotating at 1,600 rpm and a transverse load of 8 KN was employed for FSP. The material at the surface of the substrate is subjected to stirring and mixing around the pin as the tool rotates, this leads to reinforcement of the surface with Al_2O_3 particles. Due to friction between the plate and the tool, the temperature rises, and as the tool traversing speed increases, a lower reaction temperature was observed. Hence, for increasing the distribution of reinforcement particles on the substrate, a lower traversing speed was found suitable. The dispersion of particles led to increase in the microhardness by 1.7 times of the base metal (Britto, Jeevahan, Saikiran, & Mageshwaran, 2016).

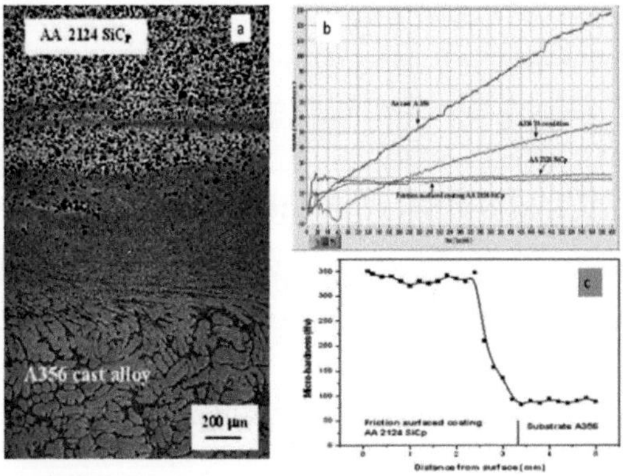

FIGURE 13.2 (a) Microstructure of coated cross-section; (b) wear rate comparisons; and (c) microhardness profile.

Source: Reprinted w ith permission from Tyagi et al., 2019. Copyright 2009. Taylor & Francis

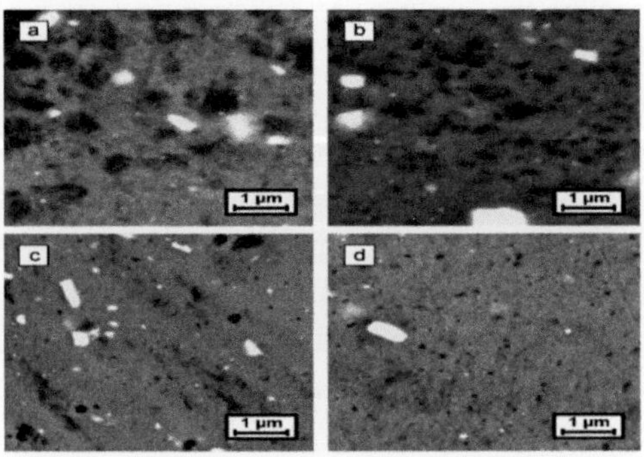

FIGURE 13.3 Dispersion of Al_2O_3 particles on the surface of composite layers produced by (a) one; (b) two; (c) three; and (d) four FSP passes.

Source: Reprinted with permission from Shafiei-Zarghani et al., 2011. Copyright 2011 Elsevier.

Friction Surfacing is finding a growing potential in the area of surface engineering. This method bears the advantage that complete melting of the part does not take place, dissimilar metals can be easily deposited as compared to fusion-based methods. Due to the application of large transverse load, the substrate requires heavy clamping, making it difficult to apply the process for thin workpieces. The process is not suitable for *in-situ* fabrication and repair works.

13.4 HARDFACING USING LASER-BASED TECHNIQUES

Man et al. (2001) did laser surface alloying of NiCrSiB powder on Al6061 using Nd-YAG laser of high power. NiCrSiB was mixed with Polyvinyl Alcohol for binding and coated on surface which was further roughened by sand-blasting. By selecting suitable parameters, NiCrSiB was effectively alloyed on Al6061 surface with efficient bonding among the deposited layer and base material. The surface consisted of intermetallic compounds (IMCs) of Ni-Al with fine dendritic structure (Figure 13.4(a)). A significantly high surface hardness of HV 900 was recorded (Figure 13.4(b)). Along with the depth of the surface, an uneven dispersion of Ni content was observed. Moreover, alloyed surface exhibited improved corrosion resistance, however, no change in the pitting potential was seen when compared to Al6061 (Man, Zhang, Yue, & Cheng, 2001).

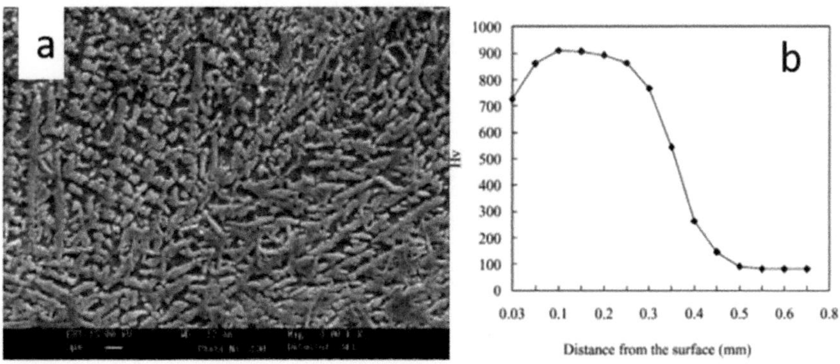

FIGURE 13.4 (a) Dendritic growth in laser surface-alloyed layer; and (b) variation of hardness along with the melt depth of the specimen.

Source: Reprinted with permission from Man et al., 2001. Copyright 2001. Elsevier

Duborg et al. (2005) used CW Nd: YAG laser cladding for producing MMC coating on Al (0, 12, 30 and 40 wt.% Si) substrate using TiC particles in the range of 0–30 vol.%. During the adhesive wear analysis, alloys with Hypoeutectic composition (Si content ≤12 wt.%) with no TiC reinforcement exhibited high wear accompanied by heavy plastic flow and considerable wear marks (Figure 13.5(a) and (b)). With the inclusion of TiC, slight wear was noticed. For alloys with hypereutectic composition (Si content ≥30%) with no TiC addition, rate of wear was comparable to that of Al–12Si/30 vol.% TiC layer (Figure 13.6(a)) and was mild. TiC reinforcement led to accelerated wear and damaged the surfaces (Figure 13.6(b)–(d)) of alloys with hypereutectic composition because of surface fatigue (Dubourg, Ursescu, Hlawka, & Cornet, 2005).

FIGURE 13.5 SEM micrographs of wear tracks: (a) Al-12Si; and (b) Al-40Si.

Source: Reprinted with permission from Dubourg et al., 2005. Copyright 2005. Elsevier

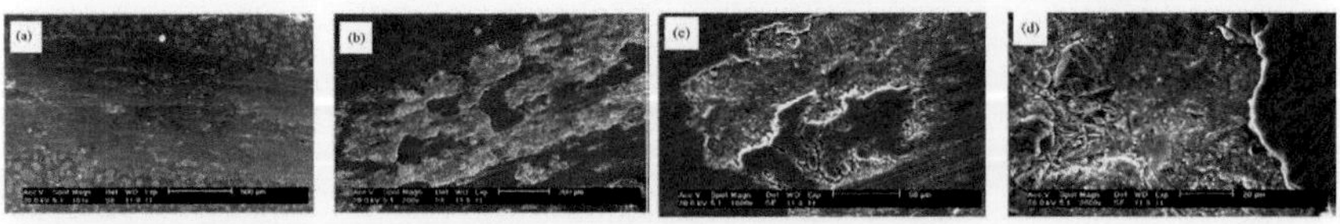

FIGURE 13.6 SEM micrographies of wear tracks after 1,600 m sliding on: (a) Al–12Si/30 vol.% TiC coating; and (b)–(d) Al–40Si/30 vol.% TiC coating.

Source: Reprinted with permission from Dubourg et al., 2005. Copyright 2005. Elsevier

Pityana (2009) did hardfacing of AA1200 alloy by developing MMC on the surface using LSA method. Composite layer was found to have evenly distributed TiC particles (Figure 13.7(a)) and XRD analysis confirmed the presence of no other phases than Al and TiC (Figure 13.7(b)). A Rofin Nd:

YAG laser was put to use for injecting the ceramic powder into the base metal. The laser power was varied from 3–4 kW, laser scan speed range was 0.8–2.0 m/min. The feeding rate of powder was restricted to 2–5 g/min. Micro-hardness of the layers was found to rise from 20 HV to 400 HV (Pityana, 2009). Further, Nath et al. (2012) used the LSA technique using a 5 kW CW-Nd:YAG laser for enhancing wear resistance by alloying the Al surface with WC+Co+NiCr. Layer obtained was found to have fine grained Al metal with the distribution of WC, W_2C, Al_4C_3, Al_9Co_2, Al_3Ni, $Cr_{23}C_6$, and Co_6W_6C. There was a marked increase in microhardness (200 HV to 650 HV) in comparison to 22 HV of base metal, however, gradual reduction in microhardness when measured from the surface to the base metal along the depth which was attributed to the change in carbide content and size of grains along the depth. Considerable increase in wear resistance was observed as can be seen from the wear spot in the modified surface and base metal (Figure 13.8(a) and (b)) (Nath, Pityana, & Dutta, 2012).

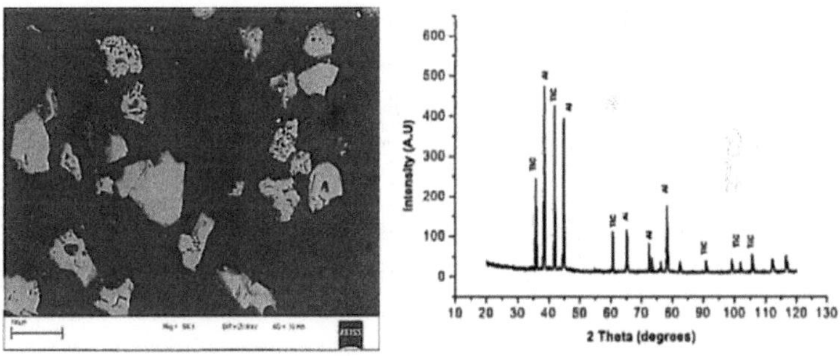

FIGURE 13.7 (a) SEM micrograph with TiC particles on Al surface; and (b) XRD analysis of the laser alloyed surface at 3 kW and 0.9 m/min laser scanning speed.

Source: Modified from Pityana, 2009

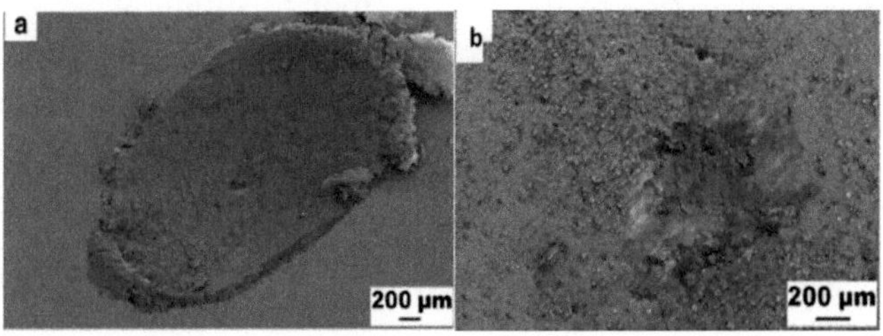

FIGURE 13.8 SEM micrographs of worn-out surface of (a) Al; and (b) laser surface alloyed Al.

Source: Reprinted with permission from Nath et al., 2012. Copyright 2012. Elsevier

Several researchers have employed laser cladding and LSA for hardfacing of Al-alloys with Nd-YAG laser as the most commonly used one. Although the method possesses high precision, however, it is highly expensive and requires a controlled environment as well as skilled operator. It is also difficult and risky to operate.

13.5 OTHER METHODS OF HARDFACING

Apachitei et al. (2000) developed autocatalytic nickel-phosphorus (Ni-P) coatings to improve the wear of an Al-alloy base material. Abrasive wear resistance was considerably improved by reducing phosphorus content and simultaneous deposition of SiC particles in the Ni-P matrix. The coatings exhibited greater wear resistance than particle-free coatings, irrespective of the phosphorus content. This was due to the entrapped SiC particles having considerably high hardness and wear resistance. A further improvement in the abrasion resistance and hardness of coatings with Phosphorous content in low and medium range was achieved by heat treating them at 220°C with 1 h holding time. For Coatings with high phosphorus content, the heat treatment was not useful (Apachitei & Duszczyk, 2000). In another study, Chen Bing-Quan et al. (2002) developed Al fillers containing Fe, Ni and Si as reinforcements for hardfacing of Al-Si alloys used for hardfacing pistons and cylinders. Filler metal showed exceptional weldability. Microstructures of the filler metal in as-cast and as-welded state were found to have uniform distribution of reinforcement and the matrix hard phase consisted of Al-Si-Fe-Ni. Mechanical and wear resistance properties of filler metal were considerably high at room temperature as well as at high temperatures. Further, it was observed that the deposited metal exhibited enhanced impact wear resistance at 220°C. At room temperature, the metal deposited had a comparable resistance to sliding wear with the use of lubricant to that of a hyper-eutectic Al-Si alloy with 1% Ni and 27% Si (Chen, Chai, & Peng, 2003). Also, Sun (2004) fabricated a rutile (TiO_2) coating on AA6061 alloy by first depositing a 2 microns layer of pure Ti using DC magnetron sputtering on the base metal followed by thermal oxidation. The coating obtained had a layered structure, consisting of a rutile-TiO_2 layer on the top, α-Ti layer with dissolved 'O' and 'N' in the middle and a TiAl layer in the intermediate region. The morphology of as deposited Ti coating is shown in Figure 13.9(a) and for oxidized coating produced at 550°C for 5 hours and 11 hours is shown in Figure 13.9(b) and (c). Hardness of oxidized surface was found to decrease with increasing penetration depth, this was because the rutile zone and coating system with layered structure had small thickness. The hardness of the oxidized surface was found to be higher than as-deposited surface up to the maximum depth of indentation, signifying better load bearing capacity of the oxidized coating. Wear resistance of coatings was found to increase with increase in the duration of oxidation. Oxidation is reported to enhance the adhesion of coating to the base metal which was attributed to the interfacial reactions that happened during oxidation, leading to metallurgical bonding between the coating and substrate by the formation of Ti–Al intermetallic and a solid solution of Ti in Al. X-ray diffraction (Figure 13.10) shows the evolution of rutile-TiO_2 phase (R), oxygen-diffused Ti (α-Ti(O)) and Ti–Al IMCs in the coating after thermal oxidation (Sun, 2004).

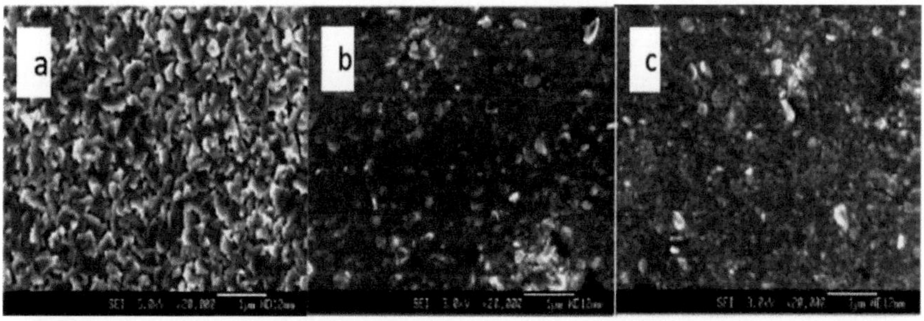

FIGURE 13.9 SEM micrographs showing the surface morphology of (a) as-deposited Ti coating; (b) oxidized coating produced at 550°C for (b) 5 h; and (c) 11 h.

Source: Reprinted with permission from Sun, 2004. Copyright 2004. Elsevier

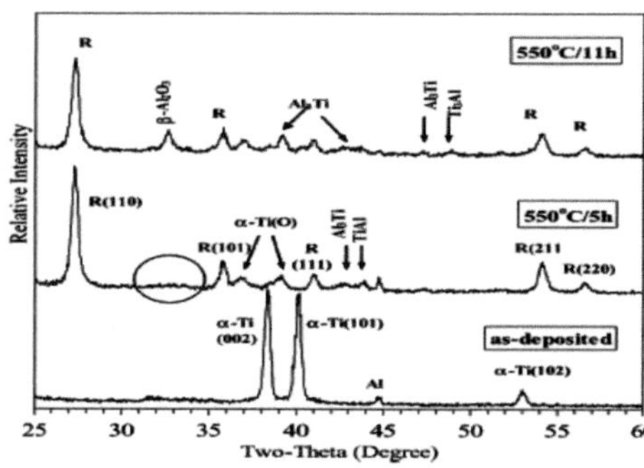

FIGURE 13.10 X-ray diffraction pattern from as-deposited and oxidized Ti coatings.

Source: Reprinted with permission from Sun, 2004. Copyright 2004. Elsevier

Josep et al. (2005) investigated the capability of high velocity oxy-fuel (HVOF) thermal spraying for improving wear behavior of 40CrMnNiMo steel, Al-7022 and metal matrix composite with Al-6061 Al alloy reinforced with 22 vol.% of Al_2O_3 particles. Tribological and mechanical characteristics of $CrC_{75}(NiCr_{20})_{25}$ and WC_{86} Co_{10} Cr_4 thermal sprayed HVOF coatings applied on the base materials were compared and it was observed that in all cases, WC–CoCr powder coatings had considerably higher hardness than CrC–NiCr coatings. Figure 13.11(a) and (b) shows the microstructure of the coatings on steel substrate consisting of metallic binders Co–Cr and Ni–Cr and dispersed carbide phases WC and CrC visible in dark and light colors. Figure 13.12(a) and (b) show the cross-sections of WC86Co10Cr4 coating on $A6061/(Al_2O_3)_p$ substrate and the interface coating-substrate with homogeneous distribution of Al_2O_3 in Al matrix. Some cracked alumina particles were seen, which was attributed to the residual stresses in the material due to plastic deformation of Al matrix during extrusion. It was established by microstructural analysis that HVOF spraying method can be successfully used for producing coatings having high density with better control over coating composition (Picas, Forn, Rilla, & Martín, 2005).

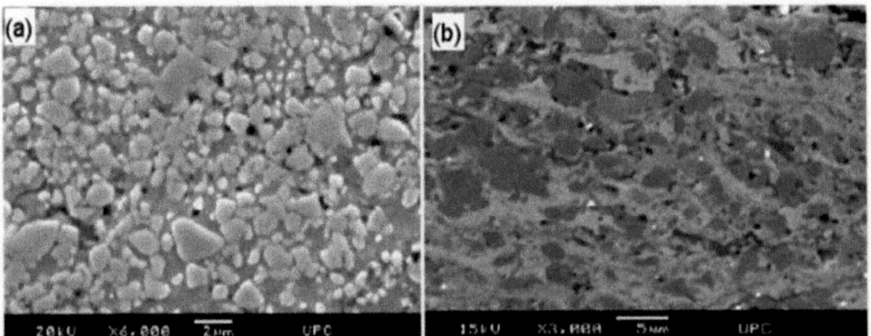

FIGURE 13.11 (a) Microstructure of $WC_{86}Co_{10}Cr_4$ coating; and (b) microstructure of $CrC_{75}NiCr_{25}$ coating on Al substrate.

Source: Reprinted with permission from Picas et al., 2005. Copyright 2005. Elsevier

Dilthey et al. (2009) enhanced the wear resistance of Al-alloys by producing fused tungsten carbide (FTC) and Titanium Carbide (TiC) composite coatings on AA5083 by plasma powder arc welding (PPAW) process. The coatings with FTC content ≤ 70 wt.% (Figure 13.13(a)) and those with 50 wt.% of TiC in the composite powders led to steady welding process and a uniform weld bead. Whereas welding with 80 wt.% FTC

(Figure 13.13(b)) and the one with 70 wt.% of TiC content was found to be unsteady and the surface of the substrate was piled up with most of the carbide particles. The addition of copper in one case and Nickel in another into the Al matrix led to development of intermetallic phases like Al-Cu and Al_3Ni, AlNi in the microstructure in respective cases which increased the hardness of the matrix. Ni addition in the AlCu and AlSiCu system led to further increase in hardness due to the formation of hard Cu_4NiAl_7 and Al–Cu–Si–Ni phases. Figure 13.14(a)–(e) shows the micrographs for the AlCuMg, AlCuSiNi and AlCuNi systems for 70% and 50% addition of FTC and TiC in the coatings. Hardness of 114HV for AlCuMg matrix and 253 HV for AlSiCuNi matrix were obtained. Improved hardness and fairly uniform dispersion of carbide particles led to higher wear resistance of the coatings (Dilthey, Kondapalli, Balashov, & Riedel, 2008).

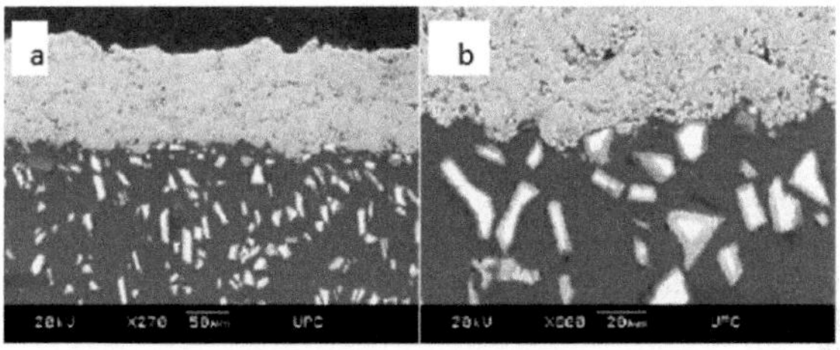

FIGURE 13.12 (a) Cross-sections of WC86Co10Cr4 coating on A6061/(Al_2O_3)p substrate; and (b) interface coating-substrate with homogeneous distribution of Al_2O_3 in Al matrix.

Source: Reprinted with permission from Picas et al., 2005. Copyright 2005. Elsevier

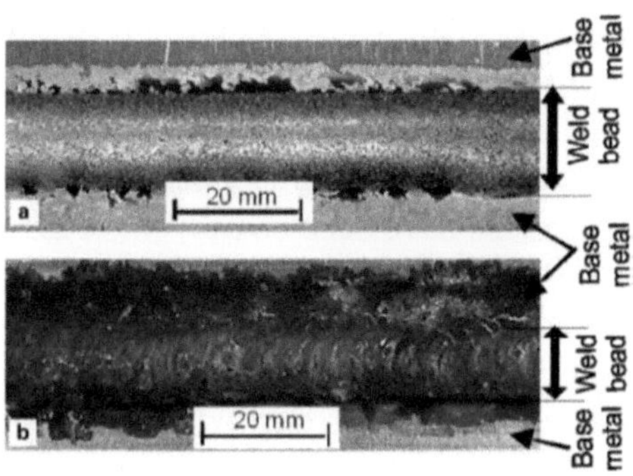

FIGURE 13.13 Surface appearance of PPA coating with AlCuMgz FTC composite powder system with a 70 wt.% FTC and b 80 wt.% FTC.

Source: Reprinted with permission from Dilthey et al., 2008. Copyright 2008. Elsevier

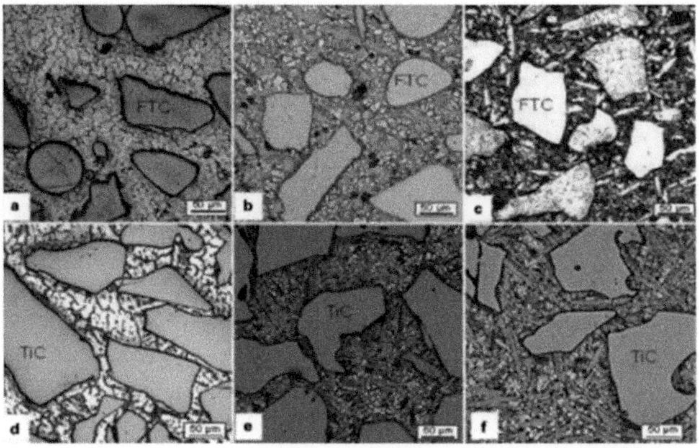

FIGURE 13.14 Microstructure of coatings with (a) AlCuMg + 70 wt.% FTC; (b) AlCuNi + 70 wt.% FTC; (c) AlSiCuNi + 70 wt.% FTC; (d) AlCuMg + 50 wt.% TiC; (e) AlCuNi + 50 wt.% TiC; and (f) AlSiCuNi + 50 wt.% TiC.

Source: Reprinted with permission from Dilthey et al., 2008. Copyright 2008. Elsevier

13.6 CONCLUSION

Wear emerges as one of the significant factors affecting the life span and maintenance cost of components operating in contact with similar or dissimilar material or medium. Hardfacing of Al alloys emerges as the most economical and fastest method for enhancing the wear resistance of the alloys. As discussed in the chapter, both conventional and non-conventional techniques can be used for hardfacing of surfaces of Al alloys. As per the results obtained in research, significant enhancement in the tribological and mechanical properties of the base material is obtained by hardfacing. Hardfacing by reinforcement of particles resulting in strengthening mechanism, development of intermetallic phases and microstructural changes leads to enhancement of wear resistance and hardness. Various methods with different mechanisms have been employed for developing composite coatings on Al alloy substrates possessing better tribological properties, and the results are different for different methods. The fusion welding methods, which involve high temperatures due to large heat input, have an adverse effect on the mechanical properties of the substrate. Close control of input parameters can, however give promising results. Methods which employ large non portable machines with limited size of fixtures such as FSP are infeasible for *in situ* modification/coating. Commercial use of such processes needs development of suitable setup through continuous research efforts. Methods using laser give high quality coatings/surfaces but they incur large processing costs due to the requirement of high-cost setup and skilled labor. The method to be adopted hence needs to be chosen with utmost care while considering cost of equipment, skill required, cost of processing, time available for processing, consumables required, end use of the product and the location of application. Although comparison of processes for particular conditions may prove one of them to be better than another one, but none of the processes can be listed as the best possible one as every process has some limitations. The need for development of highly efficient method which is economical and can be used commercially in a number of applications can be established from the review.

KEYWORDS

- **aluminum**
- **dispersion**
- **fusion welding**
- **hardfacing**
- **hardness**
- **intermetallic phases**
- **laser-based techniques**
- **reinforcement**
- **wear resistance**

REFERENCES

Ahn, D. G., (2013). Hardfacing technologies for improvement of wear characteristics of hot working tools: A review. *Int. J. Precis. Eng. Manuf.*, *14*(7), 1271–1283. https://doi.org/10.1007/s12541-013-0174-z.

Apachitei, I., & Duszczyk, J., (2000). Autocatalytic nickel coatings on Al with improved abrasive wear resistance. *Surf. Coatings Technol.*, *132*(1), 89–98. https://doi.org/10.1016/S0257-8972(00)00864-1.

Bayhan, Y., (2006). Reduction of wear via hardfacing of chisel ploughshare. *Tribol. Int.*, *39*(6), 570–574. https://doi.org/10.1016/j.triboint.2005.06.005.

Bhadauria, N., Pandey, S., & Pandey, P. M., (2019). Wear and enhancement of wear resistance: A review. *Mater. Today Proc.*, *26*(2), 2986–2991. https://doi.org/10.1016/j.matpr.2020.02.616.

Britto, J. G., Jeevahan, J., Saikiran, & Mageshwaran, G., (2016). Improving surface hardness of the Al plate by adding alumina powder by friction stir process. *Int. J. ChemTech Res.*, *9*(8), 587–593.

Chen, B. Q., Chai, C. X., & Peng, J. B., (2003). A high-Fe Al matrix welding filler metal for hardfacing Al-silicon alloys. *J. Wuhan Univ. Technol. Mater. Sci. Ed.*, *18*(1), 25–28. https://doi.org/10.1007/bf02835081.

Digambar, B., & Choudhary, D., (2014). A review paper on hardfacing processes, materials, objectives, and applications. *Iternational J. Sci. Res.*, *3*(6), 2012–2014.

Dilthey, U., Kondapalli, S., Balashov, B., & Riedel, F., (2008). Improving wear resistance of Al alloys by developing FTC and TiC based composite coatings using plasma powder arc welding process. *Surf. Eng.*, *24*(1), 75–80. https://doi.org/10.1179/174329408X277466.

Dimitrova, R., Kandeva, M., & Kamburov, V., (2017). Mechanical and tribological characteristics of TIG hardfaced dispersive layer by reinforced with particles extruded Al. *Tribol. Ind.*, *39*(1), 9–19. https://doi.org/10.24874/ti.2017.39.01.02.

Dubourg, L., Ursescu, D., Hlawka, F., & Cornet, A., (2005). Laser cladding of MMC coatings on Al substrate: Influence of composition and microstructure on mechanical properties. *Wear*, *258*(11, 12), 1745–1754. https://doi.org/10.1016/j.wear.2004.12.010.

Gandra, J., Krohn, H., Miranda, R. M., Vilaça, P., Quintino, L., & Dos, S. J. F., (2014). Friction surfacing: A review. *J. Mater. Process. Technol.*, *214*(5), 1062–1093. https://doi.org/10.1016/j.jmatprotec.2013.12.008.

Gerard, B., (2013). *Fundamentals of Hardfacing by Fusion Welding* (pp. 4–17). Welding Alloys Group.

Gualco, A., Svoboda, H. G., Surian, E. S., & Vedia, L. A. D., (2010). Effect of welding procedure on wear behavior of a modified martensitic tool steel hardfacing deposit. *Mater. Des.*, *31*(9), 4165–4173. https://doi.org/10.1016/j.matdes.2010.04.026.

Kato, K., (1992). Micro-mechanisms of wear – wear modes. *Wear*, *153*(1), 277–295. https://doi.org/10.1016/0043-1648(92)90274-C.

Kok, M., (2005). Production and mechanical properties of Al_2O_3 particle-reinforced 2024 Al alloy composites. *J. Mater. Process. Technol.*, *161*(3), 381–387. https://doi.org/10.1016/j.jmatprotec.2004.07.068.

Kuruveri, U. B., Panemangalore, D. B., Kuruveri, S. B., & John, M., (2022). Surface modification of 6xxx series Al alloys. *Coatings*, *12*(180), 1–26.

Man, H. C., Zhang, S., Yue, T. M., & Cheng, F. T., (2001). Laser surface alloying of NiCrSiB on Al6061 Al Alloy. *Surf. Coatings Technol.*, *148*(2, 3), 136–142. https://doi.org/10.1016/S0257-8972(01)01339-1.

Nath, S., Pityana, S., & Dutta, M. J., (2012). Laser surface alloying of Al with WC+Co+NiCr for improved wear resistance. *Surf. Coatings Technol.*, *206*(15), 3333–3341. https://doi.org/10.1016/j.surfcoat.2012.01.038.

Picas, J. A., Forn, A., Rilla, R., & Martín, E., (2005). HVOF thermal sprayed coatings on Al alloys and Al matrix composites. *surf. Coatings Technol.*, *200*(1-4 SPEC. ISS.), 1178–1181. https://doi.org/10.1016/j.surfcoat.2005.02.124.

Pityana, S. L., (2009). Hardfacing of Al alloy by means of metal matrix composites produced by laser surface alloying. In: 5th *Int. WLT-Conf. on Laser in Manuf. Munich* (pp. 439–443).

Pradeep, G. R. C., Ramesh, A., & Prasad, B. D., (2010). A review paper on hardfacing processes and materials. *Int. J. Eng. Sci. Technol.*, *2*(11), 6507–6510.

Reddy, G. M., Rao, K. S., & Mohandas, T., (2009). Friction surfacing: Novel technique for metal matrix composite coating on Al silicon alloy. *Surf. Eng.*, *25*(1), 25–30. https://doi.org/10.1179/174329408X298238.

Rigney, D. A., (1988). Sliding wear of metals. *Ann. Rev. Mater. Sci.*, *18*, 141–163. https://doi.org/10.1146/annurev.ms.18.080188.001041.

Selvi, S., Sankaran, S. P., & Srivatsavan, R., (2008). Comparative study of hardfacing of valve seat ring using MMAW process. *J. Mater. Process. Technol.*, *207*(1–3), 356–362. https://doi.org/10.1016/j.jmatprotec.2008.06.053.

Shafiei-Zarghani, A., Kashani-Bozorg, S. F., & Hanzaki, A. Z., (2011). Wear assessment of Al/Al_2O_3 nano-composite surface layer produced using friction stir processing. *Wear*, *270*(5, 6), 403–412. https://doi.org/10.1016/j.wear.2010.12.002.

Shibe, V., & Chawla, V., (2013). An overview of research work in hardfacing. *Mech. Confab*, *2*(3), 105–110.

Sun, Y., (2004). Tribological rutile-TiO_2 coating on Al alloy. *Appl. Surf. Sci.*, *233*(1–4), 328–335. https://doi.org/10.1016/j.apsusc.2004.03.241.

Tyagi, S. A., Bhadauria, N., Vashishtha, P., & Mishra, S., (2019). Improving surface hardness of Al 6063 alloy using hardfacing. *Mater. Today Proc.*, *26*, 1343–1346. https://doi.org/10.1016/j.matpr.2020.02.272.

Varenberg, M., (2013). Towards a unified classification of wear. *Friction*, *1*(4), 333–340. https://doi.org/10.1007/s40544-013-0027-x.

Vilaca, P., Gandra, J., & Vidal, C., (2012). Linear friction-based processing technologies for Al alloys: Surfacing, stir welding and stir channeling. *Alum. Alloy. – New Trends Fabr. Appl.* https://doi.org/10.5772/52026.

Yoshie, S., & Yasuda, K., (1995). Hardfacing of Al products using TIG arc welding methods. *Weld. Int.*, *9*(2), 94–99. https://doi.org/10.1080/09507119509548761.

CHAPTER 14

Cubosomes: An Approach to Novel Drug Delivery System (DDS)

PRATAP KALITA,[1] SWABNAM KHATANIAR,[1] RATNALI BANIA,[1] and PROBIN KR. ROY[2]

[1]*Pratiksha Institute of Pharmaceutical Sciences, Guwahati, Assam, India*

[2]*Regional Institute of Paramedical and Nursing Sciences, Zemabawk, Aizawl, Mizoram, India*

ABSTRACT

Cubosomes are crystalline nanoparticles (lipid-based) (LCNPs) which are recently attracting attention as nanocarrier in DDS among the different nanostructures. Biocompatibility and simplicity are the merits of cubosomes. Apart from these, cubosomes have the ability to increase the bioavailability of drugs. Cubosomes are having potential in controlled release of drug through functionalization. Different types of drugs owing to its different nature can be delivered by using cubosomes. LCNPs are formed by dispersion of lipid based liquid crystalline nanoparticles (LLCNPs) are prepared by dispersing the nonlamellar liquid crystalline matrix into water phase using high-energy fragmentation. These are self-assembled polar amphiphilic lipid. Cubic phase liquid crystalline nanoparticles generally prepared from aqueous dispersion of binary water like solutions and are relatively stable in presence of dispersing agent. The properties of monoolein – water-based nanoparticles or other lipid-water-based nanoparticles

Advances in Manufacturing and Processing of Materials: Characterization and Applications.
Narendra Kumar, PhD, Mohit Tyagi, PhD, Dilbagh Panchal, PhD & Ravi Pratap Singh, PhD (Eds.)

are studied with a variety of experimental method. Nowadays various methods are developed which requires minimal input of energy to ease the fabrication and application of cubosomes. Most used lipids in contriving the cubosomes are amphiphilic lipids such as monoolein which are essentially insoluble in water. This process is used to take in any lipids and hydrotropic in combination which will form cubic liquid crystalline material upon dilution. Based on the various recants reports, this chapter will emphasize on various method of preparation, characterization, evaluation techniques and application of cubosomes (Figure 14.1).

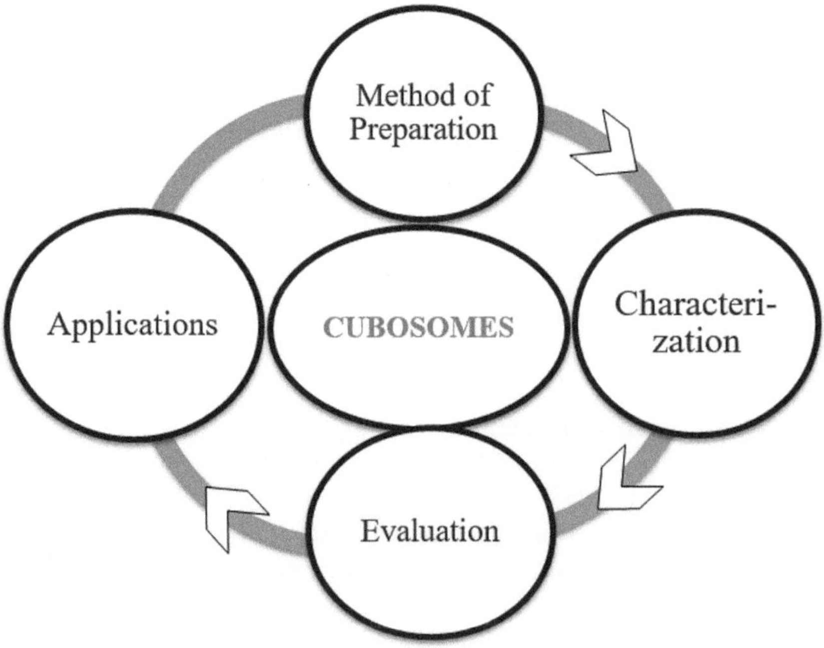

FIGURE 14.1 Graphical abstract.

14.1 INTRODUCTION

Cubosomes are lipid-based cubic phase nanoparticle are characterized by distinct nanovesicles of bi-continuous cubic structures with enhanced surface area and alike microstructure to its paternal cubic aggregates (Gaballa et al., 2020). Cubosomes term was coined by Larsson, it reflects the molecular crystallography of cubic structure. It has structural similarity to liposomes (Rao et al., 2018). LCNPs are prepared by certain Amphipathic lipids such as glycerol monooleate (GMO), phytantriol, etc. These lipids self-assemble in water to form cubosomes. Cubosomes exist as a suspension having less viscosity and they are easy to handle. Cubosomes are having potential in controlled release of drug through functionalization. They are having capability to conserve three dimensional assemblies under a variety of relevant physiological conditions.

Cubic phases in lipid-water system were recognized by Luzzati et al. (1962); and Luzzati et al. (1968) using X-ray scattering measurements. Similar conclusion was drawn by Fontell et al. (1968) about cubic phases in ternary systems such as amphiphiles, oils, and water. Similarly, biologists also started documentation of cubic symmetry in plant leaf plastid membranes (Gunning, 1965), which is now extensively acknowledged as cubic liquid crystalline phase assemblies. Various research work has been done in characterization and evaluation of cubosomes. Cubosomes loaded with doxorubicin and labeled with 177Lu has been recently used in Combined Chemo and Internal Radiotherapy for cancers. Cubosomes encapsulates doxorubicin (chemotherapeutic agents) with appropriately complexed 177Lu (radionuclide) which was a potent DDS for local therapy of various cancers (Cytryniak et al., 2020). Another research has been successfully delivered cubosomes/cubosomal gels containing beclomethasone dipropionate for the treatment of Uveitis in the eyes. The formulated cubosomes successfully enhanced residence time and ocular bioavailability (Gaballa et al., 2020, p 198).

14.1.1 ADVANTAGES

Cubosomes are having several advantages as a DDS:

1. **Drug Approaches:** Cubosomes can encapsulate different hydrophilic, hydrophobic, and amphiphilic drug moieties (Gaballa et al., 2020). Cubosomes have high loading and efficient release capacity of hydrophilic, hydrophobic, and amphiphilic drug (Flak et al., 2020). LCNPs are potential drug carriers having a great interfacial area of 400 m^2/g to encapsulate the drugs. The structure of the liquid crystalline nanoparticles consists of bi-continuous lipid

bilayer and it is surrounded by non-contacting aqueous channels. Within this biocompatible amphiphilic environment, a variety of drugs can be encapsulated, and it resembles a biological membrane (Cytryniak et al., 2020). The hydrophilic and hydrophobic drug encapsulated cubic phase get solubilized into cubic gel and release the drug in a controlled approach (Spicer et al., 2001).

2. **Bioavailability:** Cubosomes can improves the bioavailability of poorly water-soluble drugs which has an opportunistic approach to the novel DDS (Gaballa et al., 2020). Preparation Technique: Simple preparation techniques can be used to prepare cubosomes. It is less time consuming and has an ease of manufacturing approach (Gaballa et al., 2020).
3. **Structural Approach:** Cubosomes are composed of biodegradable lipids. It protects the drug which are incorporated in the cubosomes from various physical and chemical degradation. Apart of this it has high internal surface area, due to this it can load high drug contents (Gaballa et al., 2020).
4. **Drug Delivery Approach:** Cubosomes are thermodynamically stable, also shows bio-adhesion and have the capacity to encapsulate wide variety of drug moieties. It also has the capability to release the encapsulated drugs in controlled manner and regarded as a very promising carrier for administration of drugs via different routes (Zahra et al., 2016). Cubosomes are having advantages in topical administration of drugs. During penetration through a corneal and skin layer, the nanovesicles forming lipids (GMO) provides a permeation enhancer for cubosomes. It is also a potential DDSas it can be formulated for sustained release of drugs (Gaballa et al., 2020). Cubosomes are having potential in controlled release of drug through functionalization (Gaballa et al., 2020). Viscosity of cubosomes is less and it exists as a suspension. This approach makes the cubosomes much easier to handle and deliver to the suitable site.
5. **Stability Approach**: Under physiologically irrelevant conditions, it has the capability to preserve the three-dimensional assembly. Structure can be retained in all conditions. In cubic phase liquid crystalline nanoparticles, the internal cubic phases are stable in the presence of excess of water and the cubic phase can be dispersed into kinetically stable colloidal nanoparticles in the presence of a

stabilizer (Cytryniak et al., 2020). Cubic phase liquid crystalline nanoparticles, which are prepared from aqueous dispersion of two water-like solutions are quite stable in the presence of certain dispersing agent (Gaballa et al., 2020).

14.1.2 DISADVANTAGES

Apart from various advantages, few disadvantages are also there for cubosomes:

i. It has high viscosity in cubic phase, for that it imparts a great challenge in large scale production in the pharmaceutical manufacturing (Gaballa et al., 2020; Dhadwal et al., 2020).
ii. Low entrapment efficiency for water soluble drugs due to the presence high water content in its internal structure (Gaballa et al., 2020).
iii. Drug leakage may occur in preparation, preservation and transport *in vivo* (Dhadwal et al., 2020).

14.2 STRUCTURE OF CUBOSOMES

LLCNPs size varies from 10–500 nm in diameter and they look like dots square shape with slightly spherical. First dot was identified using X-ray scattering technique. Each dot corresponds to the presence of poor containing aqueous phase cubic phases in lipid water system (Bachhav et al., 2015). Cubosomes are having bicontinuous cubic structures with high internal surface area and alike microstructure to its parent aggregates (Gaballa et al., 2020). Lyotropic methods or fragmentation of bulk cubic phase gels are used to prepare LLCNPs where it preserves its structure of cubic phase and offers large surface area with low viscosity (Pan et al., 2013). The composition of cubosomes can be adjusted to fix up pore sizes or bioactive lipids can be utilized, the out corona which is composed of polymer are used for targeting at the specific site of action and they are stable under normal physiological conditions (Hanna et al., 2018).

The structure of cubosomes include bicontinuous domains of water and lipid. It separates the two internal aqueous channels. In the structure of cubosomes the surfactants are assembles itself into bilayers and twisted into 3D which are having least surface and are tightly packed like honeycomb (Bachhav et al., 2015). In non-ionic surfactant system, the cubic phases of the nanoparticles are founded to be crammed among lamellar and hexagonal liquid crystalline phase. The cubic phase region which is possesses due to the monoolein-water system it contains broad composition and temperature range. It produced reversed or inversed cubic phase (Bachhav et al., 2015, Dhadwal et al., 2020). Cubic phase nanoparticles are very unique due to its contorted lipid bilayer separating two continuous but nonintersecting water regions.

14.3 THEORIES OF SELF-ASSEMBLING OF AMPHIPHILIC LIPIDS

Certain amphiphilic surfactants are having ability to self-assemble in the aqueous media with their polar and non-polar components into highly ordered structures which resembles aggregates with some long-range order in one. The surfactant structure is available in two as well as three-dimensional form with a narrow or shorter range of atomic distance. The resultant surfactant structures which are formed due to the self-assemble of the surfactants. The surfactants are available in different form like micelle form, open and closed bilayer form and opposite micelles (Gaballa et al., 2020). There are two main theories Which will explain the mechanism of the surfactant by which it can be self-assembled into especial structure:

- Principles of opposing forces; and
- Principles of packing parameters.

14.3.1 PRINCIPLE OF OPPOSING FORCES

When amphiphilic molecules are exposed to polar solvents, it will arrange in such a way that it minimizes its free energy. Polar solvents penetrate through amphiphilic molecules, and it will expose the hydrophobic portion to aqueous environment and the hydrophobic portion of the amphiphiles are shielded from the solvent. The developement of opposing force was seen as result as a result of hydrophobic interaction between the hydrophobic hydrocarbon tails and hydrophilic head groups of the amphiphilic molecules an opposing force is developed. And these types of interaction coopete together at the interfacial area (Gaballa et al., 2020).

14.3.2 PRINCIPLE OF PACKING PARAMETERS

For a given lipid, the lipid aggregates which are formed can be best predicted by the principle of packing parameters. The critical packing parameter of the lipid was Elaborated by Israelachvilli and recently it has been well discussed by H Abdelkader. The critical packing parameter relates with the molecular shape and properties of certain lipid to its favored curvature at the lipid water interface (Gaballa et al., 2020) (Figure 14.2).

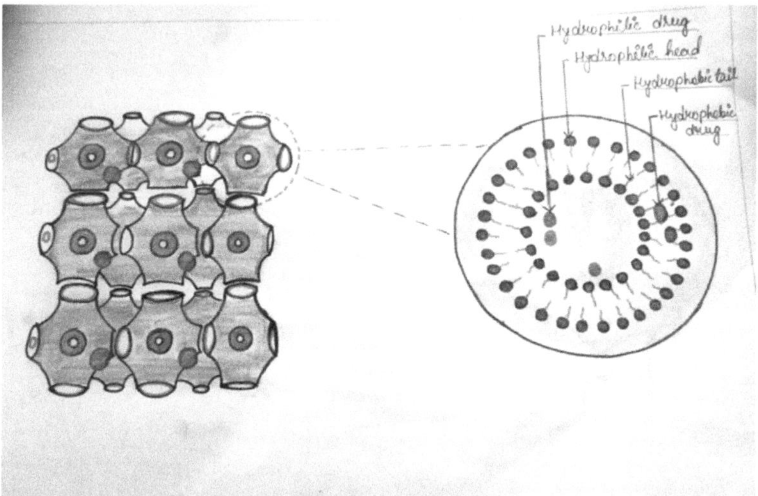

FIGURE 14.2 Internal structure of cubosomes.

14.4 TYPES OF CUBOSOMES

Cubosomes are classified into the following types:

- Liquid cubosomes precursors; and
- Powdered cubosomes precursors.

14.4.1 LIQUID CUBOSOMES PRECURSORS

Smaller and more stable cubosomes are prepared by using the hydrotropic dilution process. The growth of particles which are formed by the nucleation process is seen under crystallization and precipitation processes. Amphiphilic lipids such as monoolein is dissolved properly in a hydrotropic, such as ethanol, which will prevent it from liquid crystalline formation. Crystallization and precipitation of the Cubosomes occur spontaneously by dilution of this mixture. For the easier scale up in preparations of cubosomes, Quid precursor process is allowed, and it avoids the bulk solids handling and potentially damaging high energy processes (Bachhav et al., 2015, Dhadwal et al., 2020).

14.4.2 POWDERED CUBOSOMES PRECURSORS

Powdered cubosomes precursor are having several advantages over Liquid cubosomes precursors. It is composed of certain dehydrated surfactants with polymers. Cubosomes which are prepared from lipids are sticky and waxy solids. By the hydration of the precursor of powder, cubosomes with mean 600 nm are formed and it can be confirmed by light scattering and Cryo TEM (Bachhav et al., 2015, Dhadwal et al., 2020).

14.5 METHOD OF PREPARATION

Cubosomes are prepared by using two distinct methods:

- Top-down technique; and
- Bottom-up technique.

14.5.1 TOP-DOWN APPROACH

It is the most widely used technique for the manufacture of cubosomes in research field. This method was reported by Ljusberg-Wahren in 1996 (Naveentaj et al., 2019). In this technique bulk cubic phase is first prepared by using high-energy homogenization process and then it is processed into cubic nanoparticles (Bachhav et al., 2015). The bulk cubic phase is highly viscous, and it is prepared by mixing lipids with stabilizers. After preparation of bulk cubic phase, the mixture is dispersed into aqueous solution by using high pressure homogenization process, sonication and shearing, etc., where input of high energy is required. and resultant product will be a lyotropic liquid crystal (LLC) nanoparticle (Bachhav et al., 2015, Dhadwal et al., 2020) (Figure 14.3).

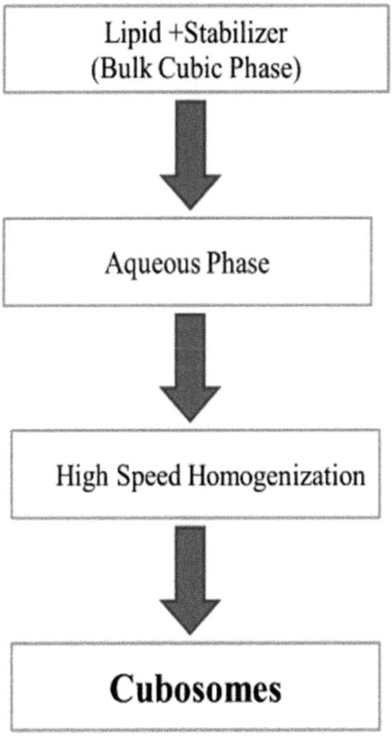

FIGURE 14.3 Top-down approach.

14.5.2 BOTTOM-UP APPROACH

Nanostructure blocks are formed in bottom-up approach and then the nanostructured block assemble them into final product. Hydrotropic gets dissolve in water insoluble lipids to create a liquid precursor. In order to prevent the formation of liquid crystals minimum energy input is required at high concentration. Inverse micellar droplets phase droplets are dispersed which will lead to the formation of cubosomes at 80°C in water, temperature is reduced until it is cooled down then the droplets get crystallized and it will lead to the formation of cubosomes (Bachhav et al., 2015; Dhadwal et al., 2020) (Figure 14.4).

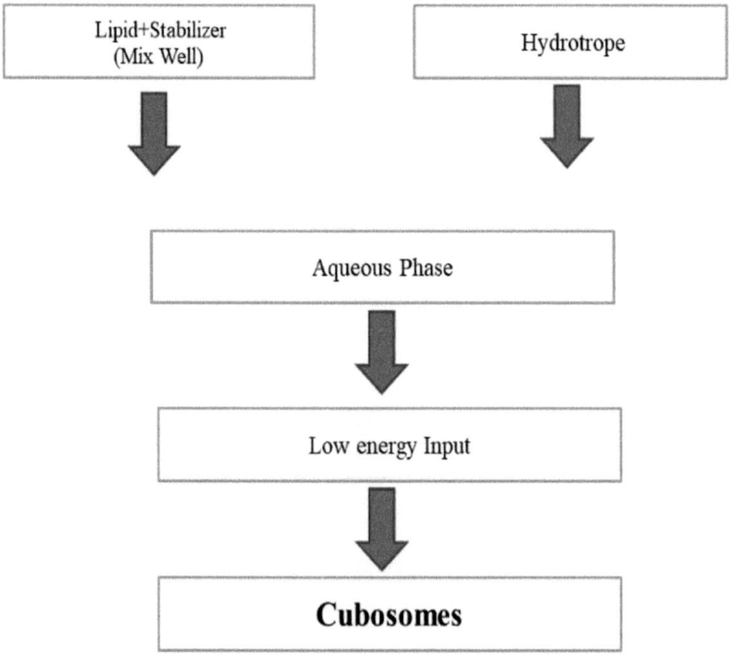

FIGURE 14.4 Bottom-up approach.

The dispersion of cubosomes is carried by two methods (Rao et al., 2018):

1. **Fabrication Method**: In this method GMO at the concentration of 5% and Poloxamer 407 at the concentration of 1% are melted together at 60°C in a hot water bath and then added required amount of drug by continuous stirring. Deionized water is added dropwise and homogenized the whole mixture by vortex. It will keep for 48 hours, and temperature will be maintained at room temperature. Now cubic gel is formed, and it is disturbed by mechanical stirring. By the use of probe sonicator crude dispersion can be fragmented at cool temperature subsequently at 20°C for 20 minutes (Rao et al., 2018).
2. **Emulsification Method:** In this method GMO at the concentration of 5% and Poloxamer 407 at the concentration of 1% are melted and mixed properly. Now 5% ethanolic solution will be added

to the melted mixture. Melted mixture is then added dropwise to deionized water. Deionized water will be preheated to 70°C before adding the melted mixture. It is now ultrasonicated for 50 minutes at the specified temperature for dispersion. Now the dispersed solution will be kept at ambient temperature (Rao et al., 2018).

14.6 EVALUATION OF CUBOSOMES

Cubosomes are evaluated for determining:

- Shape of cubosomes;
- Particle size distribution;
- Zeta potential;
- Differential scanning colorimetry;
- X-ray diffraction;
- Drug release;
- Entrapment efficiency;
- Stability studies.

14.6.1 SHAPE OF CUBOSOMES

Electron Microscopic technique is used for revealing the shape and morphology of cubosomes. Transmission electron microscope is used to photograph the detailed structure of the cubosomes whereas Scanning electron microscope is used to study the surface morphology. It provides detailed three-dimensional view of the structure (Rao et al., 2018).

14.6.2 PARTICLE SIZE DETERMINATION

Zeta sizer (Photon correlation spectroscopy) is used for determination of the particle size of cubosomes. It determined the particle size by dynamic laser light scattering. In this technique sample are diluted with suitable solvent and measured at 25°C in triplicate. Data includes PDI index, average particle size and zeta potential of the sample (Rao et al., 2018).

14.6.3 ZETA POTENTIAL

To estimate the surface charge of the nanoparticles zeta potential technique is employed to know the physical stability of nanoformulations. For a stable nanoformulation, a zeta potential value of –30 mV to +30 mV is desired. If zeta potential value is not near/close to the desired value it may results in physical instability (Joseph et al., 2019). The zeta potential values depend upon surface charge on cubosomes. The zeta potential value is affected by nature of surfactant and lipid content (Rao et al., 2018).

14.6.4 DIFFERENTIAL SCANNING COLORIMETRY (DSC)

For detection of any potential changes in the physical state of an encapsulated drug in a cubic phase gel, DSC is used by thermal scanning analysis system. Sample are heated in platinum pan at constant rate and any changes in the glass transition temperature is compared with the formulation and the standard drugs (Mohamed et al., 2015).

14.6.5 X-RAY DIFFRACTION

X-ray diffractometer is used to obtain the X-ray diffraction pattern of the sample. Data are collected using scattered angle. The diffractogram are recorded in a particular condition such as voltage 45 kV, correct 30 mA, steps 0.02° and the counting rate 0.5 s/steps at room temperature (Mohamed et al., 2015).

14.6.6 DRUG RELEASE

Cubic phases have the ability to incorporate and control the release of various molecular weight drug moieties with different polarities. Cubic phase matrix obeyed diffusion-controlled release and follow Higuchi kinetics (Bachhav et al., 2015; Dhadwal et al., 2020). *In-vitro* drug release study from cubosomes can be performed using pressure ultrafiltration (UF) and equilibrium dialysis (Rao et al., 2018). *In vitro* drug release can be evaluated by using dynamic dialysis method. Drug release can be determined by placing cubosomal dispersion in dialysis tubing after free drug get separated from drug loaded cubosomes (Mohamed et al., 2015).

14.6.7 ENTRAPMENT EFFICIENCY

UF centrifugation is used to calculate the entrapment efficiency of cubosomes (Joseph & Singhvi, 2019). The amount of drug added is analyzed by spectrophotometer (Rao et al., 2018). Some drug may be adsorbed to the UF membrane to some extent. The amount of drug adsorbed on the UF membrane is studied by filtration of known concentration of drug solution through the membrane and measuring the concentration in the ultrafiltrate (Joseph et al., 2019).

14.6.8 STABILITY STUDIES

To determine the physical stability of Cubosomes organoleptic and morphological aspects are studied in time interval. Drug Content and particle size distribution are evaluated at different time interval for the possible variations by time (Rao et al., 2018). When cubosomes are lyophilized, there is a change in particle size. With ultra-pure water, a decrease of 3.5% and with PBS buffer an increase of 16% in particle size, with an increase in polydispersion index without changing the morphological features (Barbara et al., 2021).

14.7 CHARACTERIZATION OF DRUG RELEASE FROM CUBOSOMES

To maintain the therapeutic drug concentration over a prolonged period of time, sustained release dosage form is desired. It reduces the dose frequency and produces less side effect than the conventional dosage form. Cubosomes are developed to overcome the problems and to deliver the drug in sustained manner for prolonged period of time. Cubosomes are proposed as a DDS for solubilization of poorly water-soluble drugs and also for sustained or controlled release. The release of drug is controlled by the cubic phase. During diffusion of the incorporated drug in cubosomes because of the bulk cubic phase their surface area gets enlarged and it gets divided into dispersed nanostructure as cubosomes. The properties of cubic phase such as drug loading, solubility, drug diffusion, etc., are not changed on division into cubosomes (Bachhav et al., 2015; Dhadwal et al., 2020).

14.8 APPLICATION OF CUBOSOMES

14.8.1 IN TREATMENT OF CANCER

Cubosomes performance as DDS for cancer is excellent since it is a novel lipid-based DDS (Sameh et al., 2021) and also it has the ability to transport drugs across blood brain barrier (BBB). When cubosomes are conjugated with specific ligand such as folic acid, it overexpresses folate receptor and enhance the drug delivery effectiveness. Cubosomes could be beneficial as their diagnostic agent when it is fixed with imaging and therapeutic agents. It has been reported recently that cubosomes are prepared with co delivery of two drug molecules. This type of cubosomes which are known as dual modality cubosomes have application in fluorescence-MR imaging. Cubosomes show higher specificity towards HeLa cancer cells with anticancer drug therapy (Cytryniak et al., 2020).

Cubosomes do not penetrate through the tight junction of the normal cell. But it can pass through the leaky junction of the tumor cells and release the drug at the targeted site. Thus, it kills all the cancer cells. That is why anticancer drugs are encapsulated in cubosomes and these are characterized physiologically. Anticancer drug encapsulated cubosomes can actively or passively target the cancer cells (Bachhav et al., 2015; Dhadwal et al., 2020; Rao et al., 2018). Cubosome loaded with antineoplastic agents are having bicontinuous cubic structure. Cellular imaging showed that these nanoparticles can confined in lysosomes and mitochondria for better efficacy. The nanoparticles have higher synergistic effects with least cytotoxicity (Jenni et al., 2020). Marco et al. (2020) found that cubosomes prepared with the poly(phosphoester) was found less toxic compared to prepared with Pluronic F127. The Cubosomes prepared with potentially biodegradable poly(phosphoester) represents a novel and flexible platform for fabrication of effective and smart nanocarriers. In Micelle, liposome and cubosomes, surface functionalization is considered for encapsulation of anti-cancer drugs (Mehran et al., 2021) (Figure 14.5).

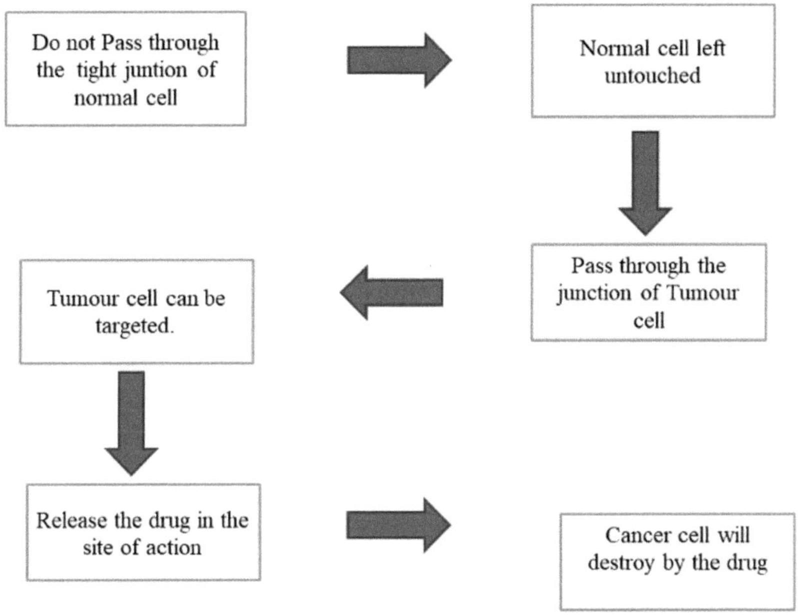

FIGURE 14.5 Action of cubosomes on tumor cell.

14.8.2 AS AN ORAL DRUG DELIVERY SYSTEM (DDS)

Cubosomes challenges oral delivery of numerous poorly water soluble, poor absorbed and drugs molecules with larger size. Cubosomes enhances the absorption of poorly water-soluble drugs due to its bio-adhesive properties via either interaction with intestinal cell membrane or induced secretion of physiological surfactants. Cubosomes incorporated the poorly water-soluble drugs within their lipid structure in the solubilized form. It prevents the drug precipitation in the Gastrointestinal tract and also improves intestinal absorption due to the mucoadhesive properties of amphiphilic lipids like GMO (Gaballa et al., 2020). It is reported that cubosomes enhances the delivery of poorly water-soluble nutrients comprising vitamins (Luzzati & Husson, 1962). Cubosomes are designed in such a way that it will form *in situ* in a controlled manner, and it help in enhancement of the effective *in vivo* distribution of drugs (Bachhav et al., 2015; Dhadwal et al., 2020).

14.8.3 AS AN INTRAVENOUS DRUG DELIVERY SYSTEM (DDS)

As comparison to the liposomes and emulsions, cubosomes have increased cargo for peptides, proteins and numerous small molecules. Cubosomes act as carrier for injections. Lipid based nanoparticles which are comprises of liquid crystal structure can solubilize the encapsulated drug and can deliver the drug to targeted site of action within the body (Rao et al., 2018).

14.8.4 AS A TOPICAL DRUG DELIVERY SYSTEM (DDS)

Topical DDS is a chemical formulation or a transport carrier which will transport the therapeutic agent to the specific area at a controlled rates that it will achieve effective concentration. The amount of drug should be adequately loaded into the carrier which depends on the interaction of the vehicle with drugs so that it will produce maximum therapeutic effects. Recent studies suggest cubosomes as a promising DDS for the treatment of skin disease (Fazilla et al., 2022). Cubosomes are showing interest as a topical and mucosal DDS of different drugs due to its bio-adhesiveness character. Due to the unique properties cubosomes can act as a potential DDS in topical application. Cubosomes facilitates controlled drug delivery to mucosal layer in an effective manner (Rao et al., 2018). Cubosomes as a topical DDS form a thin surface film at mucosal layer which consist of liquid crystal matrix which nanostructure and an optimal drug delivery profile can be achieved (Bachhav et al., 2015; Dhadwal et al., 2020).

14.8.5 AS A DRUG DELIVERY VEHICLE

Cubosomes can be act a drug delivery vehicle for numerous formulations. Many cosmetic manufacturing company are trying to use cubosomes as oil in water (o/w) emulsion stabilizers and also as pollutant absorbents in many cosmetics (Rao et al., 2018).

14.8.6 AS A CONTROLLED OR SUSTAINED DRUG DELIVERY SYSTEM (DDS)

Drugs having different physiochemical properties can be incorporated in cubosomes and the release of the drug can be controlled. Cubosomes has high internal surface area, due to this it can load high drug contents. It is also a potential DDS as it can be formulated for controlled and sustained release of drugs (Gaballa et al., 2020). Under physiologically pertinent conditions, it can maintain ability the three-dimensional structure.

14.8.7 IN THE TREATMENT OF VIRAL DISEASE

Monoglycerides which are used in the formulation of a cubosomes are having microbicidal properties. It can kill different types of disease-causing microorganisms. This monoglycerides can be used to strategically treat sexually transmitted diseases (Rao et al., 2018).

14.8.8 AS A DELIVERY VEHICLE TO THE BRAIN

Cubosome loaded with phenytoin shows a potential role in the delivery of drug moieties across the BBB. It showed required brain concentration and brain to plasma ratio (Younus et al., 2022).

14.8.9 IN THE TREATMENT OF GLAUCOMA

There is a great challenge in ophthalmological drug therapy due to the complexity of ocular barriers and which makes drugs limited or no efficacy for drugs delivered through the eye. Cubosomes loaded with latanoprost shows as a promising carrier for ophthalmological DDS. It shows effective results as compared to the conventional marketed formulation (Carolina et al., 2021).

14.9 FUTURE ASPECTS

Cubosomes are recently attracting attention as a nanocarriers for DDS. But there are a smaller number of studies are conducted in this aspect after *in vivo* administration. Various *in-vitro* investigation such as cellular response of these nanostructure are also conducted on limited number of model cell lines. From the literature it is found that most of the investigations are focused on encapsulation and biocompatible characterization of the cubosomes. In potential clinical applications exploration of these nano-structure, further *in vivo-in vitro* evaluations studies should be focused and correlation with biophysical experiments to gain further insights into their drug release behavior and entrapment efficiency along with the encapsulation behavior to the drug molecules.

The cubic phase liquid crystalline nanoparticle shows a promising aspect in delivering drug in proper manner and drug release at a constant rate for longer period of time but depending on the various parameters like route of administration, dosing frequency, mode of drug release, optimization is required for the formulation before such nanocarriers can truly realize their therapeutic potential in many diseases. Cubosomes act a nano-vehicle for loading and delivery of proteins and its derivatives along with peptides. Studies on these aspects is still on a fundamental level. So further studies are required for optimizing the field where the structural and morphological characteristic of the cubosomes nanocarriers and loading capacity, entrapment efficiency for the biomacromolecules should be diversly studied. During early stage of formulation development of cubosomes based intravenous nanomedicines, blood compatibility studies should be addressed for future aspects.

The applicability of poorly water-soluble drugs already been established as oral, ocular as well as topical drug delivery formulation but simultaneously intravenous administration of cubosomes nanocarrier is still needed to work on that. There by offering an alternative and a cost-effective opportunity for pharmaceutical science in formulation development. The rational design of smart cubosomes systems is enabled for diverse applications due to various recent advances in the formulation development and evaluation of cubosomes. Tailoring of pore sizes in lipid cubic phases, the various stabilizers which are used in the formulation of cubosomes are now available in a library, from which the cubosomes can be targeted into the different cells for studying of their several structural behaviors to understand access to the internal lipid membrane and for controlled release the systems designing are the key advances which are included in the advance studies. Key outstanding challenges are still there which would further enhance the applications of cubosomes.

Among those stabilizer membrane interaction, pore size demonstration tuning analogs in the bulk cubic phase and also mechanisms of interaction uptake in cells and smart release for the better understanding of cardiotoxicity. Cubosomes composed with bioactive lipid could be an interesting futures perspective for the novel therapeutics. As sciences developing rapidly, cubosomes are evolving as smart lipid nanoparticles and many interesting applications could be observed in the near future (Hanna et al., 2020).

14.10 CONCLUSION

Cubosomes are currently considering as a potential candidate in front of novel DDS as cubosomes nanocarriers potentially and therapeutically release the drug moieties from core at controlled rate for longer duration of time. Cubic liquid crystalline phases or cubosomes offers unique properties especially in particular interest to pharmaceutical field. Cubosomes are prepared using water and biocompatible lipids and also used various types of diseases of skin, hair, and other body parts. Applications of cubosomes include treatment of melanoma (cancer) therapy, conventional oral drug delivery to intravenous delivery. Cubosomes offers great flexibility in product development. Cubosomes prepared by dispersion techniques produces nanoscale structure similar to bulk cubic phase. Bulk cubic phase offers controlled release of drugs. Many bioactive peptides with multiple pharmacological action are available. However, due to their stability problem and difficulty in formulation, most of the peptide could be formed as dosage forms. In this perspective, cubosomes could be a game changing platform for the delivery of bioactive peptides via cubosomes. In the upcoming years, we are expecting many new applications of cubosomes specially in pharmaceutical industry.

KEYWORDS

- **characterization**
- **cubosomes**
- **internal lipid membrane**
- **melanoma**
- **nanocarriers**
- **ophthalmological drug therapy**
- **pharmaceutical industry**

REFERENCES

Adrianna, C., Ewa, N., Renata, B., Emilia, G., Zelechowska-Matysiak, K., Rafał, W., Adam, M., et al., (2020). Lipidic cubic-phase nanoparticles (cubosomes) loaded with doxorubicin and labeled with 177Lu as a potential tool for combined chemo and internal radiotherapy for cancers. *Nanomaterials, 10*, 2272. doi:10.3390/nano10112272.

Angelova, A., Angelov, B., & Drechsler, M., (2013). Protein entrapment in PEGylated lipid nanoparticles. *International Journal of Pharmac., 454*(2), 625–632.

Barbara, M., Raphael, D. D. C., Mayra, C. G. L., Bruna, R. C., & Leandro, R. S. B., (2021). Design and manufacturing of monodisperse and malleable phytantriol-based cubosomes for drug delivery applications, *Journal of Drug Delivery Science and Technology, 61.* https://doi.org/10.1016/j.jddst.2020.102149.

Carolina, D. V. B., Seyedeh, P. A., Luis, I. T., Daniela, A. Q., Watson, L., & Daniel, A. Al., (2021). Latanoprost-loaded phytantriol cubosomes for the treatment of glaucoma. *European Journal of Pharmaceutical Sciences, 160,* 105748.

Dhadwal, A., Sharma, D. R., Pandit, V., Ashawat, M. S., & Kumar, P., (2020). Cubosomes: A novel carrier for transdermal drug delivery. *Journal of Drug Delivery and Therapeutics; 10*(1), 123–130.

Dorota, K. F., Vivian, A., Grzegorz, N., Kosma, S., Michael, S., Stefan, J., & Held-Feindt, J., (2020). *AT101-Loaded Cubosomes as an Alternative for Improved Glioblastoma Therapy*. Dove Press, https://doi.org/10.2147//ijn.s265061.

Emil, J., & Gautam, S., (2019). Multifunctional nanocrystals for cancer therapy: A potential nanocarrier. In: *Nanomaterials for Drug Delivery and Therapy*.

Fazila, Z., Siti, E. A., Intan, D., Mat, A., Mohd, B., & Abdul, R., (2022). Recent advances in encapsulation of drug delivery (active substance) in cubosomes for skin diseases. *Journal of Drug Delivery Science and Technology*. https://doi.org/10.1016/j.jddst.2022.103097.

Fontell, K., Mandell, L., & Ekwall, P., (1968). Some isotropic mesophases in systems containing amphiphilic compounds. *J. Acta Chem. Scand., 22*, 3209–3223.

Gunning, B. E. S., (1965). In: Hofler, K., & Porter, K. R. J., (eds.), *The Greening Process in Plastids: The Structure of the Prolamellar Body in Protoplasma* (pp 111–139). Springer-Verlag: New York.

Hanna, M. G. B., Margaret, N. H., & Molly, M. S., (2018). *Cubosomes: The Next Generation of Smart Lipid Nanoparticles?* GDCh. https://doi.org/10.1002/anie.201804067.

Hanna, M. G. B., Margaret, N. H., & Molly, M. S., (2020). *Cubosomes: The Next Generation of Smart Lipid Nanoparticles?* PMC, doi: 10.1002/anie.201804067.

Jyoti, K. B., Bhushan, A. B., & Saudagar, R. B., (2015). Cubosomal technique: A review. *International Journal of Pharmacy & Technology*. https://www.researchgate.net/publication/283862110 (accessed on 11 September 2023).

Luzzati, V., & Husson, F., (1962). The structure of the liquid- crystalline phases of lipid–water systems. *J. of Cell Biology, 12*, 207–219.

Luzzati, V., Tardieu, A., Gulik-Krzywicki, T., Rivas, E., & Reiss-Husson, F., (1968). Structure of the cubic phases of lipid– water systems. *Nature, 220*, 485–488.

Marco, F., Stefania, B., Barbara, B., Paolo, M., Angelika, M., Frederik, R. W., & Sergio, M., (2020). Cubosomes stabilized by a polyphosphoester-analog of Pluronic F127 with reduced cytotoxicity. *Journal of Colloids and Interface Sciences, 580*, 286–297. https://doi.org/10.1016/j.jcis.2020.07.038.

Mehran, A., & Ali, N., (2021). Micro- and nanoformulations of paclitaxel based on micelles, liposomes, cubosomes, and lipid nanoparticles: Recent advances and challenges. *Drug Discovery Today*. https://doi.org/10.1016/j.drudis.2021.10.007.

Mohamed, N., Mohamed, K. G., & Ahmed, A., (2015). *In vitro* and *in vivo* evaluation of cubosomes containing 5-fluorouracil for liver targeting. *Acta Pharmaceutica Sinica B*, 79–88.

Naveentaj, S., & Indira, M. Y., (2019). A review on liquid crystalline nanoparticles (cubosomes): Emerging nanoparticulate drug carrier/ *International Journal of Current Pharmaceutical Research, 12*(1), 5–9.

Pan, X., Han, K., Peng, X., Yang, Z., Qin, L., Zhu, C., Huang, X., et al., (2013). Nanostructed cubosomes as advanced drug delivery system. *Current Pharmaceutical Design, 19*(35), 6290–6297.

Patrick, T. S., & Kristin, L. H., (2001). Novel process for producing cubic liquid crystalline nanoparticles (cubosomes). *Langmuir*, *17,* 5748–5756.

Ryan, V., Sahil, S., Purab, S., Bhargavi, K., & Dileep, K., (2022). Cubosomes in cancer drug delivery: A review. *Colloid and Interface Science Communications.* https://doi.org/10.1016/j.colcom.2021.100561.

Sadhu, V. R., Beram, N. S., & Kantamneni, P., (2018). *A Review on Cubosome: The Novel Drug Delivery System.* GSC Biological and Pharmaceutical Sciences e-ISSN: 2581-3250, CODEN (USA): GBPSC2, Article retrieved from https://doi.org/10.30574/gscbps.2018.5.1.0089.

Sameh, S., Mohamed, N. A., Saad, A. S., Mourad, N. A. E., Gobba, A. A., Shata, Abdel-Moneim, H., et al., Albendazole-loaded cubosomes interrupt the ERK1/2-HIF-1α-p300/CREB axis in mice intoxicated with diethylnitrosamine: A new paradigm in drug repurposing for the inhibition of hepatocellular carcinoma progression. *Biomedicine and Pharmacotherapy*. https://doi.org/10.1016/j.biopha.2021.112029.

Sébastien, J., Giacomo, P., Marco, F., Marianna, M., Judith, S., Yeshayahu, T., Angélique, S., et al., (2020). Multifunctional cubic liquid crystalline nanoparticles for chemo- and photodynamic synergistic cancer therapy. *Photochemical & Photobiological Sciences* (pp. 674–680). Article retrieved from https://doi.org/10.1039/C9PP00449A.

Sherif, A. G., El Garhy, O. H., & Hamdy, A., (2020). Cubosomes: Composition, preparation, and drug delivery applications. *Journal of Advanced Biomedical and Pharmaceutical Science*. http://jabps.journals.ekb.eg (accessed on 11 September 2023).

Sherif, A. G., El Garhy, O. H., Hossam, M., & Hamdy, A., (2020). Preparation and evaluation of cubosomes/cubosomal gels for ocular delivery of beclomethasone dipropionate for management of uveitis. *Pharmaceutical Research*, *37*, 198. https://link.springer.com/article/10.1007/s11095-020-02857-1 (accessed on 11 September 2023).

Yaghmur, A., & Mu, H., (2021). Recent advances in drug delivery applications of cubosomes, hexosomes, and solid lipid nanoparticles. *Acta Pharmaceutica Sinica B, 11*(4), 871–885.

Younus, M., Richard, N. P., Ben, J. B., & Shakila, B. R., (2022). Comparison of cubosomes and hexosomes for the delivery of phenytoin to the brain. *Journal of Colloid and Interface Science, 605*, 146–154.

Zahra, K., & Mehrdad, H., (2016). Cubosomes: Remarkable drug delivery potential. *Drug Discovery Today, 21*(5), 789–801. https://doi.org/10.1016/j.drudis.2016.01.004.

Index

F

G

K

L

M

O

P

S

T

U

V

W

X

Y

Z